Dynamics
of Microelectromechanical
Systems

Microsystems

Continued after index

Nicolae Lobontiu

Dynamics
of Microelectromechanical
Systems

 Springer

Nicolae Lobontiu
University of Alaska Anchorage
USA

Series Editor
Steven D. Senturia
Massachusetts Institute of Technology
Cambridge, MA
USA

ISBN 978-1-4419-4225-8   e-ISBN 978-0-387-68195-5

Printed on acid-free paper.

9 8 7 6 5 4 3 2 1

springer.com

With love to my wife, daughters, mother and father

TABLE OF CONTENTS

PREFACE

Through its objective, scope, and approach, this book offers a systematic view to the dynamics of microelectromechanical systems (MEMS). While providing an in-depth look at the main problems that involve reliable modeling, analysis, and design, the main focus of this book is the mechanical/structural micro domain, which is at the core of most MEMS. Although not designed for a specific course, the book could be used as a text at the upper-undergraduate/ graduate level, and, as such, it contains numerous fully solved examples as well as many end-of-the-chapter proposed problems whose comprehensive solutions can be accessed/downloaded from the publisher's website by the qualified instructor. At the same time, it is hoped that this book might be useful to the researchers, professionals, and academics involved with modeling/ designing mechanically based MEMS.

This text is a continuation of the book *Mechanics of Microelectromechanical Systems* by Lobontiu and Garcia (Kluwer Academic Publishers, 2004), and therefore it relies on the elements developed in its precursor, such as compliance/stiffness formulations for microcantilevers, microhinges, microbridges, and microsuspensions, as well as on the treatment given to means of actuation and sensing. However, an effort has been made to ensure that this book is self-contained as much as possible.

The material is structured into four parts (conventionally named chapters), which are briefly discussed here. Each chapter contains exposition of the theory that is necessary to developing topics specific to that part. *Chapter 1* studies the bending and torsion resonant responses of microcantilevers and microbridges by employing the distributed-parameter approach and the Rayleigh's quotient approximate method, which provides means for direct derivation of the resonant frequencies. Lumped-parameter modeling, which enables calculation of the above-mentioned resonant frequencies via the equivalent stiffness and inertia properties, is also used. Several microcantilever and microbridge configurations are analyzed, and closed-form equations are provided for the bending and torsional resonant (natural) frequencies by taking into account the number of profiles that longitudinally define the member, the number of layers in a cross-section, and the type of cross-section (either constant or variable). Designs that contain circular perforations are also analyzed together with configurations that contain externally attached matter whose quantity and position alter the main resonant frequencies.

Chapter 2 analyzes the resonant/modal response of more complex micromechanical systems by considering their components are either inertia or spring elements. The lumped-parameter modeling approach is applied to derive the free vibratory response of micromechanical systems that behave as either single degree-of-freedom (DOF) ones, or as multiple DOF systems—in case they undergo more complex vibratory motion and/or are composed of several mass elements. Lagrange's equations are employed in modeling the free response of multiple DOF micromechanical systems. Numerous examples of mass-spring microsystems undergoing linear or/and rotary resonant vibrations are presented.

Chapter 3 addresses the main mechanisms responsible for energy losses in MEMS. Quality factors and corresponding viscous damping coefficients are derived owing to fluid–structure interaction (as in squeeze- and slide-film damping), anchor (connection to substrate) losses, thermoelastic damping (TED), surface/volume losses and phonon-mediated damping.

Chapter 4 discusses MEMS by taking into account the forcing factor and therefore the forced response is analyzed. For harmonic (sinusoidal, cosinusoidal) excitation, the frequency response is modeled by quantifying the amplitude and phase shift over the excitation frequency range. The Laplace transform and the cosinusoidal transfer function approach are employed in analyzing topics such as transmissibility, coupling, mechanical-electrical analogies, as well as applications such as microgyroscopes and tuning forks. For non-harmonic excitation, the time response of MEMS is studied by means of the Laplace transform, the state-space approach and time stepping schemes. Nonlinear problems, such as those generated by large deformations are also discussed, and dedicated modeling/solution methods such as time-stepping schemes or the approximate iteration method are presented. All the solutions for the problems that appear at the ends of the chapters can be accessed at http://www.springer.com/west/home/generic/search/results?-SGWID=4-40109-22-173670220-0 by a qualified instructor.

Although many applications in this text qualify as *nano* devices, the prefix *micro* has been utilized throughout, with the understanding that both the micro and nano domains are covered by the generic denomination of *microelectromechanical systems*. Particular care has been paid to the accuracy of this text, but it is possible that unwanted errors have slipped in, and I would be extremely grateful for any related signal.

In closing, I would like to address my sincere thanks to Alex Greene, Springer Editorial Director of Engineering, for all the positive interaction, support, and profound comprehension of this project.

Anchorage, Alaska

Chapter 1

MICROCANTILEVERS AND MICROBRIDGES: BENDING AND TORSION RESONANT FREQUENCIES

1.1 INTRODUCTION

Microcantilevers and microbridges are the simplest mechanical devices that operate as standalone systems in a variety of microelectromechanical systems (MEMS) applications, such as nano-scale reading/writing in topology detection/ creation, optical detection, material properties characterization, resonant sensing, mass detection, or micro/nano electronic circuitry components such as switches or filters.

This chapter studies the bending and torsion resonant responses of microcantilevers (fixed-free flexible members) and microbridges (fixed-fixed flexible members) by mainly utilizing the distributed-parameter approach and the related Rayleigh's quotient approximate method, which enable direct derivation of the resonant frequencies. The lumped-parameter modeling, which permits separate calculation of equivalent stiffness and inertia properties en route of obtaining the above-mentioned resonant frequencies, is also used in this chapter for certain configurations.

Structurally, microcantilevers and microbridges can be identical, it is only the boundary conditions that differentiate them, and this is the reason the two members are discussed together in this chapter. The configuration of a particular microcantilever or microbridge is a combination of three features, namely: number of profiles that longitudinally define the member (there can be a single profile [geometric curve], or multiple profiles [case in which there is a series connection between various single-profile segments]), number of layers in a cross-section (there can be single-layered, homogeneous members or multi-layer [sandwich] ones), and the type of cross-section (either constant or variable). These variables are illustrated in Figure 1.1 as a three-dimensional (3D) space. Because each of the three variables can take one of two possible values, eight different configuration classes are possible by combining all possible variants (in Figure 1.1 these categories are represented by the cube's vertices). The origin of the 3D space, which is one specific design category, is defined by the parameters *SL*, *SP*, and *CCS*, and represents the subclass of microcantilevers/microbridges that is made of a single layer (*SL*). Their geometry (width) is defined by a single profile (one geometric curve—*SP*), and are of constant cross-section. This particular combination results in a homogeneous, constant cross-section member, one of the simplest and most used cantilevers/bridges. The other seven subclasses (corresponding to the remaining cube vertices in Figure 1.1) can simply be described in a similar manner.

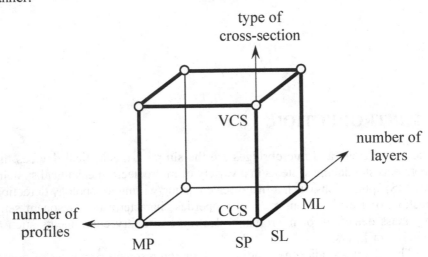

Figure 1.1 Three-dimensional space characterizing the geometric and material parameter categories that define microcantilevers/microbridges (*SL*, single layer; *ML*, multiple layer; *SP*, single profile; *MP*, multiple profile; *CCS*, constant cross-section; *VCS*, variable cross-section)

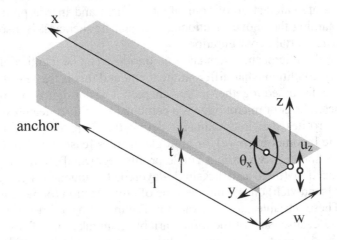

Figure 1.2 Constant cross-section microcantilever: dimensions and degrees of freedom

The assumption will be used in this chapter that variable cross-section (*VCS*) micromembers are of constant thickness and of variable width, assumption which is consistent with the usual microfabrication procedures.

Figures 1.2 and 1.3 show a microcantilever and a microbridge, respectively, both of constant rectangular cross-sections. As shown in Figure 1.2, a microcantilever is a fixed-free member, whose reference frame (which monitors the out-of-the-plane bending about the *z*-axis, and torsion about the *x*-axis) is placed at the free end where both bending and torsion deformations are maximum. A microbridge is a fixed-fixed member, as illustrated in Figure 1.3, and the reference frame can be located either at one fixed end or at its midpoint. The maximum deformations are taking place at the microbridge's midpoint.

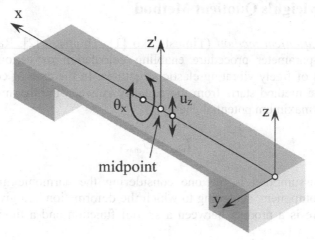

Figure 1.3 Constant cross-section microbridge: dimensions and degrees of freedom

The topic of detecting and evaluating the amount of substance that attaches to MEMS structures by monitoring the shift in the bending and torsion resonances of microcantilevers and microbridges is approached in Section 1.5.

1.2 MODAL ANALYTICAL PROCEDURES

Calculating the modal or resonant response of flexible structures can be performed by means of *analytical* and *numerical methods*. Numerical procedures, of which the *finite element method* (which is not addressed here) is the most popular, are versatile and yield precise solutions for problems that are described by partial differential equations with complex boundary conditions and geometric shapes. Although the method of choice in both academia and industry, for structures with relatively simple geometry and boundary conditions, such as microcantilevers and microbridges, the finite element method can be supplemented by simpler analytical models that are based on closed-form solutions and that offer the advantage of faster processing times.

Analytical procedures dedicated to evaluating the resonant response of elastic members comprise *distributed-parameter methods* and *lumped-parameter methods*. In a modal analysis, the distributed-parameter approach studies vibrating elastic structures by considering the time response of all points of the structure, and therefore by assuming the system's properties are distributed over the entire structure. Lumped-parameter approaches, on the other hand, consider that the system's properties are concentrated (lumped) at convenient locations and focus on the dynamic behavior at those selected locations. Small deformations of the elastic members will be assumed in this chapter, which will result in linear models.

1.2.1 Rayleigh's Quotient Method

Rayleigh's quotient method (Timoshenko [1], Thomson [2], Rao [3]) is a distributed-parameter procedure enabling calculation of various resonant frequencies of freely vibrating elastic structures. In the case of conservative systems, the method starts from the equality between the maximum kinetic energy and maximum potential energy:

$$T_{max} = U_{max} \qquad (1.1)$$

The next assumption is the one considering the harmonic motion of a vibrating component, according to which the deformation at a given point of the structure is a product between a spatial function and a time-dependent one:

$$u(x,t) = u(x)\sin(\omega t) \qquad (1.2)$$

where the deformation can be produced through bending, axial, or torsional free vibrations. The next step into Rayleigh's quotient approach is assuming a certain distribution of the elastic deformation *u (x)*. By combining all these steps yields the resonant frequency of interest. This method will be discussed in the following sections with reference to bending and torsional vibrations only.

1.2.1.1 Bending

In out-of-the-plane bending of single-component MEMS, such as the micro-cantilever and the microbridge illustrated in Figures 1.2 and 1.3, the kinetic energy is:

$$T = \frac{1}{2}\int_{V}\left[\frac{\partial u_z(x,t)}{\partial t}\right]^2 dm = \frac{\rho}{2}\int_{l} A(x)\left[\frac{\partial u_z(x,t)}{\partial t}\right]^2 dx \qquad (1.3)$$

where $u_z(x,t)$ is the deflection at an arbitrary point x on the microcantilever (microbridge) and time moment t. The member's length is l, its cross-sectional area (potentially variable) is denoted by $A(x)$, and the mass density is ρ.

The elastic potential energy stored in a bent member is:

$$U = \frac{1}{2}\int_{l} EI_y(x)\left[\frac{\partial^2 u_z(x,t)}{\partial x^2}\right]^2 dx \qquad (1.4)$$

where E is the elasticity (Young's) modulus and $I_y(x)$ is the cross-sectional moment of area with respect to the y-axis (see Figures 1.2 and 1.3). By considering the assumption:

$$u_z(x,t) = u_z(x)\sin(\omega t) \qquad (1.5)$$

the maximum kinetic energy and maximum elastic potential energy that result from Equations (1.3) and (1.4)—corresponding to values of 1 (one) for the involved sine and cosine factors—are substituted into Equation (1.1), which yields the square of the bending resonant frequency:

$$\omega_b^2 = \frac{\displaystyle\int_{l} EI_y(x)\left[\frac{\partial^2 u_z(x)}{\partial x^2}\right]^2 dx}{\displaystyle\int_{l} \rho A(x)u_z(x)^2\, dx} \qquad (1.6)$$

The deflection $u_z(x)$, which is measured at an arbitrary point along the beam and is positioned at a distance x from the origin (as already mentioned, the origin is the free end for a cantilever and either one fixed end or the midpoint for a bridge), is related to the maximum deflection u_z by means of a bending *distribution function* as:

$$u_z(x) = u_z f_b(x) \qquad (1.7)$$

By combining Equations (1.6) and (1.7), the bending resonant frequency can be reformulated as:

$$\omega_b^2 = \frac{\displaystyle\int_{l} EI_y(x)\left[\frac{d^2 f_b(x)}{dx^2}\right]^2 dx}{\displaystyle\int_{l} \rho A(x)f(x)^2\, dx} \qquad (1.8)$$

Equations (1.6) and (1.8) are two forms of Rayleigh's quotient corresponding to bending under the assumption the cross-section is variable.

1.2.1.2 Torsion

Rayleigh's quotient method can also be applied to torsion problems involving microcantilevers and microbridges. The kinetic energy of a variable rectangular cross-section rod, for instance, is expressed as:

$$T = \frac{1}{2} \int_l \frac{\rho w(x)t(x)\left[w(x)^2 + t(x)^2\right]}{12} \left[\frac{\partial \theta_x(x,t)}{\partial t}\right]^2 dx \qquad (1.9)$$

where the cross-section's width w and thickness t are assumed variable across the member's length. The torsion angle $\theta_x(x,t)$ is measured at an arbitrary abscissa x and time moment t.

The elastic potential energy is:

$$U = \frac{1}{2} \int_l GI_t(x)\left[\frac{\partial \theta_x(x,t)}{\partial x}\right]^2 dx \qquad (1.10)$$

where $I_t(x)$ is the torsion moment of area of the member's cross-section. By considering the torsional angle is defined as:

$$\theta_x(x,t) = \theta_x(x)\sin(\omega t) \qquad (1.11)$$

and by also equalizing the maximum kinetic energy to the maximum potential energy the torsion resonant frequency is calculated as:

$$\omega_t^2 = \frac{12 \int_l GI_t(x)\left[\dfrac{d\theta_x(x)}{dx}\right]^2 dx}{\int_l \rho w(x)t(x)\left[w(x)^2 + t(x)^2\right]\theta_x(x)^2 dx} \qquad (1.12)$$

The following relationship is considered relating the torsion angle at an arbitrary abscissa, $\theta_x(x)$ and the maximum (reference) torsion angle θ_x:

$$\theta_x(x) = \theta_x f_t(x) \qquad (1.13)$$

where $f_t(x)$ is the torsion distribution function. By substituting Equation (1.13) into Equation (1.12), the torsion resonant frequency becomes:

$$\omega_t^2 = \frac{12 \int_l GI_t(x) \left[\frac{df_t(x)}{dx}\right]^2 dx}{\int_l \rho w(x)t(x)\left[w(x)^2 + t(x)^2\right]f_t(x)^2 dx} \quad (1.14)$$

Again, Equations (1.12) and (1.14) express Rayleigh's quotients for torsion.

It should be mentioned that Rayleigh's quotient equations for bending and torsion give the respective resonant frequency of a non-homogeneous, variable cross-section member, irrespective of boundary conditions. The boundary conditions decide the form of the bending and torsion distribution functions, $f_b(x)$ and $f_t(x)$ over the member's length. For microcantilevers and microbridges the boundary conditions are different, and therefore the distribution functions are different as well. The distribution functions are also dependent on the abscissa origin in the case of microbridges.

1.2.2 Lumped-Parameter Method

Rayleigh's quotient method, as seen in the previous section, directly yields the resonant frequency of interest, which is sufficient when this type of response is solely needed. However, there are situations where the static or quasi-static behavior of an elastic member is also of interest, and in such cases the stiffness of that member at a specific location is necessary to use it as a connector between the applied loads and resulting deformations.

An alternative to Rayleigh's quotient distributed-parameter method to evaluating the resonant frequencies of flexible members is the lumped-parameter method, which transforms the real, distributed-parameter properties—elastic (stiffness) and inertial (mass or moment of inertia)—into equivalent, lumped-parameter ones—k_e (equivalent stiffness), m_e (equivalent mass), or J_e (equivalent mechanical moment of inertia)—which are computed separately. In doing so, one can use just the stiffness (for static applications) or both the stiffness and inertia fractions (for modal calculations), because the resonant frequency of interest is expressed as:

$$\omega_e^2 = \frac{k_e}{m_e} \quad (1.15)$$

In Equation (1.15), which is written for an elastic body whose equivalent counterpart undergoes translational motion, m_e is the mass of that body. In case the equivalent lumped body undergoes rotation, the mass of Equation (1.15) is replaced by a mechanical moment of inertia J_e. The specifics of determining the resonant frequencies corresponding to bending and torsion of microcantilevers/microbridges by the lumped-parameter approach will be discussed next.

1.2.2.1 Bending

In out-of-the-plane bending of single-component MEMS, such as micro-cantilevers or microbridges, the lumped-parameter approach substitutes the distributed-parameter flexible component by an equivalent, lumped-parameter one.

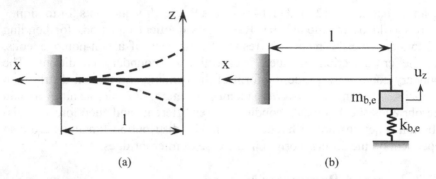

(a)   (b)

Figure 1.4 Cantilever vibrating out-of-plane: (a) real distributed-parameter system; (b) equivalent lumped-parameter, mass-spring system

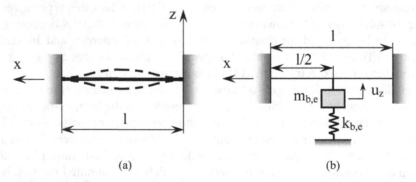

(a)   (b)

Figure 1.5 Bridge vibrating out-of-plane: (a) real distributed-parameter system; (b) equivalent lumped-parameter, mass-spring system

Figure 1.4 illustrates the lumping process for a cantilever where a mass and a linear spring are placed at the free end of a massless beam. A similar equivalence is shown in Figure 1.5 for a microbridge, where the equivalent mass and spring are placed at the midpoint. The maximum deflection is recorded at the microcantilever's free end, whereas for a microbridge the maximum deflection takes place at its midpoint. This is the reason for locating the lumped-parameter stiffness and inertia fractions at the respective points, as illustrated in Figures 1.4 and 1.5.

1.2.2.1.1 Equivalent stiffness

The stiffness of the linear spring of either Figure 1.4 or 1.5 can be found in several ways, by applying a force F_z at the point of interest (the free endpoint for the microcantilever of Figure 1.4 or the midpoint for the microbridge of Figure 1.5) and by relating it to the static deflection u_z at the same position. It is known that for linear systems, as the ones discussed here, force is proportional to deflection and the proportionality constant is the stiffness, namely:

$$F_z = k_{b,e} u_z \qquad (1.16)$$

One possibility of finding a relationship as the one of Equation (1.16) is to apply *Castigliano's first theorem*, according to which the force is expressed as:

$$F_z = \frac{\partial U_b}{\partial u_z} \qquad (1.17)$$

where U_b is the bending strain energy. The out-of-the-plane bending of cantilevers and bridges (which are beams) produces the following strain energy:

$$U_b = \frac{1}{2} \int_l EI_y(x) \left[\frac{d^2 u_z(x)}{dx^2} \right]^2 dx \qquad (1.18)$$

The deflection $u_z(x)$ is connected to the maximum deflection u_z by means of a distribution function according to Equation (1.7). Using this equation in Equation (1.17), together with Equation (1.18), results in:

$$F_z = \left\{ \int_l EI_y(x) \left[\frac{d^2 f_b(x)}{dx^2} \right]^2 dx \right\} u_z \qquad (1.19)$$

Comparing Equation (1.19) to Equation (1.16) indicates that the equivalent, lumped-parameter stiffness is:

$$k_{b,e} = \int_l EI_y(x) \left[\frac{d^2 f_b(x)}{dx^2} \right]^2 dx \qquad (1.20)$$

1.2.2.1.2 Equivalent mass

The equivalent mass $m_{b,e}$ can be found by applying Rayleigh's principle, according to which the velocity distribution over a vibrating beam is identical to the one of deflections. By also using the assumption that the kinetic energy of the distributed-parameter system is equal to the kinetic energy of the equivalent lumped-parameter mass, an equation that expresses $m_{b,e}$ can be obtained. The kinetic energy of the distributed-parameter beam is expressed by means of Equation (1.3), as it was shown in the section discussing Rayleigh's quotient method. The kinetic energy of the equivalent mass is simply:

$$T_e = \frac{1}{2} m_{b,e} \left[\frac{du_z(t)}{dt} \right]^2 \tag{1.21}$$

By equating T_e of Equation (1.21) to T of Equation (1.3), via Equation (1.7), the equivalent mass is obtained as:

$$m_{b,e} = \int_l \rho A(x) f_b(x)^2 \, dx \tag{1.22}$$

On substituting the lumped-parameter mass of Equation (1.22) and stiffness of Equation (1.20) into Equation (1.15), Equation (1.8) is obtained, which is the bending resonant frequency that has been obtained by using a distributed-parameter model and the Rayleigh quotient. This demonstrates that the lumped-parameter method and Rayleigh's quotient method yield identical results in bending.

1.2.2.2 Torsion

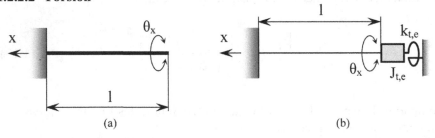

(a)   (b)

Figure 1.6 Free-fixed bar vibrating torsionally: (a) real distributed-parameter system; (b) equivalent lumped-parameter, mass-spring system

A similar approach can be followed in studying the torsion of microcantilevers and microbridges by using the lumped-parameter approach, and substitute the distributed parameters of a torsionally vibrating member by corresponding lumped-parameter ones, as illustrated in Figures 1.6 and 1.7.

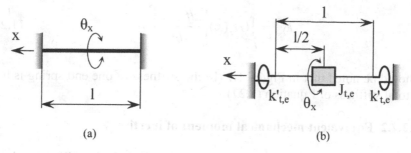

(a)   (b)

Figure 1.7 Fixed-fixed bar vibrating torsionally: (a) real distributed-parameter system; (b) equivalent lumped-parameter, mass-spring system

1.2.2.2.1 Equivalent stiffness

The torsional stiffness at the point of interest (free endpoint for the cantilever of Figure 1.6 and midpoint for the bridge of Figure 1.7) is found by means of Castigliano's first theorem as:

$$M_x = \frac{\partial U_t}{\partial \theta_x} \tag{1.23}$$

where θ_x is the maximum rotation angle and U_t is the strain energy stored in the elastic member through torsion, which is expressed as:

$$U_t = \frac{1}{2} \int_l GI_t(x) \left[\frac{d\theta_x(x)}{dx} \right]^2 dx \tag{1.24}$$

where the rotation angle at an arbitrary position on the member, $\theta_x(x)$, is expressed in terms of the maximum rotation angle θ_x and the distribution function $f_t(x)$ is given in Equation (1.13). By employing Equations (1.24) and (1.13) in Equation (1.23), the latter equation becomes:

$$M_x = \theta_x \left\{ \int_l GI_t(x) \left[\frac{df_t(x)}{dx} \right]^2 dx \right\} \tag{1.25}$$

The torsion moment is connected to the rotation angle by means of a rotary stiffness as:

$$M_x = k_{t,e} \theta_x \tag{1.26}$$

Comparison of Equations (1.25) and (1.26) yields the following lumped-parameter stiffness in torsion:

$$k_{t,e} = \int_l GI_t(x) \left[\frac{df_t(x)}{dx} \right]^2 dx \qquad (1.27)$$

It should be noted that in Figure 1.7 (b) the stiffness of one end spring is half the total stiffness of Equation (1.27).

1.2.2.2.2 Equivalent mechanical moment of inertia

Rayleigh's principle is applied again (as detailed for bending), and the kinetic energy of the distributed-parameter system, which is given in Equation (1.9), is made equal to the kinetic energy of the mass undergoing torsional vibrations, which is:

$$T_e = \frac{1}{2} J_{t,e} \left[\frac{d\theta_x}{dt} \right]^2 \qquad (1.28)$$

Equating the two energy forms—Equations (1.9) and (1.28)—results in the following lumped-parameter mechanical moment of inertia:

$$J_{t,e} = \int_l \rho J_x(x) f_t^2(x) dx \qquad (1.29)$$

Equations (1.27) and (1.29) can now be used to obtain the torsional resonant frequency, which is identical to the one of Equation (1.14). The proof was obtained that the distributed-parameter approach via Rayleigh's quotient procedure and the lumped-parameter method yield identical results in torsion.

1.3 MICROCANTILEVERS

A microcantilever, as the one shown in Figure 1.2, is a fixed-free beam and its cross-section can be variable in generally one dimension (either width or thickness). Cantilevers in which the width (cross-sectional dimension in a plane parallel to the substrate) is variable and the thickness (cross-sectional dimension that is perpendicular to the substrate) is constant are most common in MEMS. In this category, *single-profile microcantilevers* have their width defined by one curve only that can be expressed analytically, whereas *multiple-profile microcantilevers* are composed of several portions, serially connected, each segment having its width defined by a single analytical curve. Both categories will be studied in this subsection in terms of their out-of-the-plane bending and torsional resonant frequencies.

1.3.1 Single-Profile Microcantilevers

Single-profile microcantilevers have their cross-sectional dimensions (width and/or thickness for rectangular shapes, which are common configurations in MEMS) defined by a single geometric/analytic curve. As a result, the cross-section can be constant or variable, both categories being discussed next.

Constant cross-section microcantilevers, as the one sketched in Figure 1.2, are the simplest designs and regular cross-sectional shapes are rectangles and circles (e.g., the case of carbon nanotube members) that can be obtained by standard surface or bulk microfabrication procedures.

Single-profile variable cross-section microcantilevers, as mentioned previously, usually have constant thickness and their width varies following a geometric curve that is relatively simple analytically, such as a line, a circle or an ellipse segment. A generic formulation is first given in this subsection for bending as well as torsion, and then two specific configurations, the trapezoid and the circularly filleted designs, are analyzed.

1.3.1.1 Bending Distribution Functions

By using several compliances terms (in general, the compliance is the inverse of stiffness, see Lobontiu [4] and Lobontiu and Garcia [5] for more details) and the *Euler–Bernoulli model* for a relatively *long beam* (with the length at least five times larger than the largest cross-sectional dimension), the deflection at an arbitrary abscissa, $u_z(x)$, is expressed in terms of the free-end deflection u_z by means of a distribution function, as in Equation (1.7). It can be shown (see Lobontiu [4] for more information) that in case a point force F is applied at microcantilever's free end, the distribution function at the abscissa x is:

$$f_{b,F}(x) = \frac{C_{uz-Fz}(x) - xC_{uz-My}(x)}{C_{uz-Fz}} \tag{1.30}$$

where the compliances are calculated as:

$$\begin{cases} C_{uz-Fz}(x) = \int_x^l \frac{x^2}{EI_y(x)} dx \\[3mm] C_{uz-My}(x) = \int_x^l \frac{x}{EI_y(x)} dx \\[3mm] C_{uz-Fz} = \int_0^l \frac{x^2}{EI_y(x)} dx \end{cases} \tag{1.31}$$

In this case, Rayleigh's quotient of Equation (1.8) can formally be written as:

$$\omega_{b,F}^2 = \frac{\displaystyle\int_0^l EI_y(x)\left[\frac{d^2 f_{b,F}(x)}{dx^2}\right]^2 dx}{\displaystyle\int_0^l \rho A(x)\left[f_{b,F}(x)\right]^2 dx} \tag{1.32}$$

Another loading case, mentioned by Timoshenko [1], for instance, consists of a load that is distributed along the entire length of the microcantilever, case where the distribution function connecting the two deflections of Equation (1.7) can be expressed as:

$$f_{b,q}(x) = \frac{C_{uz-q}(x) - xC_{uz-Fz}(x)}{C_{uz-q}} \tag{1.33}$$

where the newly introduced compliances are calculated as:

$$\begin{cases} C_{uz-q}(x) = \displaystyle\int_x^l \frac{x^3}{EI_y(x)} dx \\[4mm] C_{uz-q} = \displaystyle\int_0^l \frac{x^3}{EI_y(x)} dx \end{cases} \tag{1.34}$$

In this case, the resonant frequency is given by:

$$\omega_{b,q}^2 = \frac{\displaystyle\int_0^l EI_y(x)\left[\frac{d^2 f_{b,q}(x)}{dx^2}\right]^2 dx}{\displaystyle\int_0^l \rho A(x)\left[f_{b,q}(x)\right]^2 dx} \tag{1.35}$$

Example 1.1
   Derive the bending distribution functions corresponding to an endpoint load and distributed load, respectively, for a constant cross-section micro-cantilever of length l.

Solution:
   The compliances of Equation (1.31) that are needed to calculate $f_{b,F}(x)$ are:

$$
\begin{cases}
C_{uz-Fz}(x) = \dfrac{l^3 - x^3}{3EI_y} \\[2ex]
C_{uz-My}(x) = \dfrac{l^2 - x^2}{2EI_y} \\[2ex]
C_{uz-Fz} = \dfrac{l^3}{3EI_y}
\end{cases}
\tag{1.36}
$$

The force-related bending distribution is obtained by substituting Equation (1.36) into Equation (1.30) as:

$$
f_{b,F}(x) = 1 - \frac{3}{2}\frac{x}{l} + \frac{1}{2}\frac{x^3}{l^3}
\tag{1.37}
$$

For a constant cross-section microcantilever, the compliances of Equation (1.34) are:

$$
\begin{cases}
C_{uz-q}(x) = \dfrac{l^4 - x^4}{4EI_y} \\[2ex]
C_{uz-q} = \dfrac{l^4}{4EI_y}
\end{cases}
\tag{1.38}
$$

By substituting Equation (1.38) into Equation (1.33), the bending distribution function corresponding to a distributed load becomes:

$$
f_{b,q}(x) = 1 - \frac{4}{3}\frac{x}{l} + \frac{1}{3}\frac{x^4}{l^4}
\tag{1.39}
$$

The following ratio can also be formulated:

$$
rf_b = \frac{f_{b,q}(x)}{f_{b,F}(x)} = \frac{3(2+c)}{2\left[3 + c(2+c)\right]}
\tag{1.40}
$$

with $c = x/l$. This ratio is plotted in Figure 1.8.

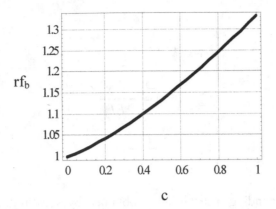

Figure 1.8 Ratio of distribution functions in terms of relative abscissa

As Figure 1.8 indicates, the two distribution functions, $f_{b,F}$ and $f_{b,q}$, are closer for smaller values of c (at locations closer to the free end) and start diverging more markedly as the relative abscissa c approaches 1 (close to the fixed end).

Example 1.2
    Calculate the bending resonant frequencies for a constant rectangular cross-section microcantilever by using the two functions $f_{b,F}$ and $f_{b,q}$ (Equations (1.37) and (1.39)).

Solution:
    In this case the cross-sectional area and moment of area are simply:

$$\begin{cases} A = wt \\ I_y = \dfrac{wt^3}{12} \end{cases} \qquad (1.41)$$

By using these values into Equations (1.32) and (1.35), the following resonant frequencies are obtained:

$$\begin{cases} \omega_{b,F} = 3.567\sqrt{\dfrac{EI_y}{ml^3}} \\ \omega_{b,q} = 3.53\sqrt{\dfrac{EI_y}{ml^3}} \end{cases} \qquad (1.42)$$

It should be noticed that the exact value of the bending-related resonant frequency is:

$$\omega_b = 3.52\sqrt{\frac{EI_y}{ml^3}} \tag{1.43}$$

which is almost the same as the prediction by the Rayleigh's quotient method when using the distribution function produced by a uniformly distributed load q. It should also be remarked that the prediction by using a point load at the free end results in a resonant frequency higher than the real one.

When the microcantilever is short (with the length less than five times the largest cross-sectional dimension), *Timoshenko's model* applies, and the two distribution functions dealt with in this section become:

$$\begin{cases} f_{b,F}^{sh} = \dfrac{C_{uz-Fz}^{sh}(x) - xC_{uz-My}(x)}{C_{uz-Fz}^{sh}} \\[3mm] f_{b,q}^{sh} = \dfrac{C_{uz-q}(x) - xC_{uz-Fz}^{sh}(x)}{C_{uz-q}} \end{cases} \tag{1.44}$$

where the shear-related compliances are:

$$\begin{cases} C_{uz-Fz}^{sh}(x) = C_{uz-Fz}(x) + \kappa\dfrac{E}{G}C_{ux-Fx}(x) \\[3mm] C_{uz-Fz}^{sh} = C_{uz-Fz} + \kappa\dfrac{E}{G}C_{ux-Fx} \end{cases} \tag{1.45}$$

as demonstrated in Lobontiu [4]. The factor κ in the equations (1.45) is dependent on the cross-sectional shape (it is equal to 5/6 for a rectangular cross-section, see Young and Budynas [6]). The axial compliances of Equation (1.45) are defined (see Lobontiu [4] for more details) as:

$$\begin{cases} C_{ux-Fx}(x) = \displaystyle\int_x^l \dfrac{dx}{EA(x)} \\[3mm] C_{ux-Fx} = \displaystyle\int_0^l \dfrac{dx}{EA(x)} \end{cases} \tag{1.46}$$

Example 1.3

Compare the bending resonant frequencies of a constant rectangular cross-section microcantilever when using the long- and short-length beam models. Also analyze the contribution of the distribution functions $f_{b,F}$ and $f_{b,q}$.

Solution:

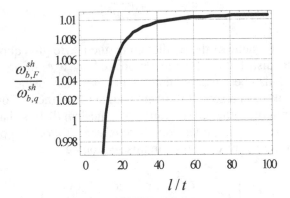

Figure 1.9 Bending frequency ratio for short-length rectangular constant cross-section cantilever

By using the distribution function corresponding to the point force loading, the following bending resonant frequency is produced:

$$\omega_{b,F}^{sh} = 20.494 G t \sqrt{\frac{E}{\rho\left(396 G^2 l^4 + 231 G E \kappa l^2 t^2 + 35 E^2 \kappa^2 t^4\right)}} \quad (1.47)$$

whereas the bending resonant frequency predicted by the distributed-force distribution function is:

$$\omega_{b,q}^{sh} = 6.481 \frac{t}{l^2} \sqrt{\frac{E\left(36 G^2 l^4 + 20 G E \kappa l^2 t^2 + 5 E^2 \kappa^2 t^4\right)}{\rho\left(1456 G^2 l^4 - 240 G E \kappa l^2 t^2 + 21 E^2 \kappa^2 t^4\right)}} \quad (1.48)$$

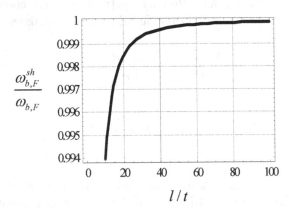

Figure 1.10 Bending frequency ratio for point force distribution function: short-length versus long-length rectangular constant cross-section cantilever

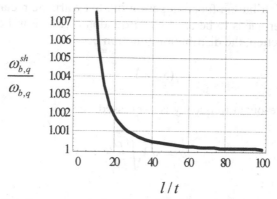

$$\frac{\omega_{b,q}^{sh}}{\omega_{b,q}}$$

Figure 1.11 Bending frequency ratio for distributed force distribution function: short-length versus long-length rectangular constant cross-section cantilever

Figures 1.9, 1.10, and 1.11 show various ratios of the bending resonant frequencies calculated according to the long- and short-beam models, as well as in terms of the two distribution functions. As Figure 1.9 indicates, the bending resonant frequency predictions by the point load and distributed load distribution functions are quite close for a length-to-thickness ratio ranging from 10 to 100. Figures 1.10 and 1.11 show that taking the shearing effects into consideration results in small errors to the long-length (Euler-Bernoulli) model of the bending resonant frequency for a constant rectangular cross-section cantilever.

1.3.1.2 Torsion Distribution Functions

In the case of microcantilevers, as shown previously, the torsion angle $\theta_x(x)$ is related to the free end torsion angle θ_x according to Equation (1.13) where the torsion-related distribution function can be calculated as:

$$f_t(x) = \frac{C_{\theta x-Mx}(x)}{C_{\theta x-Mx}} \tag{1.49}$$

with the torsion compliances being defined as:

$$\begin{cases} C_{\theta x-Mx}(x) = \int_x^l \frac{dx}{GI_t(x)} \\[2mm] C_{\theta x-Mx} = \int_0^l \frac{dx}{GI_t(x)} \end{cases} \tag{1.50}$$

The particular distribution function pertaining to a variable rectangular cross-section cantilever needs to be used into Equation (1.14). For a constant rectangular cross-section, the distribution function becomes:

$$f_t(x) = 1 - \frac{x}{l} \tag{1.51}$$

and the torsion resonant frequency is expressed as:

$$\omega_t = \sqrt{3} \sqrt{\frac{GI_t}{lJ_t}} \tag{1.52}$$

where:

$$J_t = \frac{m\left(w^2 + t^2\right)}{12} = \frac{\rho lwt\left(w^2 + t^2\right)}{12} \tag{1.53}$$

The torsional moment of inertia, I_t, depends on whether the rectangular cross-section is very thin ($t \ll w$) or thin ($t < w$). For a very thin cross-section, the torsional moment of inertia is:

$$I_t = \frac{wt^3}{3} \tag{1.54}$$

whereas for thin cross-sections the torsion moment of inertia is:

$$I_t = wt^3 \left(0.33 - 0.21\frac{t}{w}\right) \tag{1.55}$$

In the case of constant circular cross-section cantilevers, such as those consisting of carbon nanotubes, the following relationship exists between the torsional mass moment of inertia and the geometric moment of inertia:

$$J_t = \rho I_t l \tag{1.56}$$

and the torsional resonant frequency becomes:

$$\omega_t = \frac{\sqrt{3}}{l} \sqrt{\frac{G}{\rho}} \tag{1.57}$$

Because many MEMS cantilevers/bridges are of rectangular cross-section, the focus in this book will fall on this particular type of cross-section.

1.3.1.3 Solid Trapezoid Microcantilever

The solid trapezoid microcantilever (its top view is shown in Figure 1.12) has been analyzed in terms of relevant stiffnesses/compliances by Lobontiu and Garcia [5] and in terms of its resonant behavior during free bending, torsion and axial vibrations by Lobontiu [7] with the aid of the lumped-parameter approach. Rayleigh's quotient method is used here to derive the fundamental frequencies that are related to bending and torsion. The variable width $w(x)$ is calculated as:

$$w(x) = w_1 + (w_2 - w_1)\frac{x}{l} \tag{1.58}$$

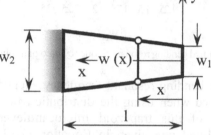

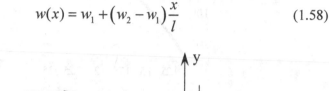

Figure 1.12 Top view of constant-thickness solid trapezoid microcantilever with defining geometry

By using the bending distribution function corresponding to a point load and a constant cross-section member, Equation (1.37), which represents a simplification because the actual bending distribution function for a variable cross-section is calculated by means of Equation (1.30), a simplified bending resonant frequency is obtained, namely:

$$\omega_{b,s} = 8.367 \frac{t}{l^2} \sqrt{\frac{E(3w_2 + w_1)}{\rho(49w_2 + 215w_1)}} \tag{1.59}$$

When considering the limit $w_2 \rightarrow w_1$, Equation (1.59) reduces to:

$$\omega_{b,s}^* = 1.03 \frac{t}{l^2} \sqrt{\frac{E}{\rho}} = 3.567 \sqrt{\frac{EI_y}{ml^3}} \tag{1.60}$$

which is the known expression of the bending resonant frequency for a constant cross-section cantilever.

Example 1.4

Compare the simplified bending resonant frequency, Equation (1.59), to the precise one that uses the actual distribution function of Equation (1.30).

Solution:

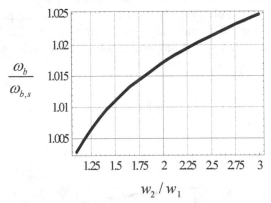

$$\frac{\omega_b}{\omega_{b,s}}$$

w_2 / w_1

Figure 1.13 Exact-to-simplified bending resonant frequency ratio of a solid trapeze microcantilever

A more complex bending resonant frequency equation, ω_b, which is not given here, is obtained when using the distribution function describing the particular geometry of this trapezoid microcantilever, Equation (1.30), instead of the simplified one given in Equation (1.37). The real (actual) bending-related distribution function is:

$$f_{b,F}(x) = \frac{w_1(w_1 + 2cx)\ln\dfrac{2cl + w_1}{2cx + w_1} + 2c(l-x)\left[c(l-x) - w_1\right]}{2cl(cl - w_1) + w_1^2 \ln\left(1 + \dfrac{cl}{w}\right)_1} \tag{1.61}$$

where $c = w_2/w_1$. The ratio of the real bending resonant frequency ω_b to the simplified one $\omega_{b,s}$ is plotted in Figure 1.13, which indicates the differences between the two models predictions are minimal.

When torsion is considered, the simplified distribution function of Equation (1.51), which corresponds to a constant cross-section cantilever, can be used to determine the resonant frequency. By following the standard procedure of Rayleigh's quotient approach, the torsional resonant frequency for a very thin cross-section (with I_t given in Equation (1.54)) is:

$$\omega_{t,s}^{v.th.} = 10.954 \frac{t}{l}\sqrt{\frac{G(w_1 + w_2)}{\rho\left[10w_1^3 + 6w_1^2 w_2 + 3w_1 w_2^2 + w_2^3 + 5t^2(w_1 + w_2)\right]}} \tag{1.62}$$

Example 1.5

Consider a trapezoid microcantilever of the type shown in Figure 1.12. Compare its torsion resonant frequency when the structure is considered a very thin structure with the one of a thin structure by using the simplified distribution function.

Solution:

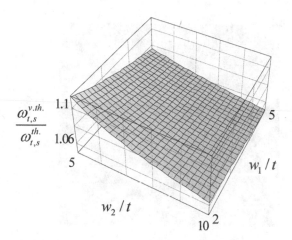

Figure 1.14 Comparison between the torsional resonant frequencies for a solid trapezoid microcantilever: very thin (*v.th.*) versus thin (*th.*) configurations

For a thin cross-section (where I_t is provided in Equation (1.55)), the torsional resonant frequency of a solid trapeze microcantilever changes to:

$$\omega_{t,s}^{th.} = 26.833 \frac{t}{l} \sqrt{\frac{G[0.165(w_1 + w_2) - 0.21t]}{\rho[10w_1^3 + 6w_1^2 w_2 + 3w_1 w_2^2 + w_2^3 + 5t^2(w_1 + w_2)]}} \qquad (1.63)$$

Figure 1.14 is the 3D plot of the ratio of the two torsion-related resonant frequencies (Equations (1.62) and (1.63)). As Figure 1.14 indicates, when the defining width parameters w_1 and w_2 are relatively large compared to the thickness t, the predictions by the two models are very similar. On the other hand, when these relationships reverse, the very thin model predicts higher values than the thin model.

1.3.1.4 Circularly Filleted Microcantilever

The circularly filleted microcantilever has been presented in terms of its relevant stiffnesses in Lobontiu and Garcia [5]. Figure 1.15 shows the picture of such a microcantilever, which is the macroscale counterpart of a micro-scale cantilever and has been produced by wire electro-discharge machining (WEDM). The geometry of a constant thickness circularly filleted design is shown in the top view of Figure 1.16. This microcantilever is defined by a single circular profile, and the variable width $w(x)$, also mentioned by Lobontiu and Garcia [5], is:

$$w(x) = w + 2\left(r - \sqrt{r^2 - x^2}\right) \tag{1.64}$$

where the parameters w, r and x are indicated in Figure 1.16.

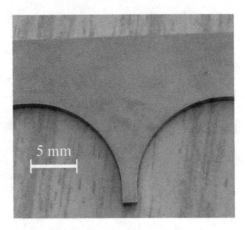

Figure 1.15 Top photograph of macro-scale circularly filleted cantilever

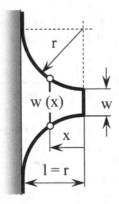

Figure 1.16 Top view of circularly filleted microcantilever with geometry

The bending resonant frequency is calculated by using Rayleigh's quotient method, and its equation is:

$$\omega_b = 1.481\frac{t}{r^2}\sqrt{\frac{E(6.575r + 8w)}{\rho(r + 16.556w)}} \tag{1.65}$$

By considering the structure is very thin, its torsion resonant frequency is:

$$\omega_t = 2\frac{t}{r}\sqrt{\frac{G(0.43r+w)}{\rho\left[0.007r^3+0.038r^2w+0.333w\left(w^2+t^2\right)+0.036r\left(3w^2+t^2\right)\right]}} \quad (1.66)$$

Example 1.6
   Compare the bending and torsional resonant frequencies of a circularly filleted microcantilever in terms of the defining geometry parameters *r*, *w* and *t*. Also establish whether the possibility exists that the two frequencies be equal.

Solution:
   One convenient way to compare the two relevant resonant frequencies of the circularly filleted microcantilever of Figure 1.16 is to analyze the torsion-to-bending resonant frequency ratio that is formed by using Equations (1.66) and (1.65); this ratio can be expressed in terms of only two non-dimensional variables, for instance r/t and w/t, and Figure 1.17 is a 3D plot illustrating this functional relationship. Figure 1.17 suggests that the frequency ratio possibly decreases towards 1 (and therefore the bending and torsion resonant frequencies are equal) when r/t assumes large values. When $r/t = 500$, for instance, by solving the equation $\omega_t/\omega_b = 1$ results in $w/t = 902.66$, which means a design which is defined by the following geometric parameters: $t = 1$ μm, $w = 902.66$ μm, and $r = 500$ μm.

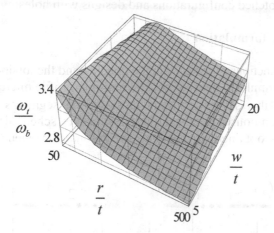

Figure 1.17 Torsional-to-bending resonant frequency ratio in terms of non-dimensional geometry parameters

1.3.2 Multiple-Profile Microcantilevers

Microcantilevers can be built up of segments having different profiles that connect continuously or stepwise. As thus, the segments can be of constant cross-section (more often rectangular) or of variable cross-section. Normally, multi-profile members are homogeneous, as they are fabricated out of a single material. In the case of layered (sandwiched) members in which the layers have different lengths, the equivalence process being applied results in multi-profile non-homogeneous members consisting of serially connected segments that have different material properties. Both cases, namely homogeneous and non-homogeneous multi-profile microcantilevers, will be studied in this section. Generic formulations are developed for both categories, where the bending and torsional resonant frequencies are derived, together with the corresponding distribution functions. Particular designs pertaining to the homogeneous category are further discussed, such as the paddle, the circularly notched configurations, as well as designs with perforations.

1.3.2.1 Homogeneous Configurations

Homogeneous multi-profile microcantilevers are configurations obtained by serially connecting several segments, each being defined as a single-profile (of either constant or variable cross-section, defined by a single geometric curve/profile), all segments being made of the same material, and by anchoring one end to the substrate (the other one being free). Generic formulations giving the transfer functions as well as the bending and torsion resonant frequencies are first developed, followed by applications to paddle configurations, circularly notched configurations and designs with holes.

1.3.2.1.1 Generic formulation

The distribution functions together with the bending and the torsional resonant frequencies are formulated here for generic multiple-profile microcantilevers.

By assuming that several constant cross-section segments are serially connected, the compound microcantilever, which is schematically shown in Figure 1.18, is obtained. In bending, the resonant frequency of such a

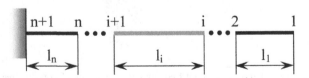

Figure 1.18 Serially compounded microcantilever

cantilever will simply be a summation of the single-profile distributed stiffness and mass properties of the individual segments, namely:

$$\omega_b^2 = \frac{E\sum\limits_{i=1}^{n}\left\{\int_{l_i} I_{y,i}(x)\left[\frac{d^2 u_{z,i}(x)}{dx^2}\right]^2 dx\right\}}{\rho\sum\limits_{i=1}^{n}\left[\int_{l_i} A_i(x)u_{z,i}^2(x)dx\right]} \tag{1.67}$$

At the same time, as it will be shown in a few subsequent examples, the deflection over a particular interval can be written as a function of the free end deflection and a distribution function corresponding to that interval as:

$$u_{z,i}(x) = f_{b,i}(x)u_z \tag{1.68}$$

Substituting Equation (1.68) into Equation (1.67) yields:

$$\omega_b^2 = \frac{E\sum\limits_{i=1}^{n}\left\{\int_{l_i} I_{y,i}(x)\left[\frac{d^2 f_{b,i}(x)}{dx^2}\right]^2 dx\right\}}{\rho\sum\limits_{i=1}^{n}\left[\int_{l_i} A_i(x)f_{b,i}(x)^2 dx\right]} \tag{1.69}$$

A similar reasoning can be applied to the torsion free vibrations of a serially compounded microcantilever, and the corresponding resonant frequency can be expressed as:

$$\omega_t^2 = \frac{12G\sum\limits_{i=1}^{n}\left\{\int_{l_i} I_{t,i}(x)\left[\frac{d\theta_{x,i}(x)}{dx}\right]^2 dx\right\}}{\rho\sum\limits_{i=1}^{n}\left[\int_{l_i} w_i(x)t_i(x)\left[w_i^2(x)+t_i^2(x)\right]\theta_{x,i}^2(x)dx\right]} \tag{1.70}$$

The angular deformation on any interval can be expressed in terms of the free end torsion angle θ_x by means of individual distribution functions as:

$$\theta_{x,i}(x) = f_{t,i}(x)\theta_x \tag{1.71}$$

Substituting Equation (1.71) into Equation (1.70) yields:

$$\omega_t^2 = \frac{12G\sum_{i=1}^{n}\left\{\int_{l_i}I_{t,i}(x)\left[\frac{df_{t,i}(x)}{dx}\right]^2 dx\right\}}{\rho\sum_{i=1}^{n}\left[\int_{l_i}w_i(x)t_i(x)\left[w_i^2(x)+t_i^2(x)\right]f_{t,i}(x)^2 dx\right]} \tag{1.72}$$

Example 1.7
    Find the bending distribution functions of a microcantilever composed of two different segments, each of constant cross-section, by considering the two segments are microfabricated of the same isotropic material.

Solution:
    The distribution functions are connected to deflections as:

$$\begin{cases} u_{z,1}(x) = f_{b,1}(x)u_z \\ u_{z,2}(x) = f_{b,2}(x)u_z \end{cases} \tag{1.73}$$

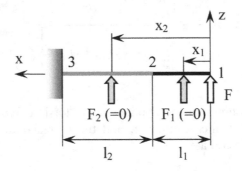

Figure 1.19 Two-segment microcantilever with point force at free end

where $u_{z,1}(x)$ is the deflection at an abscissa x_1 on the 1–2 segment, $u_{z,2}(x)$ is the deflection at an abscissa x_2 on the 2–3 segment, u_z is the deflection at the free tip, and $f_{b,1}(x)$ and $f_{b,2}(x)$ are the distribution functions on the two segments (see Figure 1.19). The tip deflection can be found by applying *Castigliano's displacement theorem* (e.g., see Lobontiu [4]), namely:

$$u_z = \frac{1}{E}\left(\int_0^{l_1}\frac{M_1}{I_{y1}}\frac{\partial M_1}{\partial F}dx + \int_{l_1}^{l_1+l_2}\frac{M_2}{I_{y2}}\frac{\partial M_2}{\partial F}dx\right) \tag{1.74}$$

where:

$$M_1 = M_2 = Fx \qquad (1.75)$$

To determine the deflection on a generic point on the first segment 1–2, a dummy load F_1 is applied at that point in addition to the point load F, and the corresponding deflection is calculated as:

$$u_{z,1} = \frac{1}{E}\left(\int_{x_1}^{l_1} \frac{M_1}{I_{y1}} \frac{\partial M_1}{\partial F_1} dx + \int_{l_1}^{l_1+l_2} \frac{M_2}{I_{y2}} \frac{\partial M_2}{\partial F_1} dx \right) \qquad (1.76)$$

where:

$$M_1 = M_2 = Fx + F_1(x - x_1) \qquad (1.77)$$

In doing so, a relationship between $u_{z,1}$ and u_z can be found of the type shown in Equation (1.7), where the distribution function corresponding to the segment 1–2 is expressed as:

$$f_{b1}(x) = a_1 + b_1 x + c_1 x^3 \qquad (1.78)$$

with:

$$\begin{cases} a_1 = 1 \\[2mm] b_1 = -\dfrac{3}{2} \dfrac{I_{y2}l_1^2 + I_{y1}l_2(2l_1 + l_2)}{I_{y2}l_1^3 + I_{y1}l_2(3l_1^2 + 3l_1 l_2 + l_2^2)} \\[4mm] c_1 = \dfrac{1}{2} \dfrac{I_{y2}}{I_{y2}l_1^3 + I_{y1}l_2(3l_1^2 + 3l_1 l_2 + l_2^2)} \end{cases} \qquad (1.79)$$

A similar relationship can be determined between $u_{z,2}$ and u_z by calculating the deflection on the 2–3 segment as produced by the tip force F and the dummy force F_2. The distribution function connecting these two deflections is:

$$f_{b2}(x) = a_2 + b_2 x + c_2 x^3 \qquad (1.80)$$

with:

$$
\begin{cases}
a_2 = \dfrac{I_{y1}\left(l_1+l_2\right)^3}{I_{y2}l_1^3 + I_{y1}l_2\left(3l_1^2 + 3l_1l_2 + l_2^2\right)} \\[4mm]
b_2 = -\dfrac{3}{2}\,\dfrac{I_{y1}\left(l_1+l_2\right)^2}{I_{y2}l_1^3 + I_{y1}l_2\left(3l_1^2 + 3l_1l_2 + l_2^2\right)} \\[4mm]
c_2 = \dfrac{1}{2}\,\dfrac{I_{y1}}{I_{y2}l_1^3 + I_{y1}l_2\left(3l_1^2 + 3l_1l_2 + l_2^2\right)}
\end{cases}
\tag{1.81}
$$

For the particular case in which the two segments have identical cross-sections, Equations (1.78) and (1.80) simplify to:

$$
f_{b1} = f_{b2} = 1 - \frac{3}{2}\frac{x}{l_1+l_2} + \frac{1}{2}\frac{x^3}{\left(l_1+l_2\right)^3}
\tag{1.82}
$$

Example 1.8
    Following the derivation of the previous Example 1.7, determine the torsion-related distribution functions of a homogeneous two-segment micro-cantilever, when each segment is of constant cross-section.

Solution:
    An approach similar to the one applied in the case of bending in Example 1.7 is used to determine the two torsion-related distribution functions. The distribution functions are connected to deflections as:

$$
\begin{cases}
\theta_{x,1}(x) = f_{t,1}(x)\theta_x \\
\theta_{x,2}(x) = f_{t,2}(x)\theta_x
\end{cases}
\tag{1.83}
$$

where $\theta_{x,1}(x)$ is the rotation at an abscissa x_1 on the 1–2 segment, $\theta_{x,2}$ is the rotation at an abscissa x_2 on the 2–3 segment, θ_x is the maximum rotation at the free end, and $f_{t,1}(x)$ and $f_{t,2}(x)$ are the distribution functions on the two segments shown in Figure 1.20. The free end rotation is determined by means of Castigliano's displacement theorem as:

$$
\theta_x = \frac{1}{G}\left(\int_0^{l_1} \frac{M_{t1}}{I_{t1}}\frac{\partial M_{t1}}{\partial M}\,dx + \int_{l_1}^{l_1+l_2} \frac{M_{t2}}{I_{t2}}\frac{\partial M_{t2}}{\partial M}\,dx \right)
\tag{1.84}
$$

where the torsion moments on the two intervals are equal, namely:

$$M_{t1} = M_{t2} = M \qquad (1.85)$$

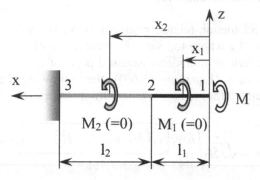

Figure 1.20 Two-segment microcantilever with point moment at free end

A dummy moment M_1 is applied at a position x_1 in addition to the point moment M to find the rotation at that point as:

$$\theta_{x,1} = \frac{1}{G}\left(\int_{x_1}^{l_1} \frac{M_{t1}}{I_{t1}} \frac{\partial M_{t1}}{\partial F} dx + \int_{l_1}^{l_1+l_2} \frac{M_{t2}}{I_{t2}} \frac{\partial M_{t2}}{\partial F} dx \right) \qquad (1.86)$$

where:

$$M_{t1} = M_{t2} = M + M_1 \qquad (1.87)$$

The ratio of $\theta_{x,1}$ to θ_x, is the torsion distribution function corresponding to the segment 1–2 and is expressed as:

$$f_{t1}(x) = 1 - \frac{I_{t2}}{I_{t2}l_1 + I_{t1}l_2} x \qquad (1.88)$$

By following a similar procedure (with the application of the dummy moment M_2 at a generic abscissa x_2 on the second interval), the distribution function for the second interval is:

$$f_{t2}(x) = \frac{I_{t1}(l_1 + l_2)}{I_{t2}l_1 + I_{t1}l_2} - \frac{I_{t1}}{I_{t2}l_1 + I_{t1}l_2} x \qquad (1.89)$$

When the two segments are also geometrically identical ($I_{t1} = I_{t2}$), Equations (1.88) and (1.89) simplify to:

$$f_t(x) = 1 - \frac{x}{l_1 + l_2}$$

(1.90)

1.3.2.1.2 Paddle microcantilever

The bending- and torsion-related resonant frequencies of a paddle micro-cantilever, as the one whose top view is sketched in Figure 1.21, are derived next from the generic formulation discussed previously. By using the generic bending formulation (Equation (1.69)) the following bending resonant frequency is obtained for the two-segment microcantilever of Figure 1.21:

$$\omega_b = \sqrt{35t}\sqrt{\frac{Ew_2\left[l_2\left(3l_1^2 + 3l_1l_2 + l_2^2\right)w_1 + l_1^3 w_2\right]}{A}}$$

(1.91)

where:

$$A = \rho[35l_1l_2^2\left(12l_1^4 + 30l_1^3l_2 + 33l_1^2l_2^2 + 18l_1l_2^3 + 4l_2^4\right)w_1^2$$

$$+l_2\left(231l_1^6 + 273l_1^5l_2 + 105l_1^4l_2^2 + 63l_1^2l_2^4 + 91l_1l_2^5 + 33l_2^6\right)w_1w_2 \quad (1.92)$$

$$+33l_1^7 w_2^2]$$

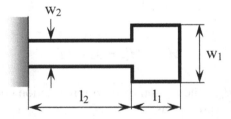

Figure 1.21 Top view of a paddle microcantilever with geometry

In calculating the bending resonant frequency by means of Equations (1.91) and (1.92), the distribution functions $f_{b1}(x)$ of Equation (1.78) and $f_{b2}(x)$ of Equation (1.80) have been used, which are the exact distribution functions. It is interesting to check how the bending resonant frequency changes when using the single approximate distribution function of Equation (1.82). By substituting Equation (1.82) into Equation (1.69) another bending resonant frequency, denoted by ω_b' and which is not given here (but which has been derived in Lobontiu [7]), is obtained and the ratio of these frequencies is plotted in Figure 1.22.

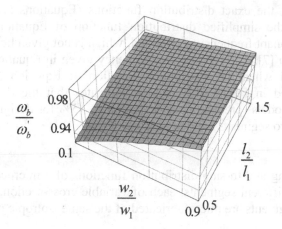

Figure 1.22 Bending resonant frequency ratio for a paddle microcantilever: exact versus approximate distribution functions

By also following the generic formulation, the torsional resonant frequency of the paddle microbridge is obtained as:

$$\omega_t = 2\sqrt{3}t \sqrt{\frac{Gw_2\left(l_1w_2+l_2w_1\right)}{\rho\left[l_1\left(3l_2^2w_1^2+3l_1l_2w_1w_2+l_1^2w_2^2\right)\left(w_1^2+t^2\right)+l_2^3w_1w_2\left(w_2^2+t^2\right)\right]}} \quad (1.93)$$

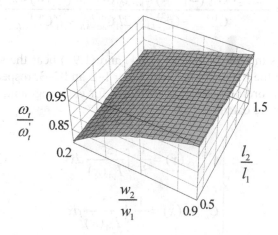

Figure 1.23 Torsion resonant frequency ratio for a paddle microcantilever: exact versus approximate distribution functions

If instead of using the exact distribution functions (Equations (1.88) and (1.89)), one uses the simplified distribution function of Equation (1.90), another torsion resonant frequency is obtained, which is not given here, but is derived in Lobontiu [7]. The ratio of the frequency given in Equation (1.93) to the one yielded when the distribution function of Equation (1.90) is employed is plotted in Figure 1.23, which indicates that the differences between the two models' predictions can amount to 15% for designs where the widths of the two segments are comparable.

Example 1.9
    Find the bending and torsion distribution functions of a microcantilever composed of two different segments, each of variable cross-section, by considering the two segments are microfabricated of the same isotropic material.

Solution:
    By following the procedure used in Example 1.7, it can be shown that the bending distribution function of the 1–2 segment in Figure 1.19 is:

$$f_{b1}(x) = \frac{C_{uz-Fz}^{(1)}(x) - x C_{uz-My}^{(1)}(x) + C_{uz-Fz}^{(2)} + (2l_1 - x) C_{uz-My}^{(2)} + l_1(l_1 - x) C_{\theta y-My}^{(2)}}{C_{uz-Fz}^{(1)} + C_{uz-Fz}^{(2)} + 2l_1 C_{uz-My}^{(2)} + l_1^2 C_{\theta y-My}^{(1)}} \tag{1.94}$$

whereas the distribution function for the 2–3 segment is:

$$f_{b2}(x) = \frac{C_{uz-Fz}^{(2)}(x) + (2l_1 - x) C_{uz-My}^{(2)}(x) + l_1(l_1 - x) C_{\theta y-My}^{(2)}(x)}{C_{uz-Fz}^{(1)} + C_{uz-Fz}^{(2)} + 2l_1 C_{uz-My}^{(2)} + l_1^2 C_{\theta y-My}^{(1)}} \tag{1.95}$$

The compliances in both Equations (1.94) and (1.95) bear the superscripts 1 and 2 indicating the two different segments, 1–2 and 2–3, respectively. They are calculated according to local coordinates for each segment, and the ones that depend on x are calculated as:

$$\begin{cases} C_{uz-Fz}^{(1)}(x) = \dfrac{1}{E} \int\limits_x^{l_1} \dfrac{x^2}{I_{y1}(x)} dx \\[4mm] C_{uz-My}^{(1)}(x) = \dfrac{1}{E} \int\limits_x^{l_1} \dfrac{x}{I_{y1}(x)} dx \end{cases} \tag{1.96}$$

for the 1–2 segment, whereas for the 2–3 segment they are:

$$
\begin{cases}
C_{uz-Fz}^{(2)}(x) = \dfrac{1}{E} \displaystyle\int_{x}^{l_2} \dfrac{x^2}{I_{y2}(x)}\, dx \\[3ex]
C_{uz-My}^{(2)}(x) = \dfrac{1}{E} \displaystyle\int_{x}^{l_2} \dfrac{x}{I_{y2}(x)}\, dx \\[3ex]
C_{\theta y-My}^{(2)}(x) = \dfrac{1}{E} \displaystyle\int_{x}^{l_2} \dfrac{dx}{I_{y2}(x)}\, dx
\end{cases}
\tag{1.97}
$$

When each of the two segments has a constant cross-section, Equations (1.94) and (1.95), together with the compliances of Equations (1.96) and (1.97) reduce to Equations (1.78) and (1.80), which define a two-segment cantilever of constant cross-section over each interval.

In torsion, the distribution function corresponding to the 1–2 segment is:

$$
f_{t1}(x) = \frac{C_{\theta x-Mx}^{(1)}(x) + C_{\theta x-Mx}^{(2)}}{C_{\theta x-Mx}^{(1)} + C_{\theta x-Mx}^{(2)}}
\tag{1.98}
$$

whereas the distribution function of the 2–3 segment is:

$$
f_{t2}(x) = \frac{C_{\theta x-Mx}^{(2)}(x)}{C_{\theta x-Mx}^{(1)} + C_{\theta x-Mx}^{(2)}}
\tag{1.99}
$$

The x-dependent torsional compliances are calculated as:

$$
\begin{cases}
C_{\theta x-Mx}^{(1)}(x) = \dfrac{1}{G} \displaystyle\int_{x}^{l_1} \dfrac{dx}{I_{t1}(x)} \\[3ex]
C_{\theta x-Mx}^{(2)}(x) = \dfrac{1}{G} \displaystyle\int_{x}^{l_2} \dfrac{dx}{I_{t2}(x)}
\end{cases}
\tag{1.100}
$$

The constant compliances of Equations (1.94), (1.95), (1.98), and (1.99) are calculated from their counterpart equations by taking a lower limit of 0 instead of x.

When the two segments are of constant cross-sections, Equation (1.98) and (1.99) change to Equations (1.88) and (1.89), respectively, of the previous Example 1.8. Furthermore, when the two segments have identical cross-sections, the two distribution functions simplify to the form given in Equation (1.90) of the same example.

1.3.2.1.3 Circularly notched microcantilever

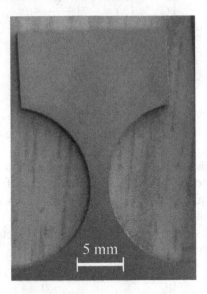

Figure 1.24 Photograph of circularly notched cantilever

The circularly notched microcantilever, a macro counterpart of it shown in Figure 1.24, is actually a paddle configuration whose thin/flexible portion is a circularly notched neck, formed of two semicircular cutouts, each or radius r. This microcantilever (also studied by Lobontiu and Garcia [5] and Lobontiu [7]) is formed of two segments, one of constant cross-section and the other one of variable width, and therefore variable cross-section. The geometry of a constant-thickness circularly notched microcantilever is shown in a top view in Figure 1.25.

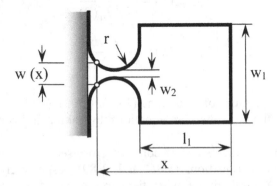

Figure 1.25 Top view of circularly notched microcantilever with geometry

The simplified bending-related distribution is considered here, namely:

$$f_b(x) = 1 - \frac{3}{2}\frac{x}{l_1 + 2r} + \frac{1}{2}\frac{x^3}{(l_1 + 2r)^3} \quad (1.101)$$

The bending resonant frequency can be expressed as:

$$\omega_b = 0.515t\sqrt{\frac{EA}{\rho B}} \quad (1.102)$$

where:

$$A = 4l_1^3 w_1 + l_1^2 r(10.3r + 24w_2) + r^3(16.88r + 32w_2) \\ + l_1 r^2(20.6r + 48w_2) \quad (1.103)$$

$$B = w_1 l_1^3(l_1^4 + 14l_1^3 r + 84l_1^2 r^2 + 280l_1 r^3 + 560r^4) \\ + l_1^2 r^5(43.46r + 610.91w_1 + 61.1w_2) + r^7(83.73r + 128w_2) \quad (1.104) \\ + l_1 r^6(120.56r + 271.51w_1 + 176.48w_2)$$

To find the torsional resonant frequency, the following simplified distribution function is used:

$$f_t(x) = 1 - \frac{x}{l_1 + 2r} \quad (1.105)$$

The torsional resonant frequency is:

$$\omega_t = 3.464t\sqrt{\frac{G(0.86r^2 + l_1 w_1 + 2rw_2)}{\rho C}} \quad (1.106)$$

with:

$$C = l_1 w_1(l_1^2 + 6l_1 r + 12r^2)(w_1^2 + t^2) \\ + r^3(4.84r^3 + 4.22rt^2 + 12.32r^2 w_2 + 8t^2 w_2 + 12.66rw_2^2 + 8w_2^3) \quad (1.107)$$

1.3.2.1.4 Constant rectangular cross-section cantilever with hole

The example of a constant cross-section cantilever with a hole in it is discussed now, and the top view of this configuration is sketched in Figure 1.26. The purpose of perforating holes in beams and plates of MEMS is twofold: one objective addresses the problem of reducing the squeeze-film damping effects in members that vibrate against the substrate, whereby holes offer the otherwise-entrapped fluid (mainly gas) an escape way, which further contributes to reducing damping; another reason consists in the possibility of modifying the stiffness and mass properties of a resonator having a prescribed shape by means of holes of various dimensions and locations. The simplest case is analyzed here, namely where a hole is perforated along the longitudinal symmetry axis of a constant rectangular cross-section microcantilever, as shown in Figure 1.26. The hole is offset by l_1 from the free edge, and the hole modifies both the mass and stiffness properties of the original constant rectangular cross-section cantilever, and therefore its bending resonant frequency. In bending, modification of the moment of inertia (which is variable along a length equal to the hole's diameter) will generate the alteration of the bending stiffness, such that the equivalent cantilever beam is made up of three different segments, as indicated in Figure 1.27. Consequently, this cantilever is a serially compounded one, and the general equation, Equation (1.69),

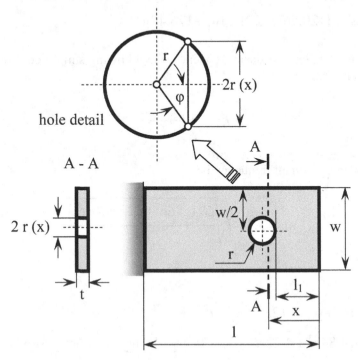

Figure 1.26 Top and section views of a rectangular cantilever with hole

which gives the bending resonant frequency, is valid. The bending-related distribution functions of the three intervals of Figure 1.27 are determined according to the methodology presented in a previous example, and is not retaken here in detail.

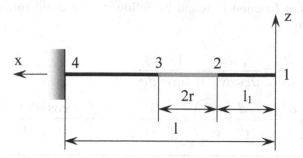

Figure 1.27 Equivalent three-segment cantilever

The hole detail of Figure 1.26 indicates that the chord that is set at a distance x can be expressed as:

$$r(x) = r \sin \varphi \qquad (1.108)$$

whereas x is:

$$x = l_1 + r(1 - \cos \varphi) \qquad (1.109)$$

and therefore:

$$dx = r \sin \varphi \, d\varphi \qquad (1.110)$$

The moment of inertia of the cantilever cross-section that is located at a distance x is:

$$I_y(x) = \frac{w(x)t^3}{12} = \frac{(w - 2r \sin \varphi)t^3}{12} \qquad (1.111)$$

whereas the cross-sectional area is:

$$A(x) = w(x)t = (w - 2r \sin \varphi)t \qquad (1.112)$$

All these relationships serve at calculating the bending distribution functions and the corresponding resonant frequencies in terms of the angular variable φ. They are quite complex, and the simpler approach in which the distribution function of a constant cross-section is used here. The distribution function corresponding to the two constant cross-section segments is the one of Equation (1.37), whereas the distribution function of the perforation segment is determined by using x of Equation (1.109). The second derivative of this last distribution function is found by following the chain rule of differentiation, namely:

$$\begin{cases} \dfrac{df_b}{dx} = \dfrac{df_b}{d\varphi}\dfrac{d\varphi}{dx} = \dfrac{1}{r\sin\varphi}\dfrac{df_b}{d\varphi} \\[3mm] \dfrac{d^2 f_b}{dx^2} = \dfrac{d}{dx}\left(\dfrac{df_b}{dx}\right) = \dfrac{1}{r^2\sin^2\varphi}\left(\dfrac{d^2 f_b}{dx^2} - \dfrac{\cos\varphi}{\sin\varphi}\dfrac{df_b}{d\varphi}\right) \end{cases} \qquad (1.113)$$

These equations are needed to carry out the integrations implied by the general formula giving the bending resonant frequency, and whereby the polar coordinate φ is used over the second segment (the one with the hole). In doing so, the resonant frequency becomes:

$$\omega_b = 23.66t\sqrt{\dfrac{EA_b}{\rho B_b}} \qquad (1.114)$$

with:

$$\begin{cases} A_b = 4l^3 w - 3\pi r^2\left(4l_1^2 + 8l_1 r + 5r^2\right) \\[2mm] B_b = 35\pi r^2[-256l^6 + 768l^5\left(l_1 + r\right) - 144l^4\left(4l_1^2 + 8l_1 r + 5r^2\right) \\[2mm] \qquad -64l^3\left(l_1 + r\right)\left(4l_1^2 + 8l_1 r + 7r^2\right) + 48l^2(8l_1^4 + 32l_1^3 r + 60l_1^2 r^2 \quad (1.115) \\[2mm] \qquad +56l_1 r^3 + 21r^4) - (64l_1^6 + 384l_1^5 r + 1200l_1^4 r^2 + 2240l_1^3 r^3 \\[2mm] \qquad +2520l_1^2 r^4 + 1584l_1 r^5 + 429r^6)] + 21121l^7 w \end{cases}$$

It can be checked that Equation (1.114), together with Equation (1.115), reduce to the first Equation (1.42), which formulates the bending resonant frequency of a constant cross-section cantilever (with no hole in it) when $r \to 0$.

Example 1.10

Compare the bending resonant frequencies of two similar rectangular cross-section microcantilevers, one with a hole in it and the other without a hole Consider $l = 100$ μm and $w = 10$ μm.

Solution:

By using the first Equation (1.42) in conjunction with Equations (1.114) and (1.115), the ratio of the bending resonant frequency of a constant rectangular cross-section microcantilever to the frequency of a similar design with a hole in it can be formed (the asterisk denotes the cantilever without a hole in it). Figure 1.28 plots this resonant frequency ratio as a function of two non-dimensional parameters that refer to the hole radius and longitudinal position.

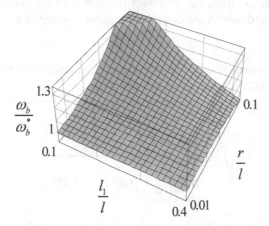

Figure 1.28 Comparison between the constant cross-section cantilever without a hole and the similar cantilever with a hole

Figure 1.28 indicates that increasing the hole radius increases the resonant frequency when the hole is placed relatively close to the free tip, whereas the opposite can be noticed when the hole is closer to the root. For the same hole radius, the bending resonant frequency is higher when the hole moves closer to the free end. Another numerical simulation looked directly at the influence of the radius dimension and hole position, as illustrated by the plot of Figure 1.29, in which the trends highlighted previously are noticed.

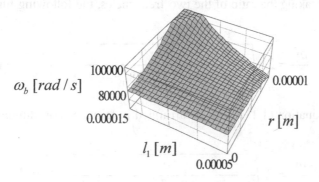

Figure 1.29 Influence of hole position and radius on the bending resonant frequency of a cantilever with a hole in it

The following numerical parameters have been used for the simulation shown in Figure 1.29: $E = 165$ GPa, $\rho = 2300$ kg/m^3 (values corresponding to polysilicon), $l = 100$ μm, $w = 10$ μm, and $t = 0.5$ μm.

Example 1.11

Calculate the bending resonant frequency of a rectangular cross-section cantilever with a hole in it, as the one of Figure 1.26 when ignoring the change in stiffness. Compare this resonant frequency to the one obtained when both the stiffness and inertia are considered altered by the hole.

Solution:

In the case in which stiffness is considered constant, the bending resonant frequency is calculated by means of the equation:

$$\left(\omega_b'\right)^2 = \frac{EI_{y1}\int_0^l \left(\frac{d^2 f_b(x)}{dx^2}\right)^2 dx}{\rho \sum_{i=1}^{3}\left(\int_{l_i} A_{y,i}(x) f_{b,i}(x)^2 dx\right)}$$ (1.116)

After carrying out the calculations of Equation (1.116), the new resonant frequency becomes:

$$\omega_b' = 47.33lt\sqrt{\frac{lw}{B_b}}$$ (1.117)

with B_b defined in Equation (1.115). Becaue the stiffness decrease through the very existence of the hole has been neglected, it is clear that Equation (1.117) yields a resonant frequency that is higher than the one produced by Equation (1.114) where both stiffness and inertia variations have been taken into account. By taking the ratio of the two frequencies, the following function is obtained:

$$\frac{\omega_b'}{\omega_b} = 2l\sqrt{\frac{lw}{4l^3 w - 3\pi r^2 \left(4l_1^2 + 8l_1 r + 5r^2\right)}}$$ (1.118)

Alternatively, Equation (1.118) can be arranged in terms of non-dimensional parameters as:

$$\frac{\omega_b'}{\omega_b} = 2\sqrt{\frac{c_3}{4c_3 - 3\pi c_2^2 \left(4c_1^2 + 8c_1 c_2 + 5c_2^2\right)}}$$ (1.119)

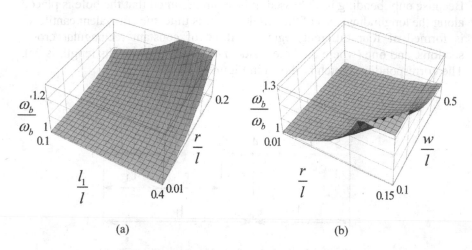

Figure 1.30 Bending resonant frequency comparison between prediction with stiffness neglected and considered for a rectangular cross-section cantilever with hole: (a) $c_3 = 0.25$; (b) $c_1 = 0.5$

where $c_1 = l_1/l$, $c_2 = r/l$, and $c_3 = w/l$. Figure 1.30 (a) shows the 3D plots of this ratio when $c_3 = 0.25$, and Figure 1.30 (b) plots the same ratio for $c_1 = 0.5$. As the figures indicate, the errors between the two models' predictions are quite negligible when the hole is placed close to the free tip, and the width-to-length ratio is larger. For the other parameter range extremities, however, the predictions can differ by as much as 30%.

1.3.2.1.5 Paddle microcantilever with hole

A paddle microcantilever with a circular perforation in the paddle region is now analyzed, by following the procedure exposed previously, and Figure 1.31 is the top view of such a paddle microcantilever.

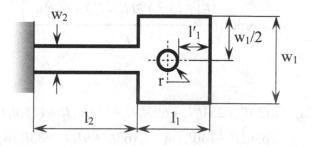

Figure 1.31 Top view of microcantilever with a hole in it and geometry

Because only bending is addressed, it is again assumed that the hole is placed along the longitudinal axis of the cantilever. This time, the equivalent cantilever is formed of four different segments, three of constant rectangular cross-sections, and one (where the hole resides) of variable width (its length is $2r$). The corresponding sketch is shown in Figure 1.32.

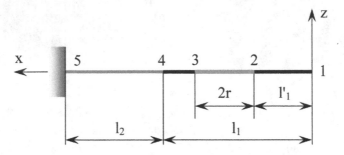

Figure 1.32 Equivalent four-segment cantilever

The resonant frequency in this case is calculated as:

$$\omega_b^2 = \frac{E\sum_{i=1}^{4}\int_{l_i} I_{yi}(x)\left(\frac{d^2 f_{bi}(x)}{dx^2}\right)^2 dx}{\rho\sum_{i=1}^{4}\int_{l_i} A_i(x) f_{bi}(x)^2 dx} \qquad (1.120)$$

The final expression of ω_b is quite complex, even when the distribution function of a constant cross-section cantilever is used instead of the exact one, and therefore the explicit ω_b is not included here. When taking $r \to 0$ (and therefore when the structure changes to a regular paddle cantilever without a hole in it), the resulting bending frequency becomes:

$$\omega_b = \sqrt{35}t\sqrt{\frac{E\left[l_2\left(3l_1^2 + 3l_1 l_2 + l_2^2\right)w_2 + l_1^3 w_1\right]}{\rho A_b}} \qquad (1.121)$$

with:

$$\begin{aligned} A_b = &\, l_1(33l_1^6 + 231l_1^5 l_2 + 693l_1^4 l_2^2 + 1155l_1^3 l_2^3 + 1155l_1^2 l_2^4 \\ &+ 630l_1 l_2^5 + 140l_2^6)w_1 + l_2^5(63l_1^2 + 91l_1 l_2 + 33l_2^2)w_2 \end{aligned} \qquad (1.122)$$

Equations (1.121) and (1.122) are slightly different from Equations (1.91) and (1.92), respectively, which expressed the bending resonant frequency of a regular paddle cantilever. It should be remembered, however, that Equations (1.91) and (1.92) have been derived by using a distribution function that varies on the two segments, whereas Equations (1.121) and (1.122) correspond to a unique distribution function; consequently, differences between the two formulations should be expected.

Example 1.12

Compare the bending resonant frequency of a paddle microcantilever with perforation with that of a similar paddle cantilever without a hole in it. Consider the particular case where $l_1 = l_2$ and $w_2 = w_1/2$.

Solution:

For the given particular conditions, the bending resonant frequency becomes:

$$\omega_b = 41\frac{t}{l_1^2}\sqrt{\frac{E}{\rho}}f(c_1,c_2,c_3) \qquad (1.123)$$

where $f(c_1, c_2, c_3)$, which is not explicitly given here, depends on the following non-dimensional parameters:

$$\begin{cases} c_1 = \dfrac{l_1'}{l_1} \\[2mm] c_2 = \dfrac{r}{l_1} \\[2mm] c_3 = \dfrac{w_1}{l_1} \end{cases} \qquad (1.124)$$

When there is no hole in the microcantilever $c_2 = 0$, and therefore Equation (1.123) simplifies to:

$$\omega_b^* = 0.19\frac{t}{l_1^2}\sqrt{\frac{E}{\rho}} \qquad (1.125)$$

Two 3D plots are drawn, one for the case where $c_3 = 0.25$, and the other one for $c_1 = 0.5$ (illustrated in Figure 1.33).

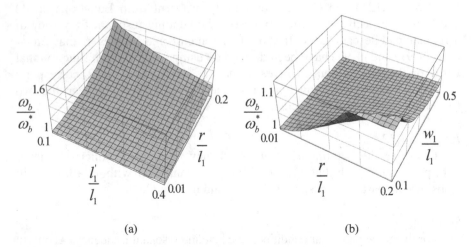

(a)   (b)

Figure 1.33 Bending resonant frequency comparison between paddle microcantilevers with and without a hole: (a) $c_3 = 0.25$; (b) $c_1 = 0.5$

It can be seen again that, relative to a paddle cantilever without a hole in it, the bending resonant frequency of a similar structure with perforation is sensitive to the hole location (better sensed are holes placed towards the free tip) and dimensions (obviously, larger holes generate larger changes in the resonant frequency; see Figure 1.33). At the same time, relatively narrower tip segments (small width-to-length ratio c_3) are capable of better sensing the presence of a perforation of specified position and dimension, as Figure 1.33 (b) indicates.

1.3.2.1.6 Hole-array microcantilever

To diminish the effect of air (fluid) damping in the case of plates that vibrate (move) about a direction perpendicular to the substrate, several holes are perforated into the moving member to enable the air, otherwise highly compressed between the plate and substrate, to escape and thus reduce the squeeze-film damping. In the majority of cases, the plates can be considered solid, the holes not altering the rigidity significantly. However, in cases in which the member thickness is relatively small, the array of holes can substantially modify the structural rigidity. At the same time, the mass is reduced by perforation, and therefore the relevant resonant frequencies are expected to modify. The example of a rectangular microcantilever with an array of identical holes perforated in it is studied here, and the bending and torsional resonant frequencies are calculated for a generic case. The assumption is made that the holes are regularly distributed over the microcantilever's area, as shown in Figure 1.34, with a constant pitch distance p between them and spaced from all the edges at a distance a.

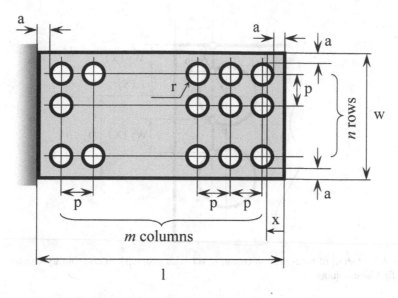

Figure 1.34 Front view of a hole-array microcantilever with geometry

There are two different strips along the *x*-direction (indicated in Figure 1.34). One contains the full width (when a line parallel to the *w* dimension is drawn) and there are *m* + 1 such areas (taking into account there are *m* hole columns). The other strip encompasses the hole zones, there are *m* such zones, and the width varies across them because the hole width varies, as also discussed in the microcantilever with one hole topic. Figure 1.35 shows the sequence of the 2*m* + 1 segments. As a series-connection resultant, the microcantilever of Figure 1.34 can be symbolized as in Figure 1.35, with 2*m* + 1 segments (*m* segments correspond to the hole regions, and *m* + 1 segments represent full-width regions).

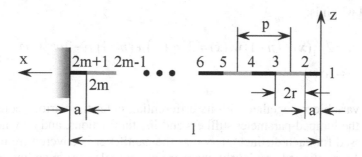

Figure 1.35 Equivalent cantilever with 2*m* + 1 segments (black, solid segment; grey, hole segment)

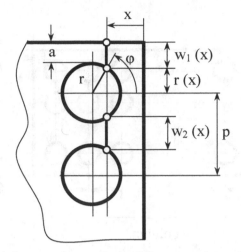

Figure 1.36 Detail of hole-array microcantilever with geometry in an arbitrary cross-section in the first hole column

In both bending (about the y-axis, which is parallel to the w dimension of Figure 1.34) and torsion, the widths $w_1(x)$ and $w_2(x)$ are important, which are indicated in Figure 1.36, because they add up to the total width $w(x)$, which is the sum of all the widths along a line parallel to w taken at a distance x from the free end through the hole strip. The aim here, as the case was with the one-hole microcantilever studied previously, is to use the polar variable instead of the Cartesian one x. The variable radius $r\,(x)$ is expressed in Equation (1.108). Figure 1.36 shows that:

$$\begin{cases} w_1(x) = a + r - r\sin\varphi \\ w_2(x) = p - 2r\sin\varphi \end{cases} \tag{1.126}$$

The total width is:

$$w(x) = 2w_1(x) + (n-1)w_2(x) = 2(a+r) + (n-1)p - 2nr\sin\varphi$$
$$= w(\varphi) \tag{1.127}$$

The variable x, together with its differential, enter the various equations defining the lumped-parameter stiffness and inertia fractions, and they need to be expressed for each distinct interval as the cantilever is covered from $x = 0$ to $x = 1$. As mentioned previously, there are m intervals corresponding to hole regions and $m + 1$ intervals for between-holes regions, overall there are $2\,m + 1$ intervals. For the hole intervals, the polar coordinate needs to be used, whereas for the between-holes regions, the Cartesian x coordinate can be used. For the j-th hole interval, the abscissa x and its differential are:

$$\begin{cases} x = a + r + (j-1)p - r\cos\varphi \\ dx = r\sin\varphi d\varphi \end{cases} \tag{1.128}$$

In bending, the following distribution function needs to be used over the j-th interval that belongs to a hole region:

$$f_{bj}(\varphi) = 1 - \frac{3}{2}\frac{a + r + (j-1)p - r\cos\varphi}{l}$$
$$+ \frac{1}{2}\frac{\left[a + r + (j-1)p - r\cos\varphi\right]^3}{l^3} \tag{1.129}$$

whereas for a non-hole area, the distribution is the regular, simplified one of Equation (1.37).

By considering the hole array microcantilever is formed of $2m + 1$ intervals, as explained previously, the lumped-parameter stiffness and mass can be determined. The equivalent stiffness is expressed as:

$$k_{b,e} = E\left\{ S_{1b}^{(k)} + S_{2b}^{(k)} + I_y\left[\int_0^a \left(\frac{d^2 f_b(x)}{dx^2}\right)^2 dx + \int_{a+2r+(m-1)p}^l \left(\frac{d^2 f_b(x)}{dx^2}\right)^2 dx \right] \right\} \tag{1.130}$$

where:

$$S_{1b}^{(k)} = r\sum_{j=1}^m \int_0^\pi I_y(\varphi)\left[\frac{d^2 f_{bj}(\varphi)}{dx^2}\right]^2 \sin\varphi d\varphi \tag{1.131}$$

and:

$$S_{2b}^{(k)} = I_y\sum_{j=2}^m \int_{a+2r+(j-2)p}^{a+(j-1)p} \left[\frac{d^2 f_b(x)}{dx^2}\right]^2 dx \tag{1.132}$$

The derivative of Equation (1.131) is calculated by means of Equation (1.113), which gives the coordinate transformation relationship. The variable moment of inertia of Equation (1.131) is:

$$I_y(\varphi) = \frac{w(\varphi)t^3}{12} \tag{1.133}$$

with $w(\varphi)$ given in Equation (1.127).

Similarly, the lumped-parameter, equivalent mass is expressed as:

$$m_{b,e} = \rho t \left\{ S_{1b}^{(m)} + S_{2b}^{(m)} + w \left[\int_0^a f_b(x)^2 \, dx + \int_{a+2r+(m-1)p}^l f_b(x)^2 \, dx \right] \right\} \quad (1.134)$$

where:

$$S_{1b}^{(m)} = r \sum_{j=1}^m \int_0^\pi w(\varphi) f_{bj}(\varphi)^2 \sin \varphi \, d\varphi \quad (1.135)$$

and:

$$S_{2b}^{(m)} = w \sum_{j=2}^m \int_{a+2r+(j-2)p}^{a+(j-1)p} f_b(x)^2 \, dx \quad (1.136)$$

The distribution function in torsion being employed over the j-th interval and corresponding to a hole region is:

$$f_{tj}(\varphi) = 1 - \frac{a+r+(j-1)p - r\cos\varphi}{l} \quad (1.137)$$

For a non-hole area, the distribution given in Equation (1.51) is used. Similar to bending, the torsion resonant frequency is expressed by means of the equivalent, lumped-parameter stiffness and mechanical moment of inertia. The equivalent stiffness is:

$$k_{t,e} = G \left\{ S_{1t}^{(k)} + S_{2t}^{(k)} + I_t \left[\int_0^a \left(\frac{df_t(x)}{dx} \right)^2 dx + \int_{a+2r+(m-1)p}^l \left(\frac{df_t(x)}{dx} \right)^2 dx \right] \right\} \quad (1.138)$$

where:

$$S_{1t}^{(k)} = r \sum_{j=1}^m \int_0^\pi I_t(\varphi) \left[\frac{df_{tj}(\varphi)}{dx} \right]^2 \sin \varphi \, d\varphi \quad (1.139)$$

and:

$$S_{2t}^{(k)} = I_t \sum_{j=2}^m \int_{a+2r+(j-2)p}^{a+(j-1)p} \left[\frac{df_t(x)}{dx} \right]^2 dx \quad (1.140)$$

Again, the derivative of Equation (1.139) is determined by means of Equation (1.113), which provides the x-φ coordinate transformation. For a very thin microcantilever, the variable mechanical moment of inertia of Equation (1.139) is:

$$I_t(\varphi) = \frac{w(\varphi)t^3}{3} \tag{1.141}$$

with $w(\varphi)$ given in Equation (1.127).

The lumped-parameter, equivalent moment of inertia corresponding to torsion is:

$$J_{t,e} = \frac{\rho t}{12}\left\{ S_{1t}^{(m)} + S_{2t}^{(m)} + w\left(w^2 + t^2\right)\left[\int_0^a f_t(x)^2\,dx + \int_{a+2r+(m-1)p}^l f_t(x)^2\,dx\right]\right\} \tag{1.142}$$

where:

$$S_{1t}^{(m)} = r\sum_{j=1}^m \int_0^\pi w(\varphi)\left[w^2(\varphi) + t^2\right]f_{tj}(\varphi)^2 \sin\varphi\,d\varphi \tag{1.143}$$

and:

$$S_{2t}^{(m)} = w\left(w^2 + t^2\right)\sum_{j=2}^m \int_{a+2r+(j-2)p}^{a+(j-1)p} f_t(x)^2\,dx \tag{1.144}$$

Example 1.13

A rectangular cross-section microcantilever 2 μm thick has $m \times n$ holes ($m = 4$, $n = 2$) each 2 μm in diameter, with a pitch $p = 4$ μm and an edge spacing $a = 5$ μm.

(a) Determine the length l and width w of this microcantilever.

(b) Find its bending and resonant frequencies and compare them to the ones corresponding to a similar blank microcantilever. Consider $E = 155$ GPa, $\mu = 0.25$, and $\rho = 2400$ kg/m³.

Solution:

(a) For a configuration as that of Figure 1.34, with regular disposition of the holes in a rectangular array, the length and width of the microcantilever are generically expressed as:

$$\begin{cases} l = 2r + 2a + (m-1)p \\ w = 2r + 2a + (n-1)p \end{cases} \tag{1.145}$$

For the numerical values of this example, $l = 24$ µm and $w = 16$ µm.

(b) Table 1.1 contains the numerical data for both the original micro-cantilever (with no holes in it) and the one perforated with $4 \times 2 = 8$ holes.

Table 1.1 Lumped-parameter model characteristics of original and hole microcantilever

	Bending			Torsion		
	Original	Altered	% Change	Original	Altered	% Change
Stiffness	358.80	338.71	5.6	1.1×10^{-7}	1.03×10^{-7}	6.4
Inertia	4.3×10^{-13}	4.15×10^{-13}	3.5	1.33×10^{-23}	1.15×10^{-23}	13.5
Resonant Frequency	2.83×10^{7}	2.85×10^{7}	0.7	9.1×10^{7}	9.4×10^{7}	3.2

In Table 1.1, all amounts are in SI units, and the resonant frequencies are in Hertz (Hz). It can be seen that torsion is more sensitive to the perforations as its resonant frequency increases by 3.2% compared to the bending resonant frequency, which only increases by 0.7%.

1.3.2.2 Non-Homogeneous Configurations

Non-homogeneous multi-profile microcantilevers consist of two or more segments, each being defined as a single-profile (of either constant or variable cross-section, defined by a unique geometric curve/profile), and each having its own material properties. For microcantilevers, one end is fixed to the substrate while the other one is free. A generic formulation giving the transfer functions as well as the bending and torsion resonant frequencies is developed here, as well as its application to a paddle configuration. Designs are possible in this category where the cross-sections of the component segments are identical, and, definitely, cases in which the segments have different cross-sections can be considered.

1.3.2.2.1 Generic formulation

By following the procedure that has been applied for homogeneous multi-profile microcantilevers, the bending resonant frequency of a series micro-cantilever which is formed of n segments of different geometries and material properties is:

$$\omega_b^2 = \frac{\sum_{i=1}^{n}\left\{ E_i \int_{l_i} I_{yi}(x)\left[\frac{d^2 f_{bi}(x)}{dx^2}\right]^2 dx \right\}}{\sum_{i=1}^{n}\left[\rho_i \int_{l_i} A_i(x) f_{bi}(x)^2 dx \right]} \qquad (1.146)$$

Determining the distribution function for each of the n intervals is done by taking the ratio of the deflection at an arbitrary point on a given interval to the maximum deflection (recorded at the free end). It can be shown by using basic mechanics o materials that the bending distribution function on an interval i is of the form:

$$f_{bi}(x) = a_{bi} + b_{bi}x + c_{bi}x^3 \qquad (1.147)$$

where:

$$\begin{cases} a_{bi} = \dfrac{a_{bi}'}{u_z} \\[2mm] b_{bi} = \dfrac{b_{bi}'}{u_z} \\[2mm] c_{bi} = \dfrac{c_{bi}'}{u_z} \end{cases} \qquad (1.148)$$

with:

$$\begin{cases} a_{bi}' = \dfrac{F_z}{3}\left[\dfrac{S_i^3}{(EI_y)_i} + \sum_{j=i+1}^{n} \dfrac{S_j^3 - S_{j-1}^3}{(EI_y)_j} \right] \\[4mm] b_{bi}' = -\dfrac{F_z}{2}\left[\dfrac{S_i^2}{(EI_y)_i} + \sum_{j=i+1}^{n} \dfrac{S_j^2 - S_{j-1}^2}{(EI_y)_j} \right] \\[4mm] c_{bi}' = \dfrac{F_z}{6(EI_y)_i} \end{cases} \qquad (1.149)$$

where u_z and F_z are the free end deflection and force and:

$$S_j = \sum_{i=1}^{j} l_i \qquad (1.150)$$

It can also be shown that:

$$u_z = \dot{a}_{b1} \tag{1.151}$$

The torsion resonant frequency of a non-homogeneous rectangular cross-section microcantilever is:

$$\omega_t^2 = \frac{12 \sum\limits_{i=1}^{n} \left\{ G_i \int\limits_{l_i} I_{ti}(x) \left[\frac{df_{ti}(x)}{dx} \right]^2 dx \right\}}{\sum\limits_{i=1}^{n} \left[\rho_i \int\limits_{l_i} w_i(x) t_i(x) \left[w_i^2(x) + t_i^2(x) \right] f_{ti}(x)^2 dx \right]} \tag{1.152}$$

By following a reasoning similar to the one applied to bending, it can be shown that for a generic interval i, the torsion-related distribution function is a first degree polynomial of the form:

$$f_{ti}(x) = a_{ti} + b_{ti}x \tag{1.153}$$

where:

$$\begin{cases} a_{ti} = \dfrac{a'_{ti}}{\theta_x} \\[4mm] b_{ti} = \dfrac{b'_{ti}}{\theta_x} \end{cases} \tag{1.154}$$

with:

$$\begin{cases} a'_{ti} = M_x \left[\dfrac{S_i}{(GI_t)_i} + \sum\limits_{j=i+1}^{n} \dfrac{l_j}{(GI_t)_j} \right] \\[4mm] b'_{ti} = -\dfrac{M_x}{(GI_t)_i} \end{cases} \tag{1.155}$$

M_x represents the x-axis torsional moment applied at the cantilever's free end, and θ_x is the resulting rotation at the same location. It can also be shown that:

$$\theta_x = \dot{a}_{t1} \tag{1.156}$$

The bending- and torsion-related distribution functions are interval-dependent, as shown previously, but a single distribution function can be used under a simplifying assumption, which is the one corresponding to a homogeneous, constant cross-section microcantilever having its length the sum of all components lengths.

In bending, formulating the individual distribution functions can be quite involved, as illustrated in the next example, which considers a two-segment microcantilever.

Example 1.14

Determine the bending distribution functions for a two-segment non-homogeneous microcantilever. Consider the case in which the segments have different cross-sections, as well as the case in which the cross-sections are geometrically identical. Compare the distribution functions with the simplified form of the distribution function corresponding to a single-segment, constant cross-section microcantilever.

Solution:

This example is similar to Example 1.7, which analyzed the homogeneous counterpart to this case, and therefore the procedure that has been fully explained in Example 1.7 is applied here. In doing so, the ratio of the deflection at an arbitrary location on the first segment (the one at the free end) and the tip (maximum) deflection is obtained, which is actually the distribution function corresponding to this segment (1–2, as shown in Figure 1.19), namely:

$$f'_{b1} = 1 + b'_1 x + c'_1 x^3 \qquad (1.157)$$

with:

$$\begin{cases} b'_1 = -\dfrac{3}{2} \dfrac{E_2 I_{y2} l_1^2 + E_1 I_{y1} l_2 \left(2l_1 + l_2\right)}{E_2 I_{y2} l_1^3 + E_1 I_{y1} l_2 \left(3l_1^2 + 3l_1 l_2 + l_2^2\right)} \\[4mm] c'_1 = \dfrac{1}{2} \dfrac{E_2 I_{y2}}{E_2 I_{y2} l_1^3 + E_1 I_{y1} l_2 \left(3l_1^2 + 3l_1 l_2 + l_2^2\right)} \end{cases} \qquad (1.158)$$

For a homogeneous two-segment microcantilever, when $E_1 = E_2$, Equation (1.158) simplifies to Equation (1.79) in Example 1.7, which gave the bending distribution function for a homogeneous two-segment cantilever.

Similarly, the following distribution function is obtained for the second (root) segment:

$$f'_{b2} = a'_2 + b'_2 x + c'_2 x^3 \qquad (1.159)$$

with:

$$
\begin{cases}
a_2' = \dfrac{E_1 I_{y1}\left(l_1+l_2\right)^3}{E_2 I_{y2}l_1^3 + E_1 I_{y1}l_2\left(3l_1^2 + 3l_1 l_2 + l_2^2\right)} \\[4ex]
b_2' = -\dfrac{3}{2}\dfrac{E_1 I_{y1}\left(l_1+l_2\right)^2}{E_2 I_{y2}l_1^3 + E_1 I_{y1}l_2\left(3l_1^2 + 3l_1 l_2 + l_2^2\right)} \\[4ex]
c_2' = \dfrac{1}{2}\dfrac{E_1 I_{y1}}{E_2 I_{y2}l_1^3 + E_1 I_{y1}l_2\left(3l_1^2 + 3l_1 l_2 + l_2^2\right)}
\end{cases}
\tag{1.160}
$$

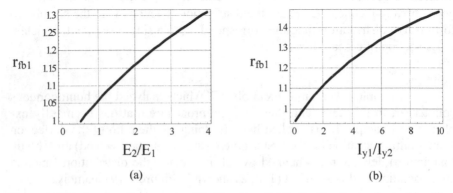

(a)   (b)

Figure 1.37 Comparison between simplified bending distribution function and actual bending distribution functions for the first (free end) component of a two-segment non-homogeneous microcantilever

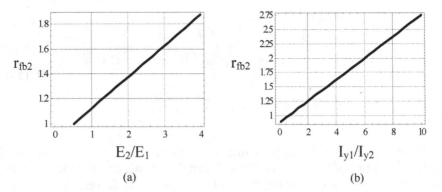

(a)   (b)

Figure 1.38 Comparison between simplified bending distribution function and actual bending distribution functions for the second (root) component of a two-segment non-homogeneous microcantilever

When $E_1 = E_2$, Equation (1.160) simplifies to Equation (1.81) of Example 1.7, as expected. For the particular case in which the two segments have identical cross-sections, and are made up of the same material, Equations (1.157) through (1.160) yield, for $l = l_1 + l_2$, the distribution function of Equation (1.37), as expected. The following bending distribution functions ratios are analyzed next:

$$\begin{cases} r_{fb1} = \dfrac{f_b(x)}{f_{b1}(x)} \\[3mm] r_{fb2} = \dfrac{f_b(x)}{f_{b2}(x)} \end{cases} \tag{1.161}$$

where the numerators in Equation (1.161) are expressed in Equations (1.157) through (1.160), and $f_b(x)$, the distribution function of a homogeneous, constant cross-section microcantilever of length $l_1 + l_2$, is given in Equation (1.37). Figures 1.37 and 1.38 show the variation of the two ratios defined in Equation (1.61). Figures 1.37 (a) and 1.38 (a) have been drawn for $I_{y2}/I_{y1} = 2$ and $x/l = 0.5$, while Figures 1.37 (b) and 1.38 (b) are plotted for $E_2/E_1 = 1.5$ and $x/l = 0.5$. As all four figures indicate, the predictions by the simplified bending distribution function are always larger than the ones by the actual distribution functions. The differences between the two models are quite large—up to a factor of 2.75, as shown in Figure 1.38 (b).

The next example will attempt to elucidate whether these differences are equally substantial when assessing the bending resonant frequency of the same structure.

Example 1.15

For the structure of Example 1.14, calculate the bending resonant frequency by using the actual distribution functions and also by using the simplified distribution function. Compare the two results.

Solution:

When using the simplified distribution function (Equation (1.37)), the bending resonant frequency is expressed as:

$$\omega_{b,s}^2 = \frac{E_1 I_{y1} \displaystyle\int_0^{l_1} \left[\dfrac{d^2 f_b(x)}{dx^2}\right]^2 dx + E_2 I_{y2} \displaystyle\int_{l_1}^{l_1+l_2} \left[\dfrac{d^2 f_b(x)}{dx^2}\right]^2 dx}{\rho_1 A_1 \displaystyle\int_0^{l_1} f_b(x)^2 dx + \rho_2 A_2 \displaystyle\int_{l_1}^{l_1+l_2} f_b(x)^2 dx} \tag{1.162}$$

where the subscript letter s indicates the simplified model.

The bending resonant frequency can also be expressed in terms of the actual distribution functions (Equations (1.157) through (1.160)) as:

$$\omega_b^2 = \frac{E_1 I_{y1} \int\limits_0^{l_1} \left[\frac{d^2 f_{b1}'(x)}{dx^2} \right]^2 dx + E_2 I_{y2} \int\limits_{l_1}^{l_1+l_2} \left[\frac{d^2 f_{b2}'(x)}{dx^2} \right]^2 dx}{\rho_1 A_1 \int\limits_0^{l_1} \left[f_{b1}'(x) \right]^2 dx + \rho_2 A_2 \int\limits_{l_1}^{l_1+l_2} \left[f_{b2}'(x) \right]^2 dx} \qquad (1.163)$$

Equations (1.162) and (1.163) enable formulating the bending resonant frequency ratio $\omega_{b,s}/\omega_b$ in terms of the following non-dimensional parameters only: E_2/E_1, ρ_2/ρ_1, l_2/l_1, and w_2/w_1. Figure 1.39 plots the frequency ratio as a function of the elasticity modulii and mass density ratios considering $l_2 = l_1$ and $w_2 = w_1$, whereas Figure 1.40 plots the frequency ratio in terms of length and width ratios when $E_2 = 1.5 \, E_1$ and $\rho_2 = \rho_1$. As Figures 1.39 and 1.40 show,

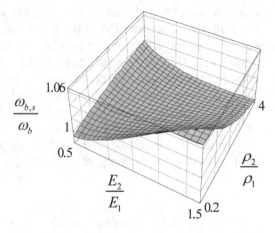

Figure 1.39 Bending resonant frequency ratio as a function of elastic modulus and mass density ratios

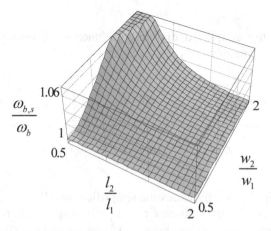

Figure 1.40 Bending resonant frequency ratio as a function of length and width ratios

the differences between the two models' predictions are less than 6% for large parameter ranges. Larger differences are recorded when the root segment is considerably shorter and wider than the free one, as well as when the root segment is much stiffer and less dense than the tip (free) segment.

As this example demonstrates, using the simplified bending distribution function of a homogeneous, constant cross-section microcantilever, instead of the individual distribution functions corresponding to each segment, is sufficiently accurate.

1.3.2.2.2 Paddle microcantilever

A paddle microcantilever is now considered as the one whose top view is sketched in Figure 1.21, and let us assume the two segments are made up of different materials. When using the approach with actual distribution functions, two such functions are needed for both bending and torsion, because the series cantilever is made up of two segments.

By applying the generic equations (Equations (1.146) through (1.151)), the resulting bending resonant frequency is expressed as:

$$\omega_b = \sqrt{35}t\sqrt{\frac{E_1E_2w_2\left[E_1l_2\left(3l_1^2 + 3l_1l_2 + l_2^2\right)w_1 + E_2l_1^3w_2\right]}{A}} \qquad (1.164)$$

with:

$$
\begin{aligned}
A = {} & 21\rho_1E_1E_2l_1^4l_2\left(11l_1^2 + 13l_1l_2 + 5l_2^2\right)w_1w_2 + 33\rho_1E_2^2l_1^7w_2^2 \\
& + E_1^2l_2^2w_1\left[35\rho_1l_1w_1\left(12l_1^4 + 30l_1^3l_2 + 33l_1^2l_2^2 + 18l_1l_2^3 + 4l_2^4\right)\right. \\
& \left. + \rho_2l_2^3w_2\left(63l_1^2 + 91l_1l_2 + 33l_2^2\right)\right]
\end{aligned}
\qquad (1.165)
$$

When $E_1 = E_2$ and $\rho_1 = \rho_2$, Equations (1.164) and (1.165) reduce to Equations (1.91) and (1.92), which were derived for a homogeneous paddle micro-cantilever.

By using the actual torsion-related distribution functions that result from the generic Equations (1.152) through (1.156), the following resonant frequency is obtained:

$$\omega_t = 2\sqrt{3}t\sqrt{\frac{G_1G_2w_2\left(G_1l_2w_1 + G_2l_1w_2\right)}{B}} \qquad (1.166)$$

with:

$$B = 3\rho_1 G_1 G_2 l_1^2 l_2 w_1 w_2 \left(w_1^2 + t^2\right) + \rho_1 G_2^2 l_1^3 w_2^2 \left(w_1^2 + t^2\right)$$
$$+ G_1^2 l_2^2 w_1 \left[3\rho_1 l_1 w_1 \left(w_1^2 + t^2\right) + \rho_2 l_2 w_2 \left(w_2^2 + t^2\right)\right] \qquad (1.167)$$

For a homogeneous paddle cantilever, Equations (1.166) and (1.167) simplify to:

$$\omega_t^* = 2\sqrt{3}t\sqrt{\frac{Gw_2\left(l_2 w_1 + l_1 w_2\right)}{B^*}} \qquad (1.168)$$

with:

$$B^* = \rho\left[3l_1 l_2 w_1 \left(w_1^2 + t^2\right)\left(w_1 l_2 + w_2 l_1\right) + l_1^3 w_2^2 \left(w_1^2 + t^2\right)\right.$$
$$\left. + l_2^3 w_1 w_2 \left(w_2^2 + t^2\right)\right] \qquad (1.169)$$

Example 1.16

For a non-homogeneous paddle microcantilever, compare the torsional resonant frequency obtained by using the actual distribution functions of Equations (1.152) through (1.156) to the one obtained by using the simplified distribution function corresponding to a homogeneous, constant cross-section member. Consider $t = 1$ μm and $w_1 = 50$ μm.

Solution:

When the simplified distribution function of Equation (1.90) is used, the torsion resonant frequency can be expressed as:

$$\omega_{t,s} = 2\sqrt{3}t\sqrt{\frac{G_1 l_1 w_1 + G_2 l_2 w_2}{\rho_1 l_1 w_1 \left(l_1^2 + 3l_1 l_2 + 3l_2^2\right)\left(w_1^2 + t^2\right) + \rho_2 l_2^3 w_2 \left(w_2^2 + t^2\right)}} \qquad (1.170)$$

For a homogeneous paddle bridge, Equation (1.170) further simplifies to:

$$\omega_{t,s}^* = 2\sqrt{3}t\sqrt{\frac{G\left(l_1 w_1 + l_2 w_2\right)}{\rho\left[l_1 w_1 \left(l_1^2 + 3l_1 l_2 + 3l_2^2\right)\left(w_1^2 + t^2\right) + l_2^3 w_2 \left(w_2^2 + t^2\right)\right]}} \qquad (1.171)$$

Equation (1.170) is now compared to Equation (1.166), which gives the torsion resonant frequency by means of the actual distribution functions through their ratio. The non-dimensional variables G_2/G_1, ρ_2/ρ_1, l_2/l_1 and w_2/w_1 are employed to draw the 3D plots of Figure 1.41. Figure 1.41 (a) has been plotted for $l_2 = l_1$ and $w_2 = 0.5\ w_1$, whereas Figure 1.41 (b) corresponds to $G_2 = 1.5\ G_1$ and $\rho_2 = 2\rho_1$.

As Figure 1.41 (a) indicates, the errors between the two models' predictions are larger in the case where the shear modulus of the root segment is considerably smaller than the one of the paddle segment; the differences in density do not appear to be so influential though. Similarly, as indicated by Figure 1.41 (b), notable differences between the two models are in place for designs with narrow roots (small w_2 compared to w_1).

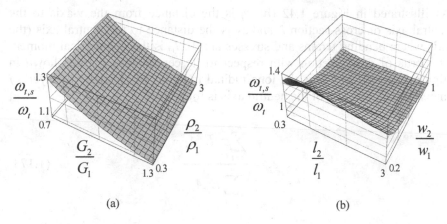

(a)   (b)

Figure 1.41 Torsion resonant frequency ratio: simplified versus actual distribution function models as a function of: (a) shear modulus and mass density ratios; (b) length and width ratios

1.3.3 Multi-Layer (Sandwich) Microcantilevers

Microcantilevers consisting of several layers are often times used in MEMS, the simplest construction being the bimorph, which is formed of a structural layer (usually thicker) and a transduction layer (thinner than the structural layer) employed for actuation or sensing. Another application, the detection of matter that deposits in a layer-like manner on homogeneous cantilevers, which results in bimorph configurations, is another important application. Although variable cross-section multi-layer cantilevers can be designed, only the constant rectangular cross-section microcantilevers will be studied in this section. It will be assumed that all layers have the same width, but two subcases will be analyzed: the simplest one, where all layers have the same length, as well as the situation in which the layers have different lengths. Again, the bending and torsion resonant frequencies will be addressed.

1.3.3.1 Equal-Length Multilayer Microcantilevers

Figure 1.42 shows two side views of a cantilever consisting of two layers only, but more than two layers can be superimposed to form a composite cross-section where all the layers are of equal lengths, in addition to having equal widths.

The equivalent rigidity of a compound cross-section formed of m layers can be found (e.g., see Lobontiu and Garcia [5]) as:

$$\left(EI_y \right)_e = \sum_{j=1}^{m} \left\{ E_j \left[I_{yj} + z_j \left(z_j - z_N \right) A_j \right] \right\}$$ (1.172)

As illustrated in Figure 1.42 (b), z_j is the distance from the y-axis to the central axis of cross-section j, and z_N is the distance to the neutral axis (the axis where bending strains and stresses are 0). I_{yj} is the geometrical moment of inertia of cross-section with respect to its central axis y_j (not shown in Figure 1.42 (b)) and E_j is the longitudinal modulus of elasticity of layer's j material. The position of the neutral axis is found as:

$$z_N = \frac{\displaystyle\sum_{j=1}^{m} z_j E_j A_j}{\displaystyle\sum_{j=1}^{m} E_j A_j}$$ (1.173)

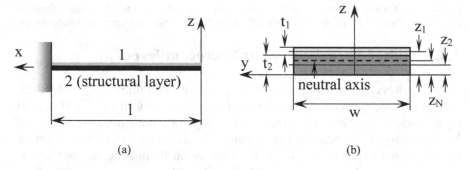

Figure 1.42 Equal-length bimorph microcantilever: (a) side view; (b) enlarged section view

The equivalent stiffness identifying the force-deflection relationship at the free end of the microcantilever is therefore:

$$k_{b,e} = \frac{3 \left(EI_y \right)_e}{l^3}$$ (1.174)

The lumped-parameter equivalent mass that is placed at the microcantilever's free end and is dynamically equivalent to the distributed-parameter system can be determined by equating the kinetic energies of the two systems, and is of the form:

$$m_{b,e} = \sum_{j=1}^{m} \left(\rho_j A_j \right) \int_0^l f_b(x)^2 \, dx = \frac{33}{140} l \sum_{j=1}^{m} \left(\rho_j A_j \right) = \frac{33}{140} m \quad (1.175)$$

where m is the total mass of the compound beam. By combining Equations (1.174) and (1.175), the bending resonant frequency is expressed as:

$$\omega_{b,e} = \sqrt{\frac{k_{b,e}}{m_{b,e}}} = 3.567 \sqrt{\frac{\left(EI_y \right)_e}{ml^3}} \quad (1.176)$$

which is very similar to the first Equation (1.42) giving the bending resonant frequency of a homogeneous, constant rectangular cross-section microcantilever.

The torsion rigidity of the compound cross-section, as shown by Lobontiu and Garcia [5], for instance, is determined as:

$$\left(GI_t \right)_e = \sum_{j=1}^{m} \left(GI_t \right)_j \quad (1.177)$$

and therefore the torsion stiffness is:

$$k_{t,e} = \frac{\left(GI_t \right)_e}{l} = \frac{\sum_{j=1}^{m} \left(GI_t \right)_j}{l} \quad (1.178)$$

It can also be shown that for thin cross-sections (see Lobontiu [7]) the mechanical moment of inertia of the compound cross-section can accurately be approximated as:

$$J_{t,e} = \frac{w}{12} \sum_{j=1}^{m} \rho_j t_j \left(w^2 + t_j^2 \right) \int_0^l f_t(x)^2 dx = \frac{wl}{36} \sum_{j=1}^{m} \rho_j t_j \left(w^2 + t_j^2 \right) = \frac{1}{3} \sum_{j=1}^{m} J_{tj} \quad (1.179)$$

where the torsion distribution function of Equation (1.51) has been used. By combining Equations (1.178) and (1.179), the torsion resonant frequency yields:

$$\omega_{t,e} = \sqrt{3} \sqrt{\frac{\left(GI_t \right)_e}{lJ_t}} \quad (1.180)$$

which resembles Equation (1.52) giving the resonant frequency of a homogeneous cantilever.

Example 1.17
An active layer is deposited uniformly over the surface of a micro-cantilever. Known are l, w, t_s (thickness of structural layer), t_a (active layer thickness, $t_a = t_s/2$), E_s, μ_s, ρ_s, and ρ_a. Evaluate the elastic properties of the active layer (its modulus of elasticity E_a and Poisson's ratio μ_a) by monitoring the bending and resonant frequencies:

    (a) Symbolically (algebraically)
    (b) Numerically (when $l = 250$ μm, $w = 25$ μm, $t_s = 1$ μm, $E_s = 160$ GPa, $\mu_s = 0.25$, $\rho_s = 2300$ kg/m³, $\rho_a = 3500$ kg/m³, $\omega_{b,e} = 1.5 \times 10^5$ rad/s, $\omega_{t,e} = 2.25 \times 10^6$ rad/s)

Solution:
(a) The compound cantilever is a bimorph that is formed of two layers, and from Equations (1.176) and (1.180) that express the bending and resonant frequencies, the unknown elasticity modulii E_a and G_a can be found as follows:

$$E_a = \frac{2\left[33\omega_{b,e}^2 l^4\left(\rho_a + 2\rho_s\right) - 560 E_s t_s^2 + \sqrt{a}\right]}{35 t_s^2} \qquad (1.181)$$

with:

$$a = 121\omega_{b,e}^4 l^8\left(\rho_a + 2\rho_s\right)^2 - 3850 E_s \omega_{b,e}^2 l^4\left(\rho_a + 2\rho_s\right)t_s^2 + 34300 E_s t_s^4 \quad (1.182)$$

$$G_a = \frac{\omega_{t,e}^2 l^2\left[\left(\rho_a + 8\rho_s\right)t_s^2 + 4\left(\rho_a + 2\rho_s\right)w^2\right] - 96 G_s t_s^2}{12 t_s^2} \qquad (1.183)$$

From the relationship between longitudinal (E) and shear (G) modulii, which involves Poisson's ratio μ, the latter one is determined as:

$$\mu_a = \frac{E_a}{2G_a} - 1 \qquad (1.184)$$

(b) For the numerical values of this problem, the following solution is obtained by using Equations (1.181) through (1.184): $E_a = 5.9 \times 10^{10}$ N/m² and $\mu_a = 0.3$.

1.3.3.2 Dissimilar-Length Multilayer Microcantilevers

A multi-layer microcantilever with dissimilar-length strata is sketched in Figure 1.43, where a bimorph is shown with the top layer (which can act as transducer) shorter than the structural layer. In such instances, the micro-cantilever can be divided in several pieces, each segment having a composite thickness.

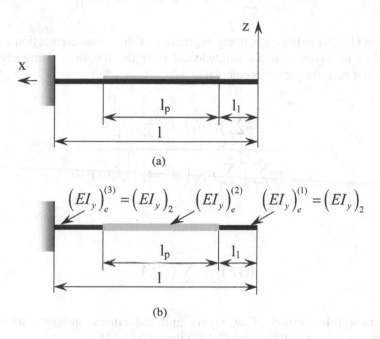

(a)

(b)

Figure 1.43 Side view of dissimilar-length bimorph

The configuration of Figure 1.43 (a), for instance, is segmented in three parts, as indicated in Figure 1.43 (b). The tip and root segments, being homogeneous, remain unaltered, whereas the mid-segment, which is composed of materials 1 (transducer *patch*) and 2 (*structure*), preserves its length l_p but gets a bending rigidity according to Equation (1.172). In doing so, the original sandwich cantilever is equivalently transformed into a similar structure serially composed of three homogeneous segments; this topic has been treated in Section 1.3.2.2 while dealing with non-homogeneous multi-profile cantilevers. The process can be applied to potentially n segments, and it can simply be shown that the bending resonant frequency is expressed as:

$$\omega_{b,e}^2 = \frac{\sum_{i=1}^{n} \int_{l_i} (EI_y)_{ei} \left[\frac{d^2 f_{bi}(x)}{dx^2} \right]^2 dx}{\sum_{i=1}^{n} \int_{l_i} \left[\sum_{j=1}^{m_i} (\rho A)_{ej} \right] f_{bi}(x)^2 dx}$$ (1.185)

with:

$$\left(EI_y\right)_{ei} = \sum_{j=1}^{m_i}\left\{E_j\left[I_{yj}+z_j\left(z_j-z_{Ni}\right)A_j\right]\right\} \tag{1.186}$$

Equation (1.186) indicates that any segment i of the series connection can be formed of m_i layers that are sandwiched over the length l_i. Similarly, the torsion resonant frequency is calculated as:

$$\omega_{t,e}^2 = \frac{12\sum_{i=1}^{n}\int_{l_i}\left(GI_t\right)_{ei}\left[\dfrac{df_{ti}(x)}{dx^2}\right]^2 dx}{tw\sum_{i=1}^{n}\int_{l_i}\left[\sum_{j=1}^{m_i}\rho_j t_j\left(w^2+t_j^2\right)\right]f_{ti}(x)^2 dx} \tag{1.187}$$

with:

$$\left(GI_t\right)_{e,i} = \sum_{j=1}^{m_i}\left(GI_t\right)_j \tag{1.188}$$

A segment i is formed of m_i layers and the corresponding distribution functions are given in Equations (1.153) through (1.156).

Example 1.18
    Determine the position of a patch of given length on a base micro-cantilever structure that would maximize the bending resonant frequency. Consider all other physical parameters defining the system are specified, namely: $l = 300$ μm, $l_p = 100$ μm, $t_1 = 1$ μm, $t_2 = 2$ μm, $E_1 = 130$ GPa, $E_2 = 165$ GPa, $\rho_1 = 3000$ kg/m³, $\rho_2 = 2400$ kg/m³.

Solution:
    By using the numerical data of this example, and the generic Equation (1.185) which gives the bending resonance of a non-homogeneous multi-profile cantilever, the bending resonant frequency is a function of only l_1, which positions the patch on the base structure. Figure 1.44 is the plot of this relationship. As shown in Figure 1.44 and as expected, moving the patch towards the cantilever root increases the stiffness and reduces the equivalent mass, the net result being an increase in the bending resonant frequency. To maximize this frequency for a given base microcantilever and patch, it is necessary to position the patch at the fixed base of the microcantilever.

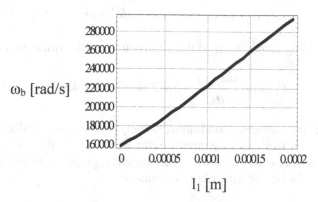

Figure 1.44 Bending resonant frequency of a patch microcantilever in terms of patch position

1.4 MICROBRIDGES

Microbridges are mechanical line-type members that are clamped at both ends to the substrate. Geometrically, a microbridge can be identical to a micro-cantilever, the only difference consisting in the change in one boundary condition between the two configurations (free end for the microcantilever versus fixed end for the microbridge). As a result of the geometric similarity between microcantilevers and microbridges, single- and multiple-profile microbridges will be studied in terms of defining their bending and torsional resonant frequencies.

1.4.1 Single-Profile Microbridges

For a microbridge consisting of a single profile (segment)—in which the cross-section dimensions are described by a single curve—the bending resonant frequency is still defined by Equation (1.8), whereas the torsion resonant frequency is calculated by means of Equation (1.14). The only difference consists in the distribution functions $f_b(x)$—in bending—and $f_t(x)$—in torsion—which will be derived in this subsection for constant and variable cross-section microbridge designs.

1.4.1.1 Constant Cross-Section

When the cross-section is constant (this is the simplest case), one can define the bending-related distribution function by relating the deflection at a generic point, $u_z(x)$, to the maximum deflection, which, in the case the load consists of a point load applied at the midpoint, will occur at the midpoint. The distribution function can be determined by seeking the out-of-the-plane deflection in the form of a four-degree polynomial, as shown by Lobontiu [7], for instance:

$$u_z(x) = a + bx + cx^2 + dx^3 + ex^4 \tag{1.189}$$

The corresponding slope, θ_y (x), is the x-derivative of the slope, namely:

$$\theta_y(x) = \frac{du_z(x)}{dx} = b + 2cx + 3dx^2 + 4ex^3 \tag{1.190}$$

By enforcing five boundary conditions, namely: 0 slope and deflection at the fixed points, as well as maximum deflection, u_z, at the midpoint, the coefficients a, b, c, d and e can be determined, together with the deflection ratio, which is the bending distribution function:

$$\frac{u_z(x)}{u_z} = f_b(x) = 16\frac{x^2}{l^2}\left(1 - \frac{x}{l}\right)^2 \tag{1.191}$$

It can be checked that this particular form of the distribution function also ensures that the slope is 0 at the midpoint.

Analyzing just half-length microcantilever instead of the full-length one, can simplify the calculations, and this is also valid for variable cross-section or multiple-profile microbridges that are symmetric with respect to their midpoint. In such cases, the midpoint needs to be guided to ensure compatibility between the full- and half-length microbridge. When x is measured from the fixed point, the distribution function can be determined by considering the following forms for deflections and slopes:

$$\begin{cases} u_z(x) = a + bx + cx^2 + dx^3 \\ \theta_y(x) = \dfrac{du_z(x)}{dx} = b + 2cx + 3dx^2 \end{cases} \tag{1.192}$$

By using four boundary conditions, namely: 0 slope and deflection at the fixed end, 0 slope at the guided end and maximum deflection at the same guided end, the four coefficients defining the deflection of Equation (1.192) can be determined, as well as the distribution function:

$$\frac{u_z(x)}{u_z} = f_b(x) = 12\frac{x^2}{l^2} - 16\frac{x^3}{l^3} \tag{1.193}$$

For a half-length microbridge, the distribution function can be found similarly by measuring the abscissa from the guided end, namely:

$$\frac{u_z(x)}{u_z} = f_b(x) = 1 - 12\frac{x^2}{l^2} + 16\frac{x^3}{l^3} \tag{1.194}$$

Example 1.19
Compare the out-of-plane and in-plane bending resonant frequencies of a constant rectangular cross-section microbridge whose length is *l*, cross-sectional dimensions are *w* (width) and *t* (thickness). The material is defined by Young's modulus E and mass density ρ.

Solution:
By using the generic method presented in this section, it can be shown that the out-of-plane (the one implying vibrations along the z-axis and bending about the y-axis) bending resonant frequency for the case in which the full-length structure is considered is:

$$\omega_{b,o-p}^{f-l} = 6.481 \frac{t}{l^2} \sqrt{\frac{E}{\rho}} \qquad (1.195)$$

whereas the in-plane (vibrations are parallel to the x-y plane) bending resonant frequency is:

$$\omega_{b,i-p}^{f-l} = 6.481 \frac{w}{l^2} \sqrt{\frac{E}{\rho}} \qquad (1.196)$$

It can be seen that the ratio of the two frequencies varies proportionally with the width-to-thickness ratio:

$$\frac{\omega_{b,i-p}^{f-l}}{\omega_{b,o-p}^{f-l}} = \frac{w}{t} \qquad (1.197)$$

When the half-length model and corresponding distribution function (Equation (1.193)) are used, the out-of-plane bending resonant frequency is:

$$\omega_{b,o-p}^{h-l} = 6.563 \frac{t}{l^2} \sqrt{\frac{E}{\rho}} \qquad (1.198)$$

which is slightly higher than the prediction by Equation (1.195). Obviously, the in-plane bending resonant frequency, according to the half-length model, is:

$$\omega_{b,i-p}^{h-l} = 6.563 \frac{w}{l^2} \sqrt{\frac{E}{\rho}} \qquad (1.199)$$

and Equation (1.197) is also valid in this case.
Similarly to bending, the torsion distribution function is determined by relating the rotation angle at a generic abscissa *x* to the maximum torsional angle (which occurs at the midpoint during free vibrations). When the full-length constant cross-section microbridge is analyzed, three rotation angles

can be specified, namely: zero rotations at the fixed ends and maximum rotation at the midpoint. It is thus sufficient to seek a distribution function in the form of a second-degree polynomial, namely:

$$\theta_x(x) = a + bx + cx^2 \tag{1.200}$$

By utilizing the boundary conditions previously mentioned, the torsion distribution function is expressed as:

$$\frac{\theta_x(x)}{\theta_x} = f_t(x) = 4\frac{x}{l} - 4\frac{x^2}{l^2} = 4\frac{x}{l}\left(1 - \frac{x}{l}\right) \tag{1.201}$$

As the case was with bending, it can be useful at times to express the torsion distribution function for only half-length the microbridge. In doing so, the resulting member can be considered as fixed-free in terms of torsion. Under these circumstances, the distribution function is sought as a first-degree polynomial because only two boundary conditions can be specified (zero rotation at the fixed point and maximum rotation angle at the free one). It can simply be shown that when the abscissa x is measured from the free end, the corresponding distribution function is:

$$\frac{\theta_x(x)}{\theta_x} = f_t(x) = 1 - 2\frac{x}{l} \tag{1.202}$$

whereas when x is measured from the fixed end, the distribution function becomes:

$$\frac{\theta_x(x)}{\theta_x} = f_t(x) = 2\frac{x}{l} \tag{1.203}$$

Example 1.20
    Calculate the torsion resonant frequencies for a very thin, constant rectangular cross-section microbridge by using both the full-length and half-length models. Compare the torsion resonant frequencies with those corresponding to bending for both models, respectively.

Solution:
    With the distribution function given in Equation (1.201), the generic Equation (1.14) gives the following torsional resonant frequency for the full-length constant rectangular cross-section microbridge:

$$\omega_t^{f-l} = 6.32\frac{t}{l}\sqrt{\frac{G}{\rho(t^2 + w^2)}} \tag{1.204}$$

where the superscript portion $f–l$ stands for full-length and the dimensions t and w are cross-sectional thickness and width, respectively whereas l is the length. In case the half-length model is considered ($h–l$ is the corresponding superscript), together with either the distribution function of Equation (1.202) or the one of Equation (1.203), the torsional resonant frequency is:

$$\omega_t^{h-l} = 6.93\frac{t}{l}\sqrt{\frac{G}{\rho\left(t^2+w^2\right)}} \qquad (1.205)$$

The full- and half-length models produce an 8.8% relative error, as Equations (1.204) and (1.205) indicate.

Comparison of the torsion and bending resonant frequencies of the constant rectangular cross-section microbridge can be performed by studying their ratio (Equations (1.204) and (1.195)), which can be set as:

$$\frac{\omega_t^{f-l}}{\omega_b^{f-l}} = 0.69\left(\frac{l}{t}\right)\frac{1}{\sqrt{\left(1+\mu\right)\left[1+\left(w/t\right)^2\right]}} \qquad (1.206)$$

where the known relationship between the elastic modulii E and G and Poisson's ratio has been taken into account, namely: $G = E/[2(1 + \mu)]$. Figure 1.45 is the 3D plot of the ratio of Equation (1.206) when the material is polysilicon ($\mu = 0.25$). For a large range of geometric parameter values, the torsional resonant frequency is higher than the bending one (a ratio larger than one is produced), but the ratio becomes less than one when, according to Equation (1.206):

$$l \le 1.62w \qquad (1.207)$$

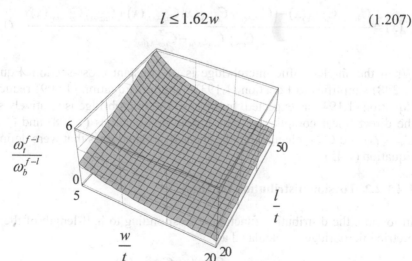

Figure 1.45 Ratio of resonant frequencies for a constant rectangular cross-section microbridge: torsion versus bending with full-length models

This is quite a severe condition, where a plate model is more likely to describe the structural behavior rather than the beam model.

1.4.1.2 Variable Cross-Section

Single-profile microbridges of constant thickness and variable width (and therefore variable cross-sections) are studied here, following a path similar to the one taken when characterizing single-profile microcantilevers. A generic formulation is first given in terms of bending and torsion followed by application to elliptically filleted and trapezoid designs.

1.4.1.2.1 Bending distribution functions

For a single-profile variable cross-section microbridge, which displays symmetry with respect to the midpoint, the bending distribution function is more convenient to be calculated for half-length microbridge by considering the midpoint is guided. The procedure follows the one depicted for a constant cross-section member, and it can be shown that when the abscissa is measured from the fixed point, the distribution function is:

$$f_b(x) = 1$$
$$+ \frac{C_{\theta y-My}C_{uz-Fz}(x) - \left(C_{uz-My} + C_{\theta y-My}x\right)C_{uz-My}(x) + C_{uz-My}C_{\theta y-My}(x)x}{C_{uz-My}^2 - C_{uz-Fz}C_{\theta y-My}} \quad (1.208)$$

When x is measured from the guided point, the distribution function is:

$$f_b(x) = \frac{C_{\theta y-My}C_{uz-Fz}(x) - \left(C_{uz-My} + C_{\theta y-My}x\right)C_{uz-My}(x) + C_{uz-My}C_{\theta y-My}(x)x}{C_{uz-Fz}C_{\theta y-My} - C_{uz-My}^2} \quad (1.209)$$

When the single-profile microbridge is of constant cross-section, Equation (1.208) simplifies to Equation (1.193), whereas Equation (1.209) reduces to Equation (1.194), as expected. In the case the microbridge is relatively short, the direct linear compliances that appear in Equations (1.208) and (1.209), $C_{uz-Fz}(x)$ and C_{uz-Fz} change into the shearing-related ones that were defined in Equation (1.45).

1.4.1.2.2 Torsion distribution function

In torsion, the distribution function corresponding to half-length of the symmetric microbridge is calculated as:

$$f_t(x) = \frac{\theta_x(x)}{\theta_x} = \frac{C_{\theta x-Mx}(x)}{C_{\theta x-Mx}} \quad (1.210)$$

where the torsion compliances are defined as:

$$
\begin{cases}
C_{\theta x-Mx}(x) = \dfrac{1}{G} \displaystyle\int_{x}^{l/2} \dfrac{dx}{I_t(x)} \\[4mm]
C_{\theta x-Mx} = \dfrac{1}{G} \displaystyle\int_{0}^{l/2} \dfrac{dx}{I_t(x)}
\end{cases}
\tag{1.211}
$$

and the torsional moment of inertia is determined depending on the cross-sectional thickness, as discussed in a previous section in this book. It should be mentioned that Equations (1.210) and (1.211) are valid irrespective of whether x is measured from the fixed or the free end of the half-length member because the compliances will change accordingly. It can also be checked that for a constant cross-section microbridge, Equations (1.210) and (1.211) simplify to Equation (1.202) when x is measured from the free (guided) end and to Equation (1.203) when x is measured from the fixed end.

1.4.1.2.3 Elliptically filleted microbridge

A right elliptically filleted microbridge, of the type introduced by Lobontiu and Garcia [5], is sketched in Figure 1.46 together with the defining geometry. It is assumed that the thickness t is constant. The out-of-the-plane bending and torsion resonant frequencies will be determined next. The full-length model is considered for bending, whereas for torsion, the half-length model is utilized.

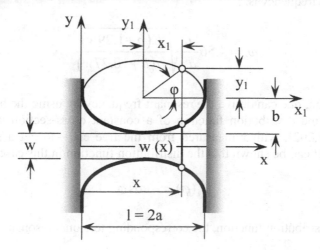

Figure 1.46 Top view of right elliptically filleted microbridge

According to Figure 1.46, the Cartesian dimensions of interest, which are the abscissa x and the variable width $w(x)$, can be expressed in terms of the polar angle φ as:

$$\begin{cases} x = a + x_1 = a + a\cos\varphi \\ w(x) = w + 2b(1 - \sin\varphi) \end{cases} \tag{1.212}$$

It therefore becomes possible to use the polar variable φ instead of the Cartesian one x, by also taking into account that:

$$dx = -a\sin\varphi d\varphi \tag{1.213}$$

and by noticing that the limits of integration for φ are 0 and π.

To keep the derivation manageable, the simplified bending distribution function (Equation (1.191)) is used. When calculating the second derivative of this distribution function, the chain rule of differentiation is used, namely:

$$\begin{cases} \dfrac{df_b}{dx} = \dfrac{df_b}{d\varphi}\dfrac{d\varphi}{dx} \\ \dfrac{d^2 f_b}{dx^2} = \dfrac{d}{dx}\left(\dfrac{df_b}{dx}\right) = \dfrac{d}{d\varphi}\left(\dfrac{df_b}{dx}\right)\dfrac{d\varphi}{dx} \end{cases} \tag{1.214}$$

where the derivative $d\varphi/dx$ can be calculated from Equation (1.213). The bending resonant frequency is:

$$\omega_b = 4.56\frac{t}{a^2}\sqrt{\frac{E(b+1.29w)}{\rho(b+10.27w)}} \tag{1.215}$$

In torsion, one can derive the resonant frequency by using the half-length model and the distribution function of a constant cross-section member— Equation (1.202), with x measured from the free end—to get a simplified prediction. It can be shown that the distribution function in this case is:

$$f_{t,s}(\varphi) = -\cos\varphi \tag{1.216}$$

With this distribution function, the corresponding torsional resonant frequency becomes:

$$\omega_{t,s} = 15.49 \frac{t}{a} \sqrt{\frac{G(0.86b+2w)}{\rho\left[0.83b^3 + 4.57b^2w + 40w(t^2+w^2) + 4.38b(t^2+3w^2)\right]}} \quad (1.217)$$

When the exact formulation is used (the generic Equations (1.210) and (1.211)), the resulting distribution function becomes:

$$f_t(\varphi) = \frac{A}{B} \quad (1.218)$$

with:

$$
\begin{cases}
A = 2\left\{ \sqrt{w(4b+w)}(\varphi-\pi) + (2b+w)[\pi \right. \\
\qquad \left. +2\arctan \dfrac{2b-(2b+w)\arctan\dfrac{\varphi}{2}}{\sqrt{w(4b+w)}}] \right\} \\
B = 2(2b+w)\left(\pi - 2\arctan\sqrt{\dfrac{w}{4b+w}} \right) - \pi\sqrt{w(4b+w)}
\end{cases}
\quad (1.219)
$$

The ratio of the torsional distribution functions given in Equations (1.218) and (1.216) is plotted in Figure 1.47, which indicates that the two functions can differ by a factor as large as 4 when the angle φ spans the $\pi/2$ to π range.

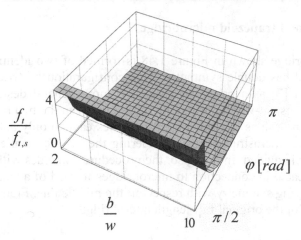

Figure 1.47 Ratio of torsional distribution functions: actual versus simplified right-elliptically filleted microbridge

However, these differences should be taken in a relative sense, because, as mentioned previously, the distribution function and its derivative are used in an averaged manner, which is achieved through the two integrations required by the equation of the resonant frequency, according to Rayleigh's procedure. However, using the distribution function of Equation (1.218) is quite difficult, and numeric integration has to be applied to determine the corresponding resonant frequencies. Table 1.2 contains the resonant frequency predictions obtained by using the simplified and exact distribution functions (Equations (1.216) and (1.218), respectively).

Table 1.2 Resonant frequencies with precise and simplified distribution functions

Case #	t [μm]	a [μm]	b [μm]	w [μm]	$\omega_{t,s}$ [MHz]	ω_t [MHz]	$\omega_{b,s}$ [MHz]	ω_b [MHz]
1	1	100	20	10	13.54	16.93	1.52	1.52
2	1.5	100	20	10	20.25	25.29	2.29	2.29
3	0.5	100	20	10	6.79	8.48	0.76	0.76
4	1	150	20	10	9.03	11.29	0.68	0.68
5	1	50	20	10	27.09	33.86	6.1	6.1
6	1	100	40	10	10.58	16.32	1.76	1.73
7	1	100	10	10	14.81	16.61	1.36	1.37
8	1	100	20	20	7.42	8.33	1.36	1.37

It can be noticed that in bending the predictions by simplified and precise distribution function models are very close, whereas in torsion the predictions by the simplified distribution function model underestimates the results yielded by the precise distribution function model.

1.4.1.2.4 Solid trapezoid microbridge

The microbridge shown in Figure 1.48 is formed of two identical trapezoid portions and has double symmetry. This configuration has been introduced by Lobontiu [7] as a particular case of a more general design having an additional constant-width portion interposed between the two mirrored trapezoid ones. In designs like this one, it becomes easy to only analyze half the length of the microstructure by considering the middle frontier line of half the member is guided. In doing so, the procedure coincides with the one that has been treated in connection to microbridges formed of a single geometric profile enjoying symmetry with respect to the middle line. Figure 1.49 shows the left half of the original full-length microbridge.

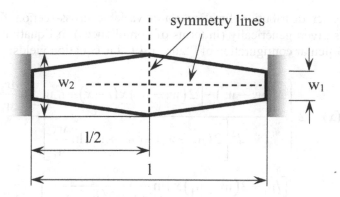

Figure 1.48 Top view of constant-thickness solid trapezoid microbridge with defining geometry

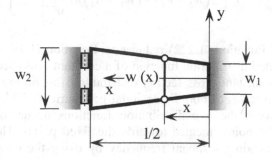

Figure 1.49 Top view of half-length solid trapezoid microbridge with guided end at middle line

The variable width $w(x)$ is calculated as:

$$w(x) = w_1 + 2\frac{w_2 - w_1}{l}x \qquad (1.220)$$

A simplified version of the bending resonant frequency is obtained when the distribution function corresponding to a constant cross-section half-length, fixed-guided beam. Equation (1.193) is used, and this frequency for the member of Figure 1.49 is:

$$\omega_{b,s} = 16.73\frac{t}{l^2}\sqrt{\frac{E\left(w_1 + w_2\right)}{\rho\left(3w_1 + w_2\right)}} \qquad (1.221)$$

In the case where $w_2 = w_1$, Equation (1.221) reduces to Equation (1.198), which gives the bending resonant frequency of a constant cross-section member.

The exact distribution function for a variable cross-section fixed-free member is given generically (in terms of compliances) in Equation (1.208). For the particular configuration of Figure 1.49, this equation yields:

$$f_b(x) = 2 \left\{ \frac{2l(w_2 - w_1) - \left[2(w_2 - w_1)x(l - x) + l^2 w_1 \right] \ln \dfrac{w_2}{w_1}}{l^2 \left[2(w_2 - w_1) - (w_2 + w_1) \ln \dfrac{w_2}{w_1} \right]} \right.$$

$$\left. + \frac{l \left[lw_1 + 2(w_2 - w_1)x \right] \ln \dfrac{lw_2}{lw_1 + 2(w_2 - w_1)}}{l^2 \left[2(w_2 - w_1) - (w_2 + w_1) \ln \dfrac{w_2}{w_1} \right]} \right\}$$

(1.222)

When $w_2 = w_1$, Equation (1.222) reduces to Equation (1.193), which gives the bending-related distribution function of a constant cross-section member. The variable-to-constant cross-section distribution functions ratio is plotted in Figure 1.50 in terms of non-dimensional geometric variables. The largest differences between the two distribution functions occur for large width ratios and for the points located towards the fixed points. However, when calculating the bending resonant frequency by using the exact distribution function of Equation (1.222) the differences between this model's predictions and the ones yielded by Equation (1.193), in which the simplified bending distribution function was used, are less prominent, as illustrated in Figure 1.51.

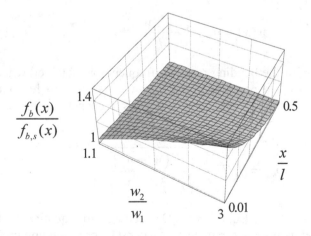

Figure 1.50 Ratio of bending distribution functions: trapezoid profile versus constant rectangular cross-section

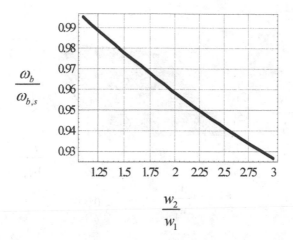

Figure 1.51 Ratio of bending resonant frequencies: trapezoid profile versus constant rectangular cross-section distribution functions

The difference between the two models results diminishes considerably, as the combined effect of having the distribution functions of each model averaged through the integration implied by elastic and inertia contributions and of deriving one particular resonant frequency as a ratio where the same distribution function enters both the numerator and denominator.

A similar procedure is applied to calculate the torsional resonant frequency by using the half-length model. When the distribution function of a constant cross-section member is used (Equation (1.203)), the torsional resonant frequency of the half-length micromember is:

$$\omega_{t,s} = 21.91\frac{t}{l}\sqrt{\frac{G\left(w_1 + w_2\right)}{\rho\left[w_1^3 + 3w_1^2 w_2 + 6w_1 w_2^2 + 10w_2^3 + 5\left(w_1 + 3w_2\right)t^2\right]}} \quad (1.223)$$

The exact distribution function, which is calculated by means of the generic Equation (1.210), becomes for the half-length doubly trapezoid microbridge:

$$f_t(x) = \frac{\ln\dfrac{lw_2}{lw_1 + 2\left(w_2 - w_1\right)x}}{\ln\dfrac{w_2}{w_1}} \quad (1.224)$$

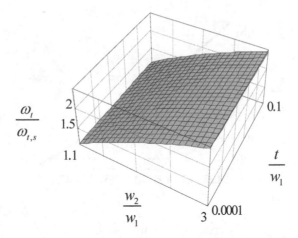

Figure 1.52 Ratio of torsional resonant frequencies: trapezoid profile versus constant rectangular cross-section distribution functions

With the aid of this function, the following torsional resonant frequency is obtained:

$$\omega_t = 22.63 \frac{t(w_2 - w_1)}{l} \sqrt{\frac{G \ln \frac{w_2}{w_1}}{\rho A}} \tag{1.225}$$

where:

$$A = (w_2 - w_1)(w_2 + w_1)(8t^2 + w_2^2 + w_1^2) - 4w_1^2$$

$$\left[2(2t^2 + w_1^2)\left(\ln \frac{w_2}{w_1} \right)^2 + (4t^2 + w_1^2) \ln \frac{w_2}{w_1} \right] \tag{1.226}$$

The 3D plot of Figure 1.52 shows the ratio of the two torsional resonant frequencies derived in Equations (1.223) and (1.225) and it can be seen that there are substantial differences between the two models' predictions.

1.4.2 Multiple-Profile Microbridges

Microbridges can be designed by combining several geometric curves, similarly to microcantilevers. Again, the particular situation in which the microbridge has double symmetry will be studied. This case comprises designs that are formed of at least three segments: two identical end segments and a middle one, all segments being symmetric with respect to the axial direction and to the direction perpendicular to it and passing to the midpoint. Both homogeneous and non-homogeneous bridge configurations will be analyzed. Bending and torsion resonant frequencies will be derived for constant, as well as variable cross-section microbridges.

1.4.2.1 Homogeneous Configurations

Homogeneous multi-profile microbridges are fabricated of the same material, and therefore have constant material and elastic properties for all the component segments. Constant and variable cross-section designs will be discussed next in terms of their bending and torsion resonant responses.

1.4.2.1.1 Constant cross-section segments

When a serially compound microbridge is formed of several constant cross-section segments, a generic model can be formulated to express the bending and torsion resonant frequencies, as shown in the following and similar to the derivation that has been presented for microcantilevers.

Generic Formulation

A serially compounded microbridge, which is formed of potentially n different segments, is sketched in Figure 1.53.

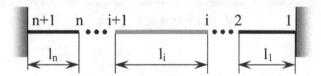

Figure 1.53 Serially compounded microbridge

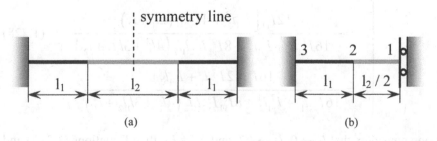

Figure 1.54 Side view of three-segment serially compounded microbridge: (a) full-length model; (b) half-length model

Equations (1.69) and (1.72), which gave the bending and torsional resonant frequencies for microcantilevers, are also valid for microbridges with different distribution functions. As mentioned here, only the particular case is analyzed of microbridges that are formed of three segments: two identical end segments (that are mirrored) and a middle one. The entire structure has two symmetry lines, as sketched in Figure 1.54 (a), and this enables the analysis of half the structure. In doing so, the midpoint is guided when bending is studied and is free when torsion is discussed.

To calculate the bending resonant frequency, the half-length model of Figure 1.54 (b) is employed. Finding the distribution functions corresponding to the two intervals, 1–2 and 2–3, has been described in detail in the subsection dedicated to microcantilevers. A similar approach is needed for microbridges and therefore only the main steps that have to be applied to derive the bending distribution functions are mentioned here, namely:

- Apply a force at the guided end and find the corresponding reaction moment at the same point.
- Express the deflection at a generic point on the 1–2 interval as produced by the tip force, as well as the maximum deflection at the guided end and determine the distribution function of the 1–2 interval as the ratio of the two deflections.
- Apply a similar procedure and determine the distribution function of the 2–3 interval.

Consequently, the distribution function corresponding to the 1–2 interval, when x is measured from the guided end in Figure 1.54 (b), is of the form:

$$f_{b1}(x) = a_1 + b_1 x^2 + c_1 x^3 \tag{1.227}$$

with:

$$
\begin{cases}
a_1 = 1 \\[2mm]
b_1 = -\dfrac{12 I_{y1}\left[I_{y1} l_2^2 + 4 I_{y2} l_1 \left(2 l_1 + l_2 \right) \right]}{16 I_{y2}^2 l_1^4 + I_{y1}^2 l_2^4 + 8 I_{y1} I_{y2} l_1 l_2 \left(4 l_1^2 + 3 l_1 l_2 + l_2^2 \right)} \\[4mm]
c_1 = \dfrac{16 I_{y1} \left(2 I_{y2} l_1 + I_{y1} l_2 \right)}{16 I_{y2}^2 l_1^4 + I_{y1}^2 l_2^4 + 8 I_{y1} I_{y2} l_1 l_2 \left(4 l_1^2 + 3 l_1 l_2 + l_2^2 \right)}
\end{cases}
\tag{1.228}
$$

If one considers that $l_2 \to 0$, $l_1 \to l/2$, and $I_{y1} = I_{y2}$, then Equations (1.227) and (1.228) reduce to Equation (1.194), which expresses the distribution function of a constant cross-section half-length microbridge.

The distribution function of the second segment 2–3 is of a similar form, namely:

$$f_{b2}(x) = a_2 + b_2 x + c_2 x^2 + d_2 x^3 \tag{1.229}$$

where:

$$
\begin{cases}
a_2 = \dfrac{I_{y2}\left(2l_1 + l_2\right)^2\left[4I_{y2}l_1\left(l_1 - l_2\right) + I_{y1}l_2\left(8l_1 + l_2\right)\right]}{16I_{y2}^2 l_1^4 + I_{y1}^2 l_2^4 + 8I_{y1}I_{y2}l_1 l_2\left(4l_1^2 + 3l_1 l_2 + l_2^2\right)} \\[4mm]
b_2 = \dfrac{24I_{y2}\left(I_{y2} - I_{y1}\right)l_1 l_2\left(2l_1 + l_2\right)}{16I_{y2}^2 l_1^4 + I_{y1}^2 l_2^4 + 8I_{y1}I_{y2}l_1 l_2\left(4l_1^2 + 3l_1 l_2 + l_2^2\right)} \\[4mm]
c_2 = -\dfrac{12I_{y2}\left[I_{y1}l_2^2 + 4I_{y2}l_1\left(l_1 + l_2\right)\right]}{16I_{y2}^2 l_1^4 + I_{y1}^2 l_2^4 + 8I_{y1}I_{y2}l_1 l_2\left(4l_1^2 + 3l_1 l_2 + l_2^2\right)} \\[4mm]
d_2 = \dfrac{16I_{y2}\left(2I_{y2}l_1 + I_{y1}l_2\right)}{16I_{y2}^2 l_1^4 + I_{y1}^2 l_2^4 + 8I_{y1}I_{y2}l_1 l_2\left(4l_1^2 + 3l_1 l_2 + l_2^2\right)}
\end{cases}
\tag{1.230}
$$

For the same limit conditions as above, namely $l_2 \to 0$, $l_1 \to l/2$, and $I_{y1} = I_{y2}$, Equations (1.229) and (1.230) simplify to Equation (1.194), which gives the distribution function of a constant cross-section half-length microbridge.

In torsion, when the half-length model is used, the middle point (which became an endpoint) is considered free, and therefore the corresponding distribution function of the 1–2 interval is:

$$f_{t1}(x) = 1 - \frac{2I_{t1}}{2I_{t2}l_1 + I_{t1}l_2}x \tag{1.231}$$

Similarly, the torsion distribution function for the 2–3 interval is:

$$f_{t2}(x) = \frac{I_{t1}\left(2l_1 + l_2\right)}{2I_{t2}l_1 + I_{t1}l_2} - \frac{2I_{t1}}{2I_{t2}l_1 + I_{t1}l_2}x \tag{1.232}$$

Both Equations (1.231) and (1.232) reduce to Equation (1.202), which gives the torsion-related distribution function of a constant cross-section microbridge by the half-length model when $l_2 \to 0$, $l_1 \to l/2$, and $I_{y1} = I_{y2}$.

Paddle Microbridge

As a direct application of the generic model developed herein, the paddle microbridge of Figure 1.55 (a) is studied now in terms of its bending and torsional resonant frequencies by using Rayleigh's quotient method applied to the half-length structure of Figure 1.55 (b). It is assumed all segments have the same thickness t.

The bending-related resonant frequency can be calculated by using the exact distribution functions of Equations (1.227) and (1.229) and this leads to quite a complicated equation. If, instead of using these distribution functions, one uses the simplified distribution function that results from Equation (1.194) by taking $l = 2l_1 + l_2$, namely:

$$f_b(x) = 1 - 12\frac{x^2}{\left(2l_1 + l_2\right)^2} + 16\frac{x^3}{\left(2l_1 + l_2\right)^3} \tag{1.234}$$

the corresponding resonant frequency can be expressed as:

$$\omega_{b,s} = 10.97t\sqrt{\frac{EA_b}{\rho B_b}} \tag{1.235}$$

where:

$$\begin{cases} A_b = \left(6l_1 + l_2\right)\left(36l_1^2 + 12l_1l_2 + 13l_2^2\right)w_1 + 27l_2\left(12l_1^2 + l_2^2\right)w_2 \\ B_b = \left(6l_1 + l_2\right)^5\left(1.24l_1^2 + 21.96l_1l_2 + l_2^2\right)w_1 \\ \quad + 5.8l_2(2240l_1^6 + 6720l_1^5l_2 + 7280l_1^4l_2^2 + 3640l_1^3l_2^3 + 1092l_1^2l_2^4 \\ \quad + 182l_1l_2^5 + 13l_2^6)w_2 \end{cases} \tag{1.236}$$

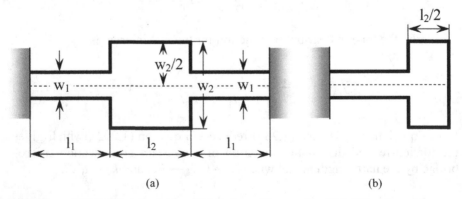

(a)   (b)

Figure 1.55 Paddle microbridge: (a) top view with geometry; (b) half-length model

When taking the following limits: $l_2 \to 0$, $l_1 \to l/2$, and $w_2 = w_1$, Equation (1.235) and Equation (1.236) reduce to Equation (1.198), which gives the bending resonant frequency of a constant rectangular cross-section bridge of length l.

The torsion resonant frequency has the following equation when using the exact distribution functions (Equations (1.231) and (1.232)):

$$\omega_t = 6.93 t w_1 \sqrt{\frac{G(2l_1 w_1 + l_2 w_2)}{\rho\left[8l_1^3 w_1^3\left(w_1^2 + t^2\right) + l_2 w_2\left(w_2^2 + t^2\right)\left(12l_1^2 w_2^2 + 6l_1 l_2 w_1 w_2 + l_2^2 w_1^2\right)\right]}} \quad (1.237)$$

Again, when taking the following limits: $l_2 \to 0$, $l_1 \to l/2$, and $w_2 = w_1$, Equation (1.237) simplifies to Equation (1.205), which expresses the torsion resonant frequency of a constant rectangular cross-section bridge of length l.

1.4.2.1.2 Variable cross-section

Serially compounded microbridges may be designed having a variable cross-section (generally, the width is considered variable and the thickness is constant, as mentioned previously) over all or just a few component segments. A generic formulation is derived in the following, which is further applied to a double trapezoid microbridge configuration.

Generic Formulation

A variable cross-section microbridge is analyzed, which is formed of two identical segments at the roots (they are mirrored with respect to the symmetry axis, as shown in Figure 1.54) and a mid-segment, which is also placed symmetrically with respect to the same symmetry line. This particular configuration is an extension of the design presented in the previous section and that consisted of three constant cross-section segments of each the root ones were identical. Figure 1.54 (b) is redrawn here (in Figure 1.56) to better emphasize the features of interest.

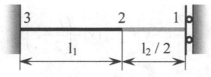

Figure 1.56 Side view of half-length model for a symmetric, variable cross-section microbridge

In bending, the half-length model of the microbridge is used because of the symmetry. The distribution functions corresponding to the two different segments are determined by using the approach detailed for the three-segment constant cross-section microbridge. It can be shown that the distribution function corresponding to the segment starting from the guided end is:

$$f_b^{(1-2)}(x) = \frac{A_{b1}(x)}{B_b} \tag{1.238}$$

where:

$$\begin{cases} A_{b1}(x) = C_{uz-Fz}^{(2-3)} + C_{uz-Fz}^{(1-2)}(x) + (l_2 - c - x)C_{uz-My}^{(2-3)} - (c+x)C_{uz-My}^{(1-2)}(x) \\ \qquad + \left(\frac{l_2}{2} - c\right)\left(\frac{l_2}{2} - x\right)C_{\theta y-My}^{(2-3)} + cxC_{\theta y-My}^{(1-2)}(x) \\ B_b = C_{uz-Fz}^{(2-3)} + C_{uz-Fz}^{(1-2)} + (l_2 - 2c)C_{uz-My}^{(2-3)} - 2cC_{uz-My}^{(1-2)} \\ \qquad + \left(\frac{l_2}{2} - c\right)^2 C_{\theta y-My}^{(2-3)} + c^2 C_{\theta y-My}^{(1-2)} \end{cases} \tag{1.239}$$

and:

$$c = \frac{C_{uz-My}^{(2-3)} + C_{uz-My}^{(1-2)} + \dfrac{l_2}{2}C_{\theta y-My}^{(2-3)}}{C_{\theta y-My}^{(2-3)} + C_{\theta y-My}^{(1-2)}} \tag{1.240}$$

Similarly, the distribution function of the second segment (the one towards the fixed point in Figure 1.56) is of the form:

$$f_b^{(2-3)}(x) = \frac{A_{b2}(x)}{B_b} \tag{1.241}$$

with:

$$A_{b2}(x) = C_{uz-Fz}^{(2-3)}(x) + \left(\frac{l_2}{2} - c - x\right)C_{uz-My}^{(2-3)}(x) - \left(\frac{l_2}{2} - c\right)xC_{\theta y-My}^{(2-3)}(x) \tag{1.242}$$

The subscript (1–2) in the equations above refer to the segment 1–2 at the guided end, whereas the superscript (2–3) denotes the root segment. All compliances that enter Equations (1.238) through (1.242) and correspond to the first segment are calculated by considering point 1 in Figure 1.56 is the origin; similarly, the compliances pertaining to the second segment are calculated with the origin at point 2. When $l_1 \rightarrow l/2$, $l_2 \rightarrow 0$, and $I_{t1} = I_{t2}$ (case where the two-segment variable cross-section bridge becomes a constant cross-section one of total length l), Equations (1.238) and (1.241) reduce to Equation (1.198), as it should be.

In torsion, the same half-length model of Figure 1.56 is used by considering now that point 1 is free (which is the appropriate assumption in terms of torsion). The distribution function corresponding to the first segment, 1–2, is:

$$f_t^{(1-2)}(x) = \frac{C_{\theta x-Mx}^{(1-2)}(x) + C_{\theta x-Mx}^{(2-3)}}{C_{\theta x-Mx}^{(1-2)} + C_{\theta x-Mx}^{(2-3)}} \qquad (1.243)$$

whereas the distribution function corresponding to the root segment is:

$$f_t^{(2-3)}(x) = \frac{C_{\theta x-Mx}^{(2-3)}(x)}{C_{\theta x-Mx}^{(1-2)} + C_{\theta x-Mx}^{(2-3)}} \qquad (1.244)$$

Again, at limit, when $l_1 \rightarrow l/2$, $l_2 \rightarrow 0$, and $I_{t1} = I_{t2}$ (conditions that render the two-segment variable cross-section bridge into a constant cross-section one of total length l), Equations (1.243) and (1.244) reduce to Equation (1.205), as expected.

Trapezoid Paddle-Type Microbridge

The configuration sketched in Figure 1.57 is an application of the generic formulation given here. Although the microbridge of Figure 1.57 (a) is formed of four segments, because of its symmetry, it is possible to analyze only half-length the structure, as shown in Figure 1.57 (b). Consequently, two segments need to be taken into account only, and therefore this particular example falls into the three-segment category presented herein.

Despite the fact that the variable cross-section segment (the trapezoid) is rather simple, the distribution functions for both bending and torsion are quite intricate, consequently the corresponding resonant frequencies are also complex and are not explicitly given here. The resonant frequencies are provided, which are calculated by means of the distribution functions defining a constant cross-section half-length bridge (Equation (1.198) for bending and Equation (1.205) for torsion).

The bending resonant frequency is:

$$\omega_b = 16.73t\sqrt{\frac{EA_b}{\rho B_b}} \qquad (1.245)$$

with:

$$A_b = 16l_1^3 w_1 + l_2 \left(12l_1^2 + l_2^2\right)\left(w_1 + w_2\right) + 4l_1 l_2^2 \left(2w_1 + w_2\right) \quad (1.246)$$

and:

$$\begin{aligned}
B_b = {} & 1664l_1^7 w_1 + 224l_1^6 l_2 \left(21w_1 + 5w_2\right) + 672l_1^5 l_2^2 \left(8w_1 + 5w_2\right) \\
& + 560l_1^4 l_2^3 \left(6w_1 + 7w_2\right) + 56l_1^3 l_2^4 \left(24w_1 + 41w_2\right) + 84l_1^2 l_2^5 \left(4w_1 + 9w_2\right) \quad (1.247) \\
& + 2l_1 l_2^6 \left(24w_1 + 67w_2\right) + l_2^7 \left(3w_1 + 10w_2\right)
\end{aligned}$$

The torsional resonant frequency is expressed as:

$$\omega_t = 21.91 t \sqrt{\frac{GA_t}{\rho B_t}} \quad (1.248)$$

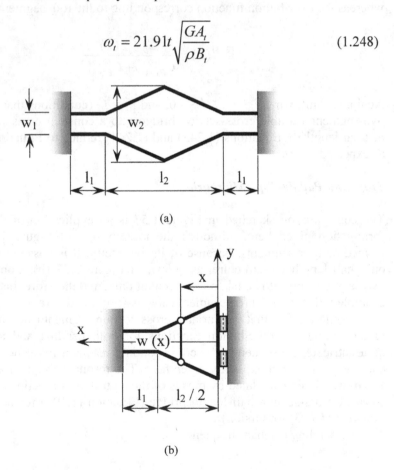

(a)

(b)

Figure 1.57 Trapezoid paddle-type microbridge with geometry: (a) full-length structure; (b) half-length structure

where:

$$A_t = 4l_1 w_1 + l_2 \left(w_1 + w_2 \right) \tag{1.249}$$

and:

$$
\begin{aligned}
B_t = {} & 160 l_1^3 w_1 \left(t^2 + w_1^2 \right) + 60 l_1^2 l_2 \left(w_1 + w_2 \right)\left(2t^2 + w_1^2 + w_2^2 \right) \\
& + l_2^3 \left[w_1^3 + 3 w_1^2 w_2 + 6 w_1 w_2^2 + 10 w_2^3 + 5t^2 \left(w_1 + 3 w_2 \right) \right] \\
& + 4 l_1 l_2^2 \left[10 t^2 \left(w_1 + 2 w_2 \right) + 3 \left(w_1^3 + 2 w_1^2 w_2 + 3 w_1 w_2^2 + 4 w_2^3 \right) \right]
\end{aligned}
\tag{1.250}
$$

A comparison is performed between the resonant frequency results obtained by means of the simplified distribution functions and those produced by using the exact distribution functions, not given here, but generically expressed by Equations (1.238), (1.241), (1.243), and (1.244). When calculating the exact-form resonant frequencies one has to consider that the variable width of the trapezoid segment (as shown in the top view of Figure 1.57 (b)) is:

$$w(x) = w_2 - 2 \frac{w_2 - w_1}{l_2} x \tag{1.251}$$

Table 1.3 comprises the results by the two models for a few values of the microbridge parameters.

Table 1.3 Resonant frequencies by means of precise and simplified distribution functions

Case	t [μm]	l_1 [μm]	l_2 [μm]	w_1 [μm]	w_2 [μm]	$\omega_{t,s}$ [MHz]	ω_t [MHz]	$\omega_{b,s}$ [MHz]	ω_b [MHz]
1	1	100	100	10	20	5.99	5.58	0.48	0.46
2	1.5	100	100	10	20	8.96	8.34	0.73	0.69
3	0.5	100	100	10	20	3.00	2.79	0.24	0.23
4	1	150	100	10	20	4.77	4.50	0.28	0.26
5	1	50	100	10	20	8.33	7.62	1.06	1.01
6	1	100	150	10	20	4.89	4.50	0.35	0.33
7	1	100	50	10	20	8.01	7.62	0.71	0.68
8	1	100	100	15	20	5.60	5.43	0.49	0.48
9	1	100	100	5	20	5.76	4.90	0.47	0.40
10	1	100	100	10	30	3.96	3.52	0.47	0.43

Rectangular Cross-Section Microbridge with Hole

Similar to the example of the rectangular cross-section cantilever with a hole, a similar perforated microbridge structure is studied here in terms of its bending resonant frequency. Figure 1.58 shows the top view of it, indicating the geometry of the structure and of the hole. The detailed view of the hole is displayed in Figure 1.26, and is not retaken. Because only bending is the matter of interest here, it is assumed the hole is placed along the longitudinal axis of symmetry, as shown in the same Figure 1.58.

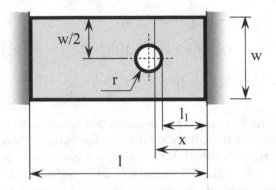

Figure 1.58 Top view of a rectangular microbridge with hole

Again, both mass and stiffness of the original constant rectangular cross-section bridge will be modified by the presence of the hole. Alteration of the bending stiffness by the hole enables consideration of a three-segment bridge, as shown in Figure 1.59. This configuration is a serially compounded one, but there is no symmetry about the structure's midpoint and the entire length has to be taken into consideration.

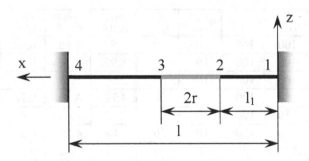

Figure 1.59 Equivalent three-segment microbridge

The distribution function of a constant cross-section bridge (Equation (1.191)), with the corresponding polar coordinate transformation (Equation (1.108)) and associated derivatives (Equation (1.113)) is used. The bending resonant frequency is expressed as:

$$\omega_b = 36.66t\sqrt{\frac{EA_b}{\rho B_b}} \tag{1.252}$$

with:

$$
\begin{cases}
A_b = 5\pi r^2[-2\left(l^2 - 6ll_1 + 6l_1^2\right)^2 + 24\left(l - 2l_1\right)\left(l^2 - 6ll_1 + 6l_1^2\right)r \\
\quad -60\left(l - 3l_1\right)\left(2l - 3l_1\right)r^2 + 252\left(l - 2l_1\right)r^3 - 189r^4] + 2l^5 w \\
B_b = -5040\pi l^4 r^2 (8l_1^4 + 32l_1^3 r + 60l_1^2 r^2 + 56l_1 r^3 + 21r^4) \\
\quad +20160\pi l^3 r^2 (l_1 + r)(8l_1^4 + 32l_1^3 r + 68l_1^2 r^2 + 72l_1 r^3 + 33r^4) \\
\quad -3780\pi l^2 r^2 (64l_1^6 + 384l_1^5 r + 1200l_1^4 r^2 + 2240l_1^3 r^3 + 2520l_1^2 r^4 \\
\quad +1584l_1 r^5 + 429r^6) + 2520\pi l r^2 (l_1 + r)(64l_1^6 + 384l_1^5 r + 1296l_1^4 r^2 \\
\quad +2624l_1^3 r^3 + 3256l_1^2 r^4 + 2288l_1 r^5 + 715r^6) - 315\pi r^2 (128l_1^8 \\
\quad +1024l_1^7 r + 4480l_1^6 r^2 + 12544l_1^5 r^3 + 23520l_1^4 r^4 \\
\quad +29568l_1^3 r^5 + 24024l_1^2 r^6 + 11440l_1 r^7 + 2431r^8) + 64l^9 w
\end{cases} \tag{1.253}
$$

When $r \to 0$, Equations (1.252) and (1.253) reduce to Equation (1.196), which gives the bending resonant frequency of a regular constant rectangular cross-section microbridge (without a hole in it).

Example 1.21

Analyze the influence of the hole position and dimension (radius) on the bending resonant frequency of a constant rectangular cross-section micro-bridge, which is defined by a length $l = 100$ μm and a width $w = 10$ μm.

Solution:

A simulation similar to the one performed for a rectangular cross-section microcantilever with a perforation is applied here, and Figure 1.60 is the 3D plot of the frequency ratio of a bridge with hole to the one without the hole in it (the asterisk [*] indicates the structure without a hole). For the parameter ranges of this example, the bending resonant frequency of the hole micro-bridge increases with larger radii and with the hole migrating towards the bridge midpoint. These trends are also highlighted in Figure 1.61, which plots the bending resonant frequency in terms of r and l_1.

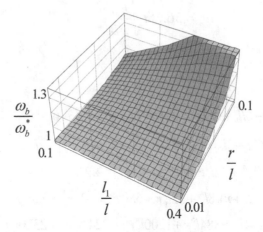

Figure 1.60 Comparison between the constant cross-section bridge without a hole and the similar bridge having a hole in it

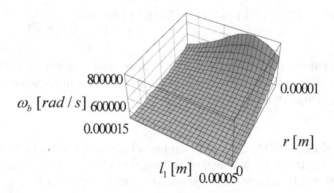

Figure 1.61 Influence of hole position and radius on the bending resonant frequency of a microbridge with a hole in it

The same numerical parameters used for the simulation in the case of a cantilever, namely: $E = 165$ GPa, $\rho = 2300$ kg/m^3 (values corresponding to polysilicon), $l = 100$ μm, $w = 10$ μm, and $t = 0.5$ μm, have been used to plot Figure 1.61.

Example 1.22

Compare the resonant frequency of a rectangular cross-section micro-bridge with perforation, as the one of Figure 1.58, when ignoring the change in stiffness, to the one obtained when both the stiffness and inertia are considered altered by the hole.

Solution:

When stiffness alteration through perforation is disregarded, the bending resonant frequency is expressed as:

$$\omega_b' = 51.85 lt \sqrt{\frac{E}{\rho}} \sqrt{\frac{lw}{B_b}} \qquad (1.254)$$

where B_b is given in Equation (1.253). Because the stiffness decrease through the very existence of the hole has been neglected, it is clear that Equation (1.254) yields a resonant frequency higher than the one where both stiffness and inertia variations have been taken into account. By taking the ratio of the two frequencies, the following function is obtained:

$$\frac{\omega_b'}{\omega_b} = \sqrt{2} l^2 \sqrt{\frac{lw}{A_b}} \qquad (1.255)$$

with A_b defined in Equation (1.253).

Equation (1.255) can be reformulated in terms of the non-dimensional parameters $c_1 = l_1/l$, $c_2 = r/lm$ and $c_3 = w/l$, which have been introduced in the example with the hole microcantilever. Figures 1.62 (a) and (b) show the 3D variations of this ratio when $c_3 = 0.25$ (Figure 1.62 (a)) and for $c_1 = 0.5$

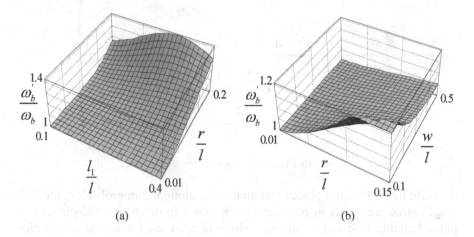

(a)   (b)

Figure 1.62 Bending resonant frequency comparison between prediction with stiffness neglected and considered for a rectangular cross-section microbridge with hole: (a) $c_3 = 0.25$; (b) $c_1 = 0.5$

(Figure 1.62 (b)), respectively. By neglecting the stiffness alteration produced through the perforation, particularly for large values of r and for small values of w, conduces to a relative increase of the bending resonant frequency ω'_b with respect to the actual ω_b, as illustrated in Figure 1.62.

Paddle Microbridge with Hole

The paddle microbridge with a hole in the middle segment, whose top view is shown in Figure 1.63, is studied now with respect to its bending resonant frequency.

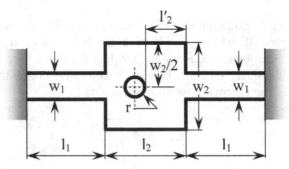

Figure 1.63 Top view of microcantilever with a hole in it and geometry

Only bending is analyzed again, and therefore it is assumed that the hole is located on the longitudinal axis. The microbridge is a serial structure comprising five different segments, as shown in Figure 1.64.

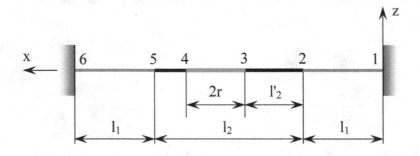

Figure 1.64 Equivalent four-segment cantilever

Because the hole is not placed symmetrically along the microbridge, the full-length structure needs to be taken into account to determine the out-of-the-plane bending resonant frequency, which is calculated with the distribution function corresponding to a constant cross-section microbridge.

The resonant frequency in this case is calculated as:

$$\omega_b^2 = \frac{E \sum_{i=1}^{5} \int_{l_i} I_{y,i}(x) \left(\frac{d^2 f_{b,i}(x)}{dx^2} \right)^2 dx}{\rho \sum_{i=1}^{5} \int_{l_i} A_i(x) f_{b,i}(x)^2 dx} \tag{1.256}$$

The expression of ω_b is complex again and is therefore not explicitly given here. As the case was with the paddle cantilever, a check is performed by taking $r \to 0$ into the resonant frequency equation and the resulting bending frequency is:

$$\omega_b = \sqrt{35}t \sqrt{\frac{EA_b}{\rho B_b}} \tag{1.257}$$

with:

$$\begin{cases} A_b = 8l_1^3 w_1 + l_2^3 w_2 + 3l_1 l_2 [l_1(3w_1 + w_2) + l_2(w_1 + w_2)] \\ B_b = 4224l_1^7 w_1 + 7l_1^6 l_2(1987w_1 + 125w_2) + 21l_1^5 l_2^2(931w_1 \\ \qquad + 125w_2) + 35l_1^4 l_2^3(425w_1 + 103w_2) + 105l_1^3 l_2^4(61w_1 + 27w_2) \\ \qquad + 42l_1^2 l_2^5(35w_1 + 31w_2) + 14l_1 l_2^6(10w_1 + 23w_2) + 33l_2^7 w_2 \end{cases} \tag{1.258}$$

Compared to Equations (1.235) and (1.236), which gave the bending resonant frequency of a paddle microbridge by using the half-length model, Equations (1.257) and (1.258) are slightly different. It should be remembered again that Equations (1.235) and (1.236) have been obtained by using different distribution functions for the two segments, whereas Equations (1.257) and (1.258) correspond to a unique distribution function.

Example 1.23
    Compare the bending resonant frequency of a paddle microbridge with hole with that of a similar bridge without a hole in it. Consider the particular case where $l_1 = l_2$ and $w_2 = 2w_1$.

<u>Solution:</u>
    For the given particular conditions, the bending resonant frequency becomes of the form:

$$\omega_b = \sqrt{35}t\sqrt{\frac{E}{\rho}}f(c_1,c_2,c_3) \qquad (1.259)$$

where $f(c_1, c_2, c_3)$, which is not explicitly given here, depends on three non-dimensional parameters, of which c_2 and c_3 have been defined in Example 1.22, and c_1 is:

$$c_1 = \frac{l_2'}{l_2} \qquad (1.260)$$

When there is no hole in the microcantilever $c_2 = 0$, and therefore Equation (1.257) simplifies to:

$$\omega_b^* = 0.19\frac{t}{l_1^2}\sqrt{\frac{E}{\rho}} \qquad (1.261)$$

Two 3D plots are drawn, one for the case in which $c_3 = 0.25$, and the other one for $c_1 = 0.5$, illustrated in Figure 1.65.

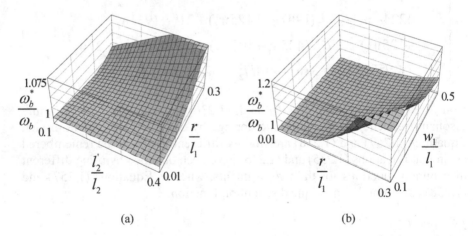

(a)   (b)

Figure 1.65 Bending resonant frequency comparison between paddle microbridges with and without a hole: (a) $c_3 = 0.25$; (b) $c_1 = 0.5$

For the set of analysis parameters, the resonant frequency of the original paddle microbridge is higher than the one of the similar configuration with a hole in it, which indicates that the change in mass (through removal of matter) is more important that the change in stiffness, and leads to an overall decrease in the resonant frequency of the hole microbridge, as Figure 1.65 illustrates. Also indicated in the same figures is the structure sensitivity to hole position and dimensions, as well as to the width dimensions.

1.4.2.2 Non-Homogeneous Configurations

In non-homogeneous microbridges, the component segments are built of various materials, which is not so common in current MEMS design. However, This situation can be encountered when analyzing microbridges having at least one portion in sandwich construction, which enables transforming the respective segment into an equivalent one, of constant rigidity. Coupled with the other segments of the microbridge that are fabricated of the same material, the resulting bridge is a non-homogeneous one. This particular example will be analyzed in Section 1.4.3.

The generic formulation for both bending and torsion will briefly be discussed, followed by an example of a non-homogeneous paddle microbridge.

1.4.2.2.1 Generic model

The generic formulation developed for non-homogeneous microcantilevers in Section 1.3.2.2 is valid for microbridges, too, when adequate distribution functions are used. It has been shown for non-homogeneous micro-cantilevers that by using the simplified distribution function instead of the precise, actual ones, accurate results are obtained in bending. On the other hand, it has been shown that sensible errors in the torsion resonant frequency can occur when the simplified distribution function is employed instead of the actual ones. Consequently, distribution functions will only be derived in torsion for microbridges, while for bending the simplified distribution functions will be used.

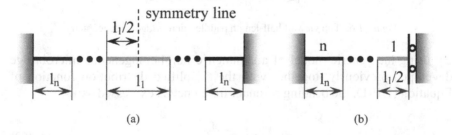

Figure 1.66 Non-homogeneous, multi-segment, serially compounded microbridge: (a) full-length model; (b) half-length model

The particular category of microbridges that have a symmetry line passing through their midpoint (as shown in Figure1.66 (a)), will be analyzed, as previously. This restriction enables studying only half the microbridge, as illustrated in Figure 1.66 (b). Also, the additional assumption is used that there is a mid-segment (the one having the length l_1 in Figure 1.66 (a)) and therefore the microbridge comprises an odd number of segments. In doing so, the model that has been presented for full-length non-homogeneous micro-cantilevers in Section 1.3.2.2 is valid for half-length microbridges as well. The only modifications consist in utilizing $l_1/2$ instead of l_1 and $l/2$ instead of l.

1.4.2.2.2 Paddle microbridge

As an example, the non-homogeneous paddle microbridge sketched in Figure 1.55 (b) is considered here, where the two end segments are fabricated of the same material, different from the one of the mid-segment. For both bending and torsion, the half-model is employed.

The geometric and material properties of the guided half segment of Figure 1.67 are assigned the subscript 1 and the root segment of the same figure is denoted by subscript 2.

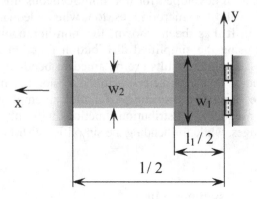

Figure 1.67 Top view of half-length paddle microbridge with geometry

By using the generic model of a symmetric non-homogeneous microbridge developed previously, together with the simplified distribution function of Equation (1.191), the bending resonant frequency is expressed as:

$$\omega_{b,e}^2 = \frac{A_b}{B_b} \tag{1.262}$$

with:

$$A_b = 560t^2 \left[E_1 l_1 \left(3l^2 - 6ll_1 + 4l_1^2 \right) w_1 + E_2 \left(l - l_1 \right) \left(l^2 - 2ll_1 + 4l_1^2 \right) w_2 \right] \tag{1.263}$$

and

$$B_b = \rho_1 w_1 \left(35l^6 l_1 - 70l^4 l_1^3 + 35l^3 l_1^4 + 63l^2 l_1^5 - 70l l_1^6 + 20l_1^7\right)$$
$$+\rho_2 w_2 \left(l - l_1\right)^5 \left(13l^2 + 30l l_1 + 20l_1^2\right)$$
(1.264)

By using the exact (actual) torsion-related distribution functions derived for non-homogeneous microcantilevers, together with the length corrections mentioned at the beginning of the non-homogeneous microbridge section, the torsion resonant frequency is:

$$\omega_{t,e}^2 = \frac{A_t}{B_t}$$
(1.265)

with:

$$A_t = 48 G_1 G_2 \left[G_1 \left(l - l_1\right) w_1 + G_2 l_1 w_2\right]$$
(1.266)

and:

$$B_t = \rho_2 G_1^2 \left(l - l_1\right)^3 w_1 w_2 \left(w_2^2 + t^2\right) + \rho_1 l_1 \left(w_1^2 + t^2\right)[3 G_2^2 \left(l - l_1\right)^2 w_1^2$$
$$+3 G_1 G_2 \left(l - l_1\right) l_1 w_1 w_2 + G_2^2 l_1^2 w_2^2]$$
(1.267)

Example 1.24

Consider a symmetric microbridge of constant rectangular cross-section over its length l, composed of three segments, of which the end ones are of the same material. Compare the bending and torsion resonant frequencies of this microbridge with the ones corresponding to a homogeneous microbridge having the same geometry and the material of the mid-segment of the non-homogeneous configuration.

Solution:

The bending resonant frequency is readily computable by means of Equation (1.146) and by using the distribution function of Equation (1.194), which describes half-length microbridge and where the length l is simply the sum of lengths of all component segments. The bending resonant frequency of a homogeneous, constant cross-section microbridge by means of the half-length model is given in Equation (1.198). Consequently, the ratio of the two frequencies can be formulated and it only depends on three non-dimensional variables, namely: E_2/E_1, ρ_2/ρ_1, and l_1/l. Figure 1.68 (a) plots the bending frequency ratio for $l_1/l = 0.4$, whereas Figure 1.68 (b) plots the same ratio for $E_2/E_1 = 0.7$ and $\rho_2/\rho_1 = 0.5$.

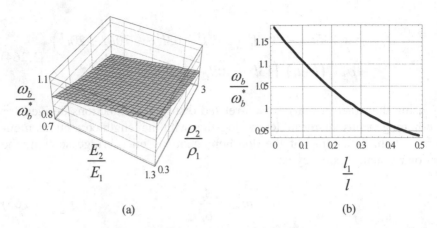

Figure 1.68 Bending resonant frequency ratio: non-homogeneous, three-segment microbridge versus homogeneous bridge as a function of: (a) elasticity modulus ratio; (b) length ratio

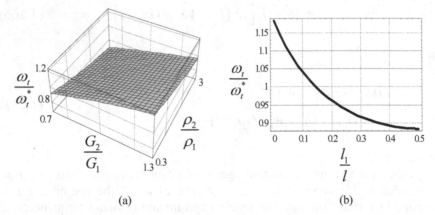

Figure 1.69 Torsion resonant frequency ratio: non-homogeneous, three-segment microbridge versus homogeneous bridge as a function of: (a) elasticity modulus ratio; (b) length ratio

A similar comparison is made between the torsion resonant frequency of the non-homogeneous, three-segment microbridge of this example (computed by means of Equations (1.231) and (1.232)), and a homogeneous counterpart (whose expression is given in Equation (1.205)). Both models refer to the half-length microbridge and Figure 1.69 illustrates this ratio in function of non-dimensional variables. Figure 1.69 (a) was plotted for $l_1/l = 0.4$ and Figure 1.69 (b) corresponds to $G_2/G_1 = 0.7$ and $\rho_2/\rho_1 = 0.5$. As Figures 1.68 (a) and 1.69 (a) indicate, the resonant frequencies (bending and torsion) of the non-homogeneous microbridge can be smaller than the corresponding resonant frequencies of the homogeneous counterpart when the side segments possess low elastic modulii and large densities. Similarly, when the length of the side segments decreases (equivalent to an increase in the length of the mid-segment for constant total length), the resonant frequencies of the non-homogeneous microbridge are smaller than the ones of the homogeneous design.

1.4.3 Multi-Layer (Sandwich) Microbridges

As the case was with microcantilevers, microbridges can be designed by superimposing (attaching) several layers. The process can involve layers that have the same length or layers that have different lengths. The two cases will be briefly analyzed next.

1.4.3.1 Equal-Length Multilayer Microbridges

A microbridge with equal-length layers can be obtained from the cor-responding microcantilever of Figure 1.42, by anchoring its the free end, and thus, obtaining the fixed-fixed boundary conditions. The generic derivation that has been presented for equal-length multi-layer microcantilevers remains valid for bridges, the only alteration being the different bending and torsion distribution functions that have to be used. The lumped-parameter method will be used next by considering the full-length bridge.

For a microbridge of length l with layers having equal lengths and equal widths, the stiffness corresponding to the midpoint, which expresses the force-deflection relationship is, similar to a homogeneous bridge:

$$k_{b,e} = \frac{192\left(EI_y\right)_e}{l^3} \tag{1.268}$$

where the equivalent bending rigidity $(EI_y)_e$ is calculated by means of Equation (1.172). The equivalent lumped-parameter bending mass is determined as:

$$m_{b,e} = \sum_{j=1}^{m}\left[\rho_j w t_j \int_0^l f_b(x)^2 \, dx\right] = \frac{128}{315}m \tag{1.269}$$

where m is the total mass of the composite beam, and the bending dis-tribution function $f_b(x)$ is given in Equation (1.191). By combining Equations (1.268) and (1.269), the bending resonant frequency becomes:

$$\omega_{b,e} = 21.737\sqrt{\frac{\left(EI_y\right)_e}{ml^3}} \tag{1.270}$$

By comparing Equation (1.270) to the first Equation (1.176) it can be seen that the bending resonant frequency of a microbridge is approximately six times larger than the one of a cantilever having the same geometry and material parameters.

Similar to a homogeneous microbridge, the mid-point torsion stiffness of a sandwiched microbridge is:

$$k_{t,e} = \frac{4(GI_t)_e}{l} \tag{1.271}$$

where the equivalent torsion rigidity is calculated by means of Equation (1.177). The equivalent mechanical moment of inertia corresponding to very thin layers is computed as:

$$J_{t,e} = \frac{w}{12} \sum_{j=1}^{m} \left[\rho_j t_j \left(w^2 + t_j^2 \right) \int_0^l f_t(x)^2 \, dx \right] = \frac{8}{15} J_t \tag{1.272}$$

where $f_t(x)$ is the torsion distribution function (Equation (1.201)) and J_t is the total mechanical moment of inertia, which is approximated to:

$$J_t = \sum_{j=1}^{m} J_{t,j} = \sum_{j=1}^{m} \frac{m_j \left(w^2 + t_j^2 \right)}{12} \tag{1.273}$$

where m_j is the mass of the j-th layer. The torsion resonant frequency is obtained by means of Equations (1.271) and (1.272) as:

$$\omega_{t,e} = 2.739 \sqrt{\frac{(GI_t)_e}{J_t l}} \tag{1.274}$$

Again, comparing the torsion resonant frequency of a microbridge to that of a microcantilever with identical geometry and inertia properties (Equation (1.180)), reveals the former is approximately 1.58 larger than the latter.

Example 1.25

For a two-component sandwich microbridge (bimorph), determine the thickness of the upper layer that would result in a maximum separation between the torsional and bending resonant frequencies. Consider $E_1 = 180$

GPa, E_2 = 165 GPa, ρ_1 = 3500 kg/m^3, ρ_2 = 2300 kg/m^3, l = 350 μm, w = 50 μm, t_2 = 1 μm.

Solution:

Equations (1.274) and (1.270) are used to set the torsion-to-bending resonant frequency ratio. By utilizing the numerical data of this problem, the frequency ratio can be expressed as a function of the two layers thickness ratio, and Figure 1.70 is a 2D plot of this relationship.

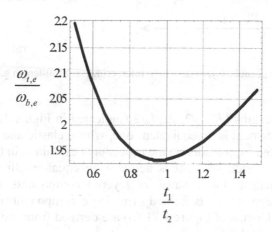

Figure 1.70 Torsion-to-bending torsion frequency ratio as a function of the thickness ratio for a bimorph microbridge

As Figure 1.70 indicates, the resonant frequency ratio has a minimum of about 1.9 for t_1/t_2 = 0.87, whereas the 2.2 maximum is reached at approximately t_1/t_2 = 0.5. The limit to the resonant frequency ratio is:

$$\max\left(\frac{\omega_{t,e}}{\omega_{b,e}}\right) = \lim_{t_1/t_2 \to 0}\left(\frac{\omega_{t,e}}{\omega_{b,e}}\right) = 3.864 \qquad (1.275)$$

which is also the value of the same frequency ratio for a homogeneous microbridge corresponding to the numerical values of this example.

1.4.3.2 Dissimilar-Length Multilayer Microbridges

Sandwich microbridges can be fabricated with layers having dissimilar lengths, as sketched in Figure 1.71 (a), where m layers are shown (the top layer is denoted by 1). When the layers are disposed symmetrically with respect to the microbridge midpoint, half-length model can be used to simplify calculating the bending and torsion resonant frequencies.

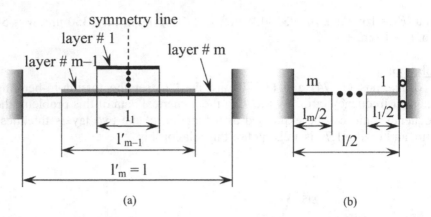

Figure 1.71 Side view of multi-layer (sandwich) microbridge: (a) full-length model; (b) half-length model

Over each of the lengths $l_1/2$, $l_2/2$, ..., $l_m/2$ (indicated in Figure 1.71 (b)), the respective beam segment is an equivalent one, whose elastic and inertia properties are calculated by considering the procedures explained in the previous subsection, because each segment is actually an equal-length sandwiched portion. On segment 1, for instance, m layered components are stacked, whereas the last segment, m, is formed of a single component (as shown in Figure 1.71). The lengths of Figure 1.71 (b) are derived from the actual ones of Figure 1.71 (a) as follows:

$$l_i = l_i' - l_{i-1}' \tag{1.276}$$

with $i = 1, 2, ..., m$. This particular notation has been used here to enable using the procedure exposed at non-homogeneous multi-profile microbridges.

In bending, the resonant frequency of the half-model shown in Figure 1.71 (b) is calculated as:

$$\omega_{b,e}^2 = \frac{\sum_{i=1}^{m}\left\{ \sum_{j=1}^{m-i+1}\left(E_j \left[I_{yj} + \left(z_j - z_{Ni} \right)^2 A_j \right] \right) \int_{l_{iu}} \left(\frac{d^2 f_b(x)}{dx^2} \right)^2 dx \right\}}{w \sum_{i=1}^{m}\left[\sum_{j=1}^{m-i+1}\left(\rho_j t_j \right) \int_{l_i} f_b(x)^2 dx \right]} \tag{1.277}$$

In Equation (1.277), a unique bending distribution function has been used, namely the one given in Equation (1.194), instead of individual distribution functions, as it was shown that the errors introduced by using the simplified distribution function of Equation (1.194) are minor.

The torsion resonant frequency is calculated as:

$$\omega_{t,e}^2 = \frac{12\sum_{i=1}^{m}\left[\sum_{j=1}^{m-i+1}\left(G_j I_{tj}\right)\int_{l_i}\left(\frac{df_{ti}(x)}{dx}\right)^2 dx\right]}{w\sum_{i=1}^{m}\left[\sum_{j=1}^{m-i+1}\left(\rho_j t_j\left(w^2+t_j^2\right)\right)\int_{l_i}f_{ti}(x)^2 dx\right]}$$ (1.278)

where the distribution functions $f_{t,i}(x)$ are calculated for each of the m intervals, by using the generic form of Equation (1.153) in which the coefficients of the first-degree polynomial are:

$$a_{ti} = \left[\frac{S_i}{\left(GI_t\right)_i} + \sum_{j=i+1}^{m}\frac{l_j}{2\left(GI_t\right)_j}\right] / \sum_{j=1}^{m}\frac{l_j}{2\left(GI_t\right)_j}$$ (1.279)

and

$$b_{ti} = -\left[\frac{1}{\left(GI_t\right)_i}\right] / \sum_{j=1}^{m}\frac{l_j}{2\left(GI_t\right)_j}$$ (1.280)

The sum S_i of Equation (1.279) was defined in Equation (1.150).

Example 1.26
A symmetric three-layer microbridge as the one sketched in Figure 1.72 (a) has its outer layers deposited of the same material and also of equal length and equal thickness. Study the bending resonant frequency of the structure in terms of the elastic and inertia properties of the materials by considering the middle (structural) layer is defined by: $l = 200$ μm, $t_2 = 1$ μm, $E_2 = 165$ GPa, and $\rho_2 = 2300$ kg/m^3.

Solution:
By using the generic Equation (1.277) for the case where $m = 3$ (three layers), as well as the specifications and numerical data of the problem, the bending resonant frequency of the trimorph can be expressed as a function of the following two ratios: E_1/E_2 and ρ_1/ρ_2 and Figure 1.73 is the 3D plot showing this relationship. As Figure 1.73 indicates, the bending resonant frequency increases for added layers with higher elastic modulii and lower mass densities.

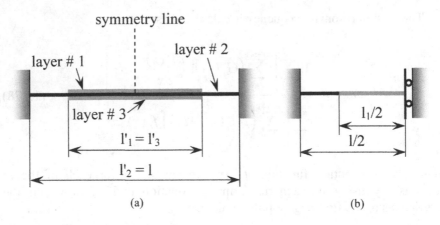

Figure 1.72 Side view of three-layer (trimorph) microbridge: (a) full-length model; (b) half-length model

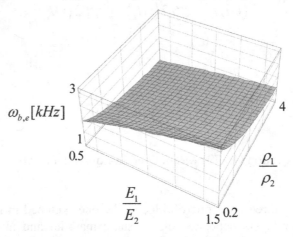

Figure 1.73 Bending resonant frequency of a three-layer (trimorph) microbridge with identical outer layers as a function of material properties

1.5 MASS DETECTION BY THE RESONANT FREQUENCY SHIFT METHOD

1.5.1 Method

The bending or torsion resonant response of microcantilevers and microbridges is used to detect extraneous matter that attaches to the resonant device. Mass addition changes the original resonant frequency of the microdevice and the shift in frequency is used as a metric for the quantity of added mass.

A simple qualitative analysis can be performed for a beam (cantilever or bridge) to determine the effect of added mass. When a mass Δm attaches to the beam, the new resonant frequency can be expressed, by using the lumped-parameter approach as:

$$\omega_b = \sqrt{\frac{k_b}{m_b + \Delta m}} \tag{1.281}$$

As shown a bit later, the mass Δm_b is a weighted value of the real mass Δm (weighting is due to various positions the added mass can assume on a beam), but, as a first approximation, it can be considered in Equation (1.281) that $\Delta m_b = \Delta m$ with sufficient accuracy. By also considering that the altered resonant frequency diminishes by a quantity $\Delta\omega$ and that the stiffness can be expressed as: $k_b = m_b\,(\omega_{b,0})^2$, Equation (1.281) gives the quantity of the deposited mass as:

$$\Delta m = \left[\left(\frac{\omega_{b,0}}{\omega_{b,0} - \Delta\omega} \right)^2 - 1 \right] m_b \tag{1.282}$$

which indicates that smaller masses can be detected by smaller frequency increments. Any detection equipment possesses a threshold value (in our case, the frequency), below which it is no longer accurate or even responsive. Clearly, such a value $\Delta\omega_{min}$ will generate, through Equation (1.282), a minimum mass Δm that can be detected. By using the non-dimensional frequency and mass fractions (Lobontiu [7] and Lobontiu et al. [8]):

$$\begin{cases} f_m = \dfrac{\Delta m}{m_b} \\[4mm] f_\omega = \dfrac{\Delta\omega}{\omega_b} \end{cases} \tag{1.283}$$

Equation (1.282) permits expressing f_m as a function of f_ω as:

$$f_m = \frac{1}{(1 - f_\omega)^2} - 1 \tag{1.284}$$

It is clear from Equation (1.284) that zero added mass ($f_m = 0$) produces no change in the resonant frequency ($f_\omega = 0$). Figure 1.74 also shows the quasi-linear proportionality between the two fractions. While all this discussion pertained to beams and bending, a similar reasoning could be developed for torsional resonators, where the added mass in positions that are off the structure's longitudinal axis produces a change in the mechanical moment of inertia, which decreases the original resonant frequency.

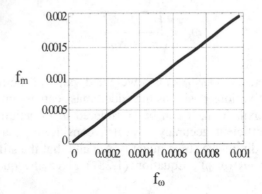

Figure 1.74 Mass fraction in terms of frequency fraction (Equation (1.284))

1.5.2 Microcantilever Support

Mass deposition will be first analyzed by studying the changes in the bending and torsion resonant frequencies of cantilevers. Constant and variable cross-section designs will be discussed in this subsection.

1.5.2.1 Constant Cross-Section

The simplest assumption will only be used here, namely that the mass attaches to the microcantilever in a point-like manner, as shown in Figure 1.75. Mass that deposits in a layer-like fashion over a portion of the cantilever length and over the whole width is a problem amenable to non-homogeneous structures (both cantilevers and bridges), and that problem has thoroughly been discussed in this chapter. In going with this assumption, the only modification will be in the total mass—when bending is taken into account, or in the mechanical moment of inertia—when torsion is discussed. It will be considered that there is no alteration in the structural stiffness, and therefore the original strain energy expression (Equation (1.4)) will be used.

The mass is offset by a quantity l_1 from the free end (as shown in Figure 1.75) about the longitudinal direction of the microcantilever and the bending effect of this change to the original resonant frequency can be quantified by evaluating the altered kinetic energy due to mass attachment, the energy being:

$$T = \frac{\rho w t}{2} \int_0^l \left[\frac{\partial u_z(x,t)}{\partial t} \right]^2 dx + \frac{\Delta m}{2} \left[\frac{\partial u_z(l_1,t)}{\partial t} \right]^2 \qquad (1.285)$$

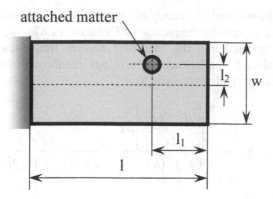

Figure 1.75 Top view of a rectangular cantilever with point-like deposited mass

By considering Equation (1.5), which separates the space and time variables, as well as Equation (1.7), which gives the deflection distribution in terms of the tip deflection, the maximum energy resulting from Equation (1.285) can be assessed to be:

$$T_{max} = \left\{ \frac{\rho wt}{2} \int_0^l f_b(x)^2 \, dx + \frac{\Delta m}{2} f_b(l_1)^2 \right\} \omega^2 u_z^2 \qquad (1.286)$$

By equating the maximum kinetic energy of Equation (1.286) to the maximum potential energy resulting from Equation (1.4), the bending resonant frequency becomes:

$$\omega_b^2 = \frac{EI_y \int_0^l \left[\dfrac{d^2 f_b(x)}{dx^2} \right]^2 dx}{\rho wt \int_0^l f_b(x)^2 \, dx + \Delta m f_b(l_1)^2} \qquad (1.287)$$

with the distribution function $f_b(x)$ being given in Equation (1.37). By taking into account the relationship between the regular and circular frequency ($f = 2\pi \omega$), Equation (1.287) allows expressing the attached mass as:

$$\Delta m = g_b(l_1) \frac{\Delta f_b \left(2 f_{b,0} - \Delta f_b \right)}{f_{b,0}^2 \left(f_{b,0} - \Delta f_b \right)^2} \qquad (1.288)$$

where $f_{b,0}$ is the original frequency of the cantilever, expressible by means of its corresponding circular frequency of Equation (2.87), Δf_b is the difference between the original and the modified natural frequencies (the modified frequency is always smaller than the original one through the mass increase), and the function $g_b(l_1)$ is expressed as:

$$g_b(l_1) = \frac{EI_y \int_0^l \left[\frac{d^2 f_b(x)}{dx^2} \right] dx}{4\pi^2 f_b^2(l_1)} = \frac{El^3 t^3 w}{4\pi^2 (l - l_1)^4 (2l + l_1)^2} \qquad (1.289)$$

The last side of Equation (1.289) used the particular values of the transfer function and the geometry of rectangular cross-section.

Equation (1.288) can also express the change in frequency as a result of a known deposited mass as:

$$\Delta f_b = \frac{f_{b,0} \left\{ \sqrt{g_b(l_1) \left[g_b(l_1) + \Delta m f_{b,0}^2 \right]} - \left(g_b(l_1) + \Delta m f_{b,0}^2 \right) \right\}}{g_b(l_1) + \Delta m f_{b,0}^2} \qquad (1.290)$$

Example 1.27

Analyze the mass detection sensitivity of a microcantilever with $l = 100$ μm, $w = 20$ μm, and $t = 1$ μm to the landing position of the mass and to the frequency shift. Consider the microcantilever is fabricated of polysilicon with $E = 1.5 \times 10^{11}$ N/m^2, and $\rho = 2300$ kg/m^3.

Solution:

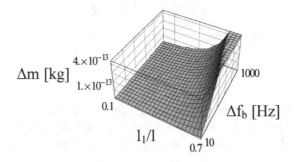

Figure 1.76 Detected mass in terms of position and frequency change

Equations (1.288) and (1.289), together with the first Equation (1.42), which gives the original bending resonant frequency (with no mass attached to it), are used with the numerical data of the problem and the plots of Figures 1.76 and 1.77 are obtained, which show the influence of l_1 and Δf_b on the quantity of deposited mass, Δm. As shown in Figures 1.76 and 1.77, more mass can be detected when attached closer to the root of the cantilever for the same change in frequency (when the l_1/l ratio increases). A larger mass can also be detected when the frequency change increases.

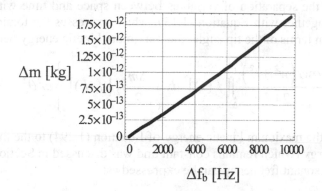

Δm [kg]

Δf_b [Hz]

Figure 1.77 Detected mass in terms of frequency change (for $l_1/l = 0.5$)

While Δm depends on Δf_b quasi-linearly (a better representation is given in Figure 1.77), the influence of deposited matter position on the micro-cantilever is highly nonlinear, particularly for values of l_1 that approach l. This can better be seen if the sensitivity of Δm to Δf_b and l_1/l is studied. The differential of Δm can be expressed as:

$$\Delta m = \frac{\partial(\Delta m)}{\partial(\Delta f_b)} d(\Delta f_b) + \frac{\partial(\Delta m)}{\partial(c_l)} dc_l \qquad (1.291)$$

where $c = l_1/l$. The partial derivatives in Equation (1.291) are the sensitivities of Δm to Δf and l_1/l, respectively. Their equations are not included but the sensitivities are proportional to:

$$\begin{cases} \dfrac{\partial(\Delta m)}{\partial(\Delta f_b)} \sim \Delta f_b \\[4mm] \dfrac{\partial(\Delta m)}{\partial(c_l)} \sim c_l^7 \end{cases} \qquad (1.292)$$

which indicates the strong nonlinear character of the partial derivative of Δm in terms of c_l.

The offset mass (which is shown at a distance l_2 from one side in Figure 1.75) changes the inertia properties of the microcantilever in torsion and therefore its torsion resonant response. The new (altered) kinetic energy due to mass attachment is:

$$T = \frac{\rho w t \left(w^2 + t^2\right)}{24} \int_0^l \left[\frac{\partial \theta_x(x,t)}{\partial t}\right]^2 dx + \frac{\Delta m}{2} l_2^2 \left[\frac{\partial \theta_x(l_1,t)}{\partial t}\right]^2 \quad (1.293)$$

By applying the separation of variables between space and time with Equation (1.11) together with Equation (1.13), which expresses the torsion angle distribution in terms of the tip angle, the maximum kinetic energy becomes:

$$T_{max} = \left\{\frac{\rho w t \left(w^2 + t^2\right)}{24} \int_0^l \left[f_t(x)\right]^2 dx + \frac{\Delta m}{2} l_2^2 \left[f_t(l_1)\right]^2 \right\} \omega^2 \theta_x^2 \quad (1.294)$$

By equating the maximum kinetic energy of Equation (1.294) to the maximum potential energy (which remains constant and was discussed in Section 1.2.1), the torsion resonant frequency can be expressed as:

$$\omega_t^2 = \frac{12 G I_t \int_0^l \left[\frac{df_t(x)}{dx}\right]^2 dx}{\rho w t \left(w^2 + t^2\right) \int_0^l \left[f_t(x)\right]^2 dx + 12 \Delta m l_2^2 \left[f_t(l_1)\right]^2} \quad (1.295)$$

The distribution function $f_t(x)$ of a constant cross-section microcantilever is given in Equation (1.51). By following a procedure similar to the one detailed in the case of bending, the attached mass can be formulated from Equation (1.295) as:

$$\Delta m = g_t(l_1, l_2) \frac{\Delta f \left(2 f_{t,0} - \Delta f_t\right)}{f_{t0}^2 \left(f_{t,0} - \Delta f_t\right)^2} \quad (1.296)$$

where $f_{t,0}$ is the original torsional frequency of the cantilever, expressed generically through its corresponding circular frequency of Equation (1.14), Δf_t is the difference between the original and the modified natural frequencies, and the function $g_t(l_1)$ is:

$$g_t(l_1, l_2) = \frac{G I_t \int_0^l \left[\frac{df_t(x)}{dx}\right]^2 dx}{4 \pi^2 l_2^2 f_t(l_1)^2} = \frac{G I_t l}{4 \pi^2 l_2^2 \left(l - l_1\right)^2} \quad (1.297)$$

The last side of Equation (1.297) took into account the distribution function pertaining to a constant cross-section microcantilever subjected to free torsional vibrations, and I_t is expressed differently depending on the relationship between w and t.

Similar to the bending case, the change in torsional resonant frequency can be expressed in terms of the deposited mass as:

$$\Delta f_t = \frac{f_{t,0}\left\{\sqrt{g_t(l_1)\left[g_t(l_1)+\Delta m f_{t,0}^2\right]}-\left(g_t(l_1)+\Delta m f_{t,0}^2\right)\right\}}{g_t(l_1)+\Delta m f_{t,0}^2} \quad (1.298)$$

Example 1.28 simply analyzes the relationship between bending and torsional resonant frequencies of a cantilever without no mass attachment as a preparation for the next example, Example 1.29, which compares the same resonant frequencies when mass is deposited on the microcantilever.

Example 1.28

Compare the bending and torsional resonant frequencies of a constant cross-section microcantilever.

Solution:

By using Equations (1.42) and (1.51) through (1.54), which give the resonant frequencies in bending and torsion, the following ratio can be formulated:

$$\frac{\omega_t}{\omega_b} = 2.379 \frac{l}{\sqrt{(1+\mu)(w^2+t^2)}} \quad (1.299)$$

where μ is Poisson's ratio expressing the connection between the longitudinal (Young's) modulus and the shear modulus, namely: $G = E/[2(1+\mu)]$. Figure 1.78 plots the resonant frequency ratio as a function of two non-dimensional parameters by considering $\mu = 0.25$ for a polysilicon material. As Figure 1.78

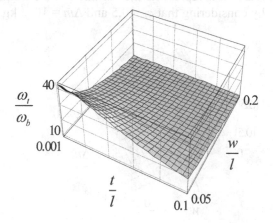

Figure 1.78 Torsion-to-bending frequency ratio in terms of geometry parameters

shows it, the torsional resonant frequency is 10 to 40 times larger than the bending resonant frequency for the geometric parameter ranges. As cantilevers become relatively thicker and wider, the torsional and bending resonant frequencies get closer.

Example 1.29

Consider mass attaches in a point-like manner on a constant cross-section microcantilever at $l_1 = l_2$ (as shown in Figure 1.75). Compare the altered bending and torsion resonant frequencies.

Solution:

By using Equation (1.287), the modified bending resonant frequency becomes:

$$\omega_b^2 = \frac{35EI^3t^3w}{35\Delta m\left(1-l_1\right)^4\left(2l+l_1\right)^2 + 33\rho l^7 tw} \tag{1.300}$$

Similarly, Equation (1.295) gives the altered resonant frequency as:

$$\omega_t^2 = \frac{12Glt^3w}{36\Delta m\left(1-l_1\right)^2 l_1^2 + \rho l^3 tw\left(w^2 + t^2\right)} \tag{1.301}$$

The following non-dimensional variables have been considered in the numerical results presentation that follows: $c_l = l_1/w$, $c_t = t/l$, and $c_w = w/l$. The plot of Figure 1.79 shows the torsion-to-bending altered resonant frequency ratio as a function of c_l and Δm by considering that $c_t = 0.05$ and $c_w = 0.2$. For the parameter ranges that have been chosen for this example, the altered torsion resonant frequency is 10.5 to 11.25 times larger than the bending one. The next plot, shown in Figure 1.80, represents the variation of the frequency ratio in terms of c_t and c_w, by considering that $c_l = 0.5$ and $\Delta m = 10^{-12}$ kg.

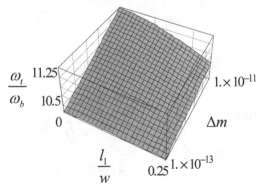

Figure 1.79 Torsion-to-bending altered resonant frequencies ($c_t = t/l = 0.05$, $c_w = w/l = 0.2$)

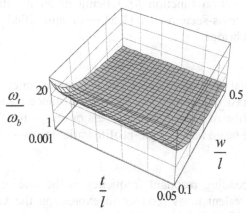

Figure 1.80 Torsion-to-bending resonant frequencies ($c_l = l_1/w = 0.25$, $\Delta m = 10^{-12}$ kg)

The frequency ratio ranges between a value of 2 (relatively thick and wide cantilever configurations) and 20 (for very thin and narrow cantilevers), as illustrated in Figure 1.80.

1.5.2.2 Single-Profile Variable Cross-Section Microcantilever

For single-profile variable cross-section microcantilevers, the formulation is similar to the one derived for constant-cross section microcantilevers that was just presented. The only difference consists in the geometric properties being functions of x. As thus, the altered bending resonant frequency becomes:

$$\omega_b^2 = \frac{E \int_0^l I_y(x) \left[\frac{d^2 f_b(x)}{dx^2} \right]^2 dx}{\rho \int_0^l A(x) f_b(x)^2 \, dx + \Delta m f_b(l_1)^2} \tag{1.302}$$

The bending distribution function $f_b(x)$ is given in Equation (1.30) in its precise form and in Equation (1.37) under the simplifying assumption that the cross-section is constant.

Similarly, the altered torsion resonant frequency is calculated as:

$$\omega_t^2 = \frac{12G \int_0^l I_t(x) \left[\frac{df_t(x)}{dx} \right]^2 dx}{\rho t \int_0^l w(x) \left(w(x)^2 + t^2 \right) f_t(x)^2 \, dx + 12\Delta m l_2^2 f_t(l_1)^2} \tag{1.303}$$

with the torsion distribution function $f_t(x)$ being given in either Equation (1.49)—for variable cross-section—or (1.51)—in simplified form—for a constant cross-section bridge.

Example 1.30
   Determine the total equivalent mass for a resonant microcantilever or microhinge when a mass of Δm attaches on it at a distance l_1 measured from one end of the beam (the free one for cantilevers or one of the fixed ends for bridges). Consider the beam has a single-profile variable cross-section.

Solution:
   For beams, the bending resonant frequency is the one of interest and finding the total equivalent mass reduces to expressing the kinetic energy after mass addition. Definitely:

$$T = T_0 + \Delta T \tag{1.304}$$

where T_0 is the original kinetic energy of the beam, which can be calculated by means of Equation (1.3) and ΔT is the kinetic energy produced by the added mass, which can be expressed as:

$$\Delta T = \frac{1}{2} \Delta m \left[\frac{du_z(l_1,t)}{dt} \right]^2 \tag{1.305}$$

The time- and space-dependent deflection $u_z(x,t)$ is discretized in both time (by means of Equation (1.5)) and space (by means of the bending distribution function $f_b(x)$; Equation (1.7)), as:

$$u_z(x,t) = u_z f_b(x) \sin(\omega t) \tag{1.306}$$

where u_z is the maximum deflection (free end for cantilevers and midpoint for symmetric bridges). By using Equation (1.306) to express T_0 and ΔT, the time-independent maximum kinetic energy is:

$$T_{max} = \frac{1}{2} \omega^2 u_z^2 \left[\rho \int_l A(x) f_b(x)^2 \, dx + \Delta m f_b(l_1)^2 \right] \tag{1.307}$$

   It has been shown in Section 1.2.2 when introducing the lumped-parameter method that the first term in the bracket represents the original equivalent mass of the beam, reduced at the point of interest (where the maximum deflection is recorded). The second term in the same bracket of Equation (1.307) contains Δm, which is moderated/weighted by the bending distribution function corresponding to the landing position of the added mass. Consequently, the total equivalent mass is:

$$m_b = m_{b,0} + \Delta m_b \tag{1.308}$$

with:

$$\begin{cases} m_{b,0} = \rho \int_l A(x) f_b(x)^2 \, dx \\ \Delta m_b = \Delta m f_b(l_1)^2 \end{cases} \tag{1.309}$$

Example 1.31
Express the mass that attaches to a resonant single-profile microcantilever in terms of the frequency shift, the original beam mass and resonant frequency, by also considering the landing position of the additional mass. Study the errors that are set when using the original quantity Δm instead of the weighted one Δm_b of Equation (1.309).

Solution:
In the brief discussion at the beginning of the section on mass detection by resonant microdevices, it has been shown that a simplification in carrying out an analysis is to consider that the total mass (after external matter attachment) is simply the sum of the original beam mass $m_{b,0}$ and the unweighted additional mass Δm. In reality, the total (final) mass contains the weighted added mass Δm_b of Equation (1.309). Consequently, Equation (1.182) transforms to:

$$\Delta m = \left[\left(\frac{\omega_{b,0}}{\omega_{b,0} - \Delta \omega} \right)^2 - 1 \right] \frac{m_b}{f_b(l_1)^2} \tag{1.310}$$

The same consideration changes Equation (1.284) into:

$$f_m = \left[\frac{1}{(1 - f_\omega)^2} - 1 \right] / f_b(l_1)^2 \tag{1.311}$$

The errors produced when utilizing Δm instead of Δm_b are quantified by considering the following ratio resulting from Equation (1.309):

$$r_{\Delta m} = \frac{\Delta m_b}{\Delta m} = f_b(l_1)^2 \tag{1.312}$$

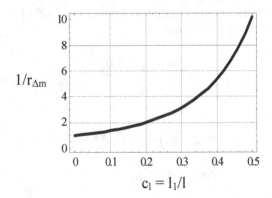

$c_l = l_1/l$

Figure 1.81 Error ratio between weighted and actual added mass as a function of landing position

For a single-profile microcantilever, the simplified bending distribution function is given in Equation (1.37) and therefore the error ratio of Equation (1.312) becomes:

$$r_{\Delta m} = \left(1 - \frac{3}{2}\frac{l_1}{l} + \frac{1}{2}\frac{l_1^3}{l^3}\right)^2 \tag{1.313}$$

By using the connection $l_1 = c\,l$, Figure 1.81 plots the inverse of the error ratio as a function of c_l.

The differences between the two mass fractions appear large, and the difference grows larger as the landing position moves toward the cantilever's root. It should be considered, however, that the added mass enters square-rooted into the bending resonant frequency (and this diminishes the differences). In addition, the added mass is many times just a very small fraction of the resonator's mass, and therefore the bending resonant frequency is expected to be less sensitive to the way of considering the added mass (original or weighted). To highlight these two aspects, one can construct a ratio between the altered resonant frequency that takes the weighted added mass and the one that simplifies the approach and considers the actual added mass. This ratio is:

$$\frac{\omega_b}{\omega_b'} = \sqrt{\frac{k_b}{m_b + \Delta m_b}} \times \sqrt{\frac{m_b + \Delta m}{k_b}} = \sqrt{\frac{1 + f_m}{1 + f_m f_b(l_1)^2}} \tag{1.314}$$

where f_m is the mass ratio defined in Equation (1.311). Since f_b (l_1) can be expressed in terms of $c_l = l/l_1$, the frequency ratio of Equation (1.314) depends only on f_m and on c_l and Figure 1.82 illustrates this relationship.

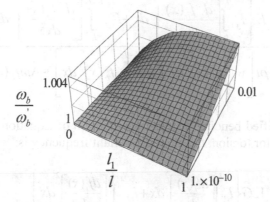

Figure 1.82 Resonant frequency ratio: Equation (1.314) as a function of landing position of added mass and mass fraction

As Figure 1.82 indicates, the differences between the two resonant frequencies are small, even for large mass fractions and locations of this additional mass that are very close to the microcantilever root.

1.5.2.3 Paddle Microcantilever

A paddle microcantilever is considered now in connection to mass detection of point-like attached matter. It is assumed the particle lands at distances a and b, as sketched in Figure 1.83.

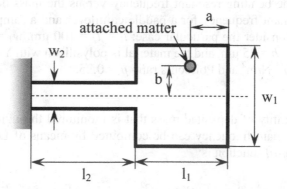

Figure 1.83 Top view of a paddle microcantilever with point-like attached matter

By following again Rayleigh's procedure, applied this time to the two-segment paddle configuration, the altered bending frequency can be expressed as:

$$\omega_b^2 = \frac{E\left\{I_{y1}\int_0^{l_1}\left[\frac{d^2 f_b(x)}{dx^2}\right]^2 dx + I_{y2}\int_{l_1}^{l_1+l_2}\left[\frac{d^2 f_b(x)}{dx^2}\right]^2 dx\right\}}{\rho t\left[w_1\int_0^{l_1} f_b(x)^2 dx + w_2\int_{l_1}^{l_1+l_2} f_b(x)^2 dx\right] + \Delta m f_b(a)^2} \tag{1.315}$$

where the simplified bending distribution function of Equation (1.37) can be used. Similarly, for torsion, the altered resonant frequency is:

$$\omega_t^2 = \frac{12G\left\{I_{t1}\int_0^{l_1}\left[\frac{df_t(x)}{dx}\right]^2 dx + I_{t2}\int_{l_1}^{l_1+l_2}\left[\frac{df_t(x)}{dx}\right]^2 dx\right\}}{\rho t\left[w_1\left(w_1^2+t^2\right)\int_0^{l_1} f_t(x)^2 dx + w_2\left(w_2^2+t^2\right)\int_{l_1}^{l_1+l_2} f_t(x)^2 dx\right] + 12\Delta m b^2 f_t(a)^2} \tag{1.316}$$

where the torsion distribution function can be the simplified one of Equation (1.51). It can be seen that whether the modified bending frequency depends only on the parameter a, the torsion resonant frequency depends on both landing parameters, a and b.

Example 1.32
    Determine the mass that can be determined through monitoring the change in the bending resonant frequency versus the mass detected through torsion resonant frequency for a paddle cantilever and a sampling frequency of 50 Hz. Consider the particular case: $l_1 = l_2 = 100$ μm, $w_1 = 4w_2 = 20$ μm, $t = 1$ μm, $a = b = 25$ μm, and the material is polysilicon with Young's modulus $E = 1.5 \times 10^{11}$ N/m^2 and Poisson's ratio $\mu = 0.25$.

Solution:
    The quantity of deposited mass that is monitored through shifting of the bending resonant frequency can be computed by means of Equation (1.288) where the $g_b(a)$ function is:

$$g_b(a) = \frac{E\left\{I_{y1}\int_0^{l_1}\left[\frac{d^2 f_b(x)}{dx^2}\right]^2 dx + I_{y2}\int_{l_1}^{l_1+l_2}\left[\frac{d^2 f_b(x)}{dx^2}\right]^2 dx\right\}}{4\pi^2 f_b(a)^2} \tag{1.317}$$

After performing the calculations, Equation (1.317) becomes:

$$g_b(a) = \frac{Et^3 \left[l_1^3 w_1 + l_2 w_2 \left(3l_1^2 + 3l_1 l_2 + l_2^2 \right) \right]}{4\pi^2 \left(l_1 + l_2 - a \right)^4 \left[2\left(l_1 + l_2 \right) + a \right]^2}$$

(1.318)

For torsion, the added mass is expressed by means of the generic Equation (1.296) where:

$$g_t(a,b) = \frac{G \left\{ I_{t1} \int_0^{l_1} \left[\frac{df_t(x)}{dx} \right]^2 dx + I_{t2} \int_{l_1}^{l_1+l_2} \left[\frac{df_t(x)}{dx} \right]^2 dx \right\}}{4\pi^2 b^2 f_t(a)^2}$$

(1.319)

After calculations are carried out, Equation (1.319) becomes:

$$g_t(a,b) = \frac{Gt^3 \left(l_1 w_1 + l_2 w_2 \right)}{12\pi^2 b^2 \left(l_1 + l_2 - a \right)^2}$$

(1.320)

Before computing the actual masses detected by the resonant frequency shifts, the functions of Equations (1.318) and (1.320) are compared in terms of the landing parameters a and b by means of their ratio, $r_g = g_t\ (a,b)\ /\ g_b\ (a)$.

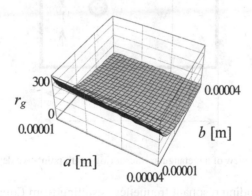

Figure 1.84   Torsion-to-bending ratio of functions g_t to g_b

As Figure 1.84 suggests, for small values of b (when the torsion produced by the additional mass is relatively small), the ratio between the torsion and the bending influence functions of Equations (1.320) and (1.318) is quite large.

As b increases, this ratio reduces significantly for any given value of a. This trend can also be applied to the ratio of the modified frequencies. For the numerical values of this example, it is found that the mass detected through bending is 1.61×10^{-14} kg (for an original resonant frequency of 19,732 Hz), whereas the mass detected through torsion is 3.16×10^{-17} kg—almost 500 times smaller (at a resonant frequency of 593,545 Hz).

1.5.3 Microbridge Support

Mass deposition can also be detected by monitoring the bending and torsion resonant shifts of microbridges. Constant cross-section and paddle configurations will be analyzed next in conjunction with mass attachment.

1.5.3.1 Constant Rectangular Cross-Section Microbridge

A constant rectangular cross-section microbridge is now studied to determine its altered resonant frequencies in bending and torsion when mass attaches in a point-like manner on it, as shown in Figure 1.85. The formulations in both bending and torsion are identical to the ones developed for the constant rectangular cross-section cantilever, only the distribution functions are the ones pertaining to a bridge—Equation (1.191) for bending and Equation (1.201) for torsion, respectively.

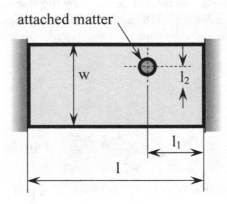

Figure 1.85 Top view of a rectangular microbridge with point-like deposited mass

The altered bending resonant frequency resulting from Equation (1.287) is:

$$\omega_b^2 = \frac{Ewt^3}{l^3 \left[\dfrac{m}{42} + 15 \left(\dfrac{l_1}{l} \right)^4 \left(1 - \dfrac{l_1}{l} \right)^4 \Delta m \right]} \tag{1.321}$$

where $m = \rho lwt$ is the mass of the bridge. The attached mass can be expressed in the form of Equation (1.288) by taking into account the original bending resonant frequency of a bridge and the change in frequency occurred due to mass deposition. The function g_b (l_1) of Equation (1.289) is here:

$$g_b(l_1) = \frac{Elwt^3}{60\pi^2 l_1^4 \left(1 - \dfrac{l_1}{l}\right)^4} \tag{1.322}$$

The altered torsion resonant frequency is obtained from Equation (1.295) as:

$$\omega_t^2 = \frac{40Glwt^3}{l^2\left(w^2 + t^2\right)m + 360 l_1^2 l_2^2 \left(1 - \dfrac{l_1}{l}\right)^2 \Delta m} \tag{1.323}$$

Equation (1.323), too, can be reformulated in the form of Equation (1.296) to express the quantity of deposited mass in terms of the original resonant frequency and the change in frequency. In this case, the function g_t (l_1, l_2) is:

$$g_t(l_1, l_2) = \frac{Glwt^3}{36\pi^2 l_1^2 l_2^2 \left(1 - \dfrac{l_1}{l}\right)^2} \tag{1.324}$$

Example 1.33

Evaluate the landing position (l_1 and l_2) of a point-like mass that attaches on a microbridge knowing the shifts in bending and torsion resonant frequencies Δf_b and Δf_t. Known are also the geometry and material properties of the microbridge. (a) Make the computations algebraically (symbolically); (b) Evaluate numerically the unknowns for $\Delta f_b = 500$ Hz, $\Delta f_t = 1500$ Hz, $\Delta m = 10^{-12}$ kg, $l = 200$ μm, $w = 40$ μm, $t = 2$ μm, $E = 1.65 \times 10^{11}$ N/m², $\mu = 0.28$, $\rho = 2400$ kg/m³.

Solution:

(a) The deposited mass can be expressed for both bending (by using Equations (1.288) and (1.289), and torsion) by means of Equations (1.296) and (1.297), namely:

$$
\begin{cases}
\Delta m = \dfrac{c_b \Delta f_b \left(2f_{b,0} - \Delta f_b\right)}{l_1^4 \left(1 - \dfrac{l_1}{l}\right)^4 f_{b,0}^2 \left(f_{b,0} - \Delta f_b\right)^2} \\[4ex]
\Delta m = \dfrac{c_t \Delta f_t \left(2f_{t,0} - \Delta f_t\right)}{l_1^2 l_2^2 \left(1 - \dfrac{l_1}{l}\right)^2 f_{t,0}^2 \left(f_{t,0} - \Delta f_t\right)^2}
\end{cases}
\tag{1.325}
$$

where:

$$
\begin{cases}
c_b = \dfrac{Elwt^3}{60\pi^2} \\[3ex]
c_t = \dfrac{Glwt^3}{36\pi^2}
\end{cases}
\tag{1.326}
$$

The first Equation (1.325) can be solved for l_1, namely:

$$
l_1 = \frac{l}{2} \pm \sqrt{\left(\frac{l}{2}\right)^2 - c_{b1}}
\tag{1.327}
$$

with:

$$
c_{b1} = l_4 \sqrt{\frac{c_b \Delta f_b \left(2f_{b,0} - \Delta f_b\right)}{\Delta m f_{b,0}^2 \left(f_{b,0} - \Delta f_b\right)^2}}
\tag{1.328}
$$

As shown in Equation (1.327), two solutions can be valid for l_1 because both values are in the $0 \rightarrow l$ interval where l_1 can range.

With l_1 determined, the second Equation (1.325) is used to find l_2, namely:

$$
l_2 = \frac{l}{f_{t,0}\left(f_{t,0} - \Delta f_t\right)l_1\left(l - l_1\right)} \times \sqrt{\frac{c_t \Delta f_t \left(2f_{t,0} - \Delta f_t\right)}{\Delta m}}
\tag{1.329}
$$

(b) The numerical solution for the algorithm presented at point (a) gives the following two values for l_1, according to Equation (1.327): $l_{1,1} = 25$ μm and $l_{1,2} = 175$ μm, and it can be seen that both are valid, because the bridge's length is 200 μm. It can also be seen that the second position is symmetric to the first one with respect to the microbridge midpoint, and this makes complete sense. Moreover, one single value results for l_2 when using either $l_{1,1}$ or $l_{1,2}$ given above, namely: $l_{2,1} = l_{2,2} = 5.7$ μm, but, because torsion is insensitive to one or the other semi-regions determined by the longitudinal axis, the landing position may also be determined by a value of 5.7 μm measured from the axis in the direction opposed to the one indicated in Figure 1.85. As a result, the mass can attach in four different positions, namely: P_1 ($l_1 = 25$ μm, $l_2 = 5.7$ μm), P_2 ($l_1 = 175$ μm, $l_2 = 5.7$ μm), P_3 ($l_1 = 25$ μm, $l_2 = -5.7$ μm), and P_4 ($l_1 = 175$ μm, $l_2 = -5.7$ μm).

1.5.3.2 Paddle Microbridge

The paddle microbridge can also be used to detect mass attachment on the paddle area particularly (which is larger). Figure 1.86 shows the top view and the landing position of a point-like mass on a paddle microbridge.

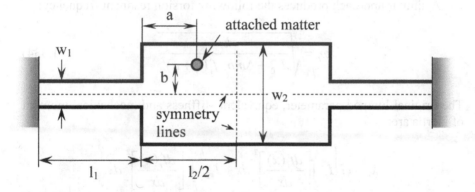

Figure 1.86 Top view of a paddle microbridge with point-like attached matter

In the case of bending, and by using the lumped-parameter approach, the resonant frequency resulting after mass attachment is:

$$\omega_b = \sqrt{\frac{k_b}{m_{b,0} + \Delta m f_b (l_1 + a)^2}} \qquad (1.330)$$

The origin of measuring the x-coordinate is at the root on the left in Figure 1.86. The original lumped-parameter stiffness and mass are:

$$k_b = E\left\{ I_{y1} \int_0^{l_1} \left[\frac{d^2 f_b(x)}{dx^2} \right]^2 dx + I_{y2} \int_{l_1}^{l_1+l_2} \left[\frac{d^2 f_b(x)}{dx^2} \right]^2 dx \right.$$
$$\left. + I_{y1} \int_{l_1+l_2}^{2l_1+l_2} \left[\frac{d^2 f_b(x)}{dx^2} \right]^2 dx \right\} \tag{1.331}$$

$$m_{b,0} = \rho t \left[w_1 \int_0^{l_1} f_b(x)^2 dx + w_2 \int_{l_1}^{l_1+l_2} f_b(x)^2 dx + w_1 \int_{l_1+l_2}^{2l_1+l_2} f_b(x)^2 dx \right] \tag{1.332}$$

and the distribution function is the simplified one (corresponding to a constant cross-section bridge):

$$f_b(x) = \left[4 \frac{x}{2l_1+l_2} \left(1 - \frac{x}{2l_1+l_2} \right) \right]^2 \tag{1.333}$$

A similar approach produces the following torsion resonant frequency:

$$\omega_t = \sqrt{\frac{k_t}{J_{t,0} + \Delta m b^2 f_t(l_1 + a)^2}} \tag{1.334}$$

The original lumped-parameter, equivalent stiffness and mechanical moment of inertia are:

$$k_t = G\left\{ I_{t1} \int_0^{l_1} \left[\frac{df_t(x)}{dx} \right]^2 dx + I_{t2} \int_{l_1}^{l_1+l_2} \left[\frac{df_t(x)}{dx} \right]^2 dx \right.$$
$$\left. + I_{t1} \int_{l_1+l_2}^{2l_1+l_2} \left[\frac{df_t(x)}{dx} \right]^2 dx \right\} \tag{1.335}$$

$$J_{t,0} = \frac{\rho t}{12} \left[w_1 \left(w_1^2 + t^2 \right) \int_0^{l_1} f_t(x)^2 dx + w_2 \left(w_2^2 + t^2 \right) \int_{l_1}^{l_1+l_2} f_t(x)^2 dx \right.$$
$$\left. + w_1 \left(w_1^2 + t^2 \right) \int_{l_1+l_2}^{2l_1+l_2} f_t(x)^2 dx \right] \tag{1.336}$$

Example 1.34
    Compare the resonant bending and torsion performance of sensing the mass of substance deposited in a point-like manner on a paddle nanobridge in

terms of the landing parameters a and b for a microcantilever whose dimensions are $l_1 = l_2 = 10$ μm, $w_1 = 500$ nm, $w_2 = 2$ μm, $t = 80$ nm. Consider the material parameters are $E = 1.55 \times 10^{11}$ N/m^2, $\mu = 0.25$, $\rho = 2400$ kg/m^3. Also determine the quantities determined by each resonance for a resonant frequency variation of $\Delta f = 100$ Hz.

Solution:
Equations (1.330) and (1.334) enable formulating the amount of mass detected through bending and torsion as:

$$\Delta m_b = \frac{1}{f_b(l_1+a)^2}\left[\frac{k_b}{(\omega_b - \Delta\omega)^2} - m_{b,0}\right] \qquad (1.337)$$

$$\Delta m_t = \frac{1}{b^2 f_t(l_1+a)^2}\left[\frac{k_t}{(\omega_t - \Delta\omega)^2} - J_{t,0}\right] \qquad (1.338)$$

The ratio $r_m = \Delta m_b/\Delta m_t$ is plotted in Figure 1.87 as a function of the parameters a and b of Figure 1.1.

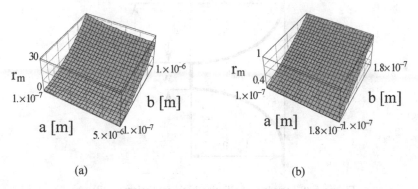

(a)   (b)

Figure 1.87 Bending versus torsion mass detection: (a) full-range for position parameters; (b) reduced-range for position parameters

As Figure 1.87 indicates, the amount of mass detected by the bending resonant shift is up to 30 times higher than the amount of mass detected by the torsional resonant shift, except for a small range of the position parameters in an area located towards the microbridge axis. As b increases, the equivalent inertia due to torsion also increases and the corresponding torsional resonant frequency decreases, and this makes the ratio r_m increase, as shown in Figure 1.87.

Problems

Problem 1.1

The elastic modulus E and Poisson's ratio μ of an unknown material need to be determined. Propose an experimental method to evaluate these amounts that would employ the resonant response of a constant rectangular cross-section microcantilever.

Problem 1.2

Calculate the bending resonant frequency of a constant rectangular cross-section microbridge by using a distribution function that results from applying a distributed load over the whole microbridge length. Compare the result with the regular equation produced when considering the distribution function generated by a point force applied at the midpoint.

Problem 1.3

Using Rayleigh's quotient method, derive the bending resonant frequency of the elliptically filleted microcantilever whose top view is shown in Figure 1.88. The elliptical semi-axes are a and b, as shown in the figure.

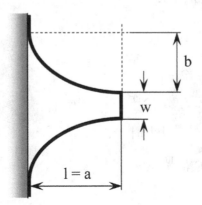

Figure 1.88 Top view of elliptically filleted microcantilever

Problem 1.4

Derive the torsional resonant frequency of the elliptically filleted micro-cantilever of Figure 1.88.

Problem 1.5

Compare the resonant frequencies (in both bending and torsion) of a circularly filleted microcantilever to those of an elliptically filleted one considering both configurations have the same length l, minimum width w and thickness t.

Problem 1.6

Design a constant cross-section microcantilever capable of discerning a minimum mass of $\Delta m = 10^{-12}$ kg when the detecting equipment has a sensitivity of $\Delta f = 10$ Hz.

Problem 1.7

Evaluate the total equivalent mechanical moment of inertia for a torsion-resonant microcantilever when a mass of Δm attaches on it at a distance l_2 measured from the longitudinal axis. Consider the beam has a single-profile variable cross-section.

Problem 1.8

Evaluate the mass that attaches to a resonant single-profile microbridge in terms of the frequency shift, the original beam mass, resonant frequency, and position of the additional mass. Also study the errors that are produced in calculating the bending resonant frequency when using the original quantity Δm instead of the weighted one Δm_b.

Problem 1.9

Two parallelepiped microcantilevers have the same geometric envelope defined by $l = 240$ μm, $w = 60$ μm, and $t = 1$ μm. Holes are perforated in the cantilevers: 6×2 in one of them and 12×4 in the other one. The hole radius is $r = 1.5$ μm and the edge distance is $a = 6$ μm.

  (a) Determine the pitch distance p for each configuration.
  (b) Evaluate the equivalent stiffness, inertia and resonant frequency (bending and torsion) for each configuration. Known are: $E = 165$ GPa, $\mu = 0.25$, $\rho = 2350$ kg/m^3.

Problem 1.10

A parallelepiped microcantilever made up of polysilicon ($E = 150$ GPa, $\mu = 0.28$) is functionalized locally to enable orderly deposition of an unknown substance at several locations (to increase the quantity of attached mass), as shown in the Figure 1.89. Known are the bending resonant frequency shift $\Delta \omega = 2\pi \times 500$ rad/s, also $l = 40$ μm, $w = 20$ μm, $t = 0.5$ μm, $p = 10$ μm. The radius of a small circular spot is measured and is equal to $r = 0.3$ μm and the thickness of a deposed mass is 0.2 μm. Considering point-like mass deposition, find the density of the unknown substance.

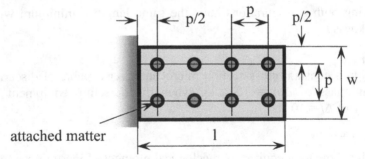

Figure 1.89 Top view of a rectangular microcantilever with point-like deposited mass in eight spots

Problem 1.11

Determine the optimum length and location of a patch on a constant rectangular cross-section microcantilever to obtain the best separation between the bending and the torsion resonant frequencies. The patch and the microcantilever have the same width w. Known are: $l = 200$ μm, $t_1 = 0.8$ μm, $t_2 = 1$ μm, $E_1 = 140$ GPa, $E_2 = 165$ GPa, $\rho_1 = 2500$ kg/m³, $\rho_2 = 2300$ kg/m³.

Problem 1.12

For a non-homogeneous paddle microcantilever, compare the bending resonant frequency obtained by using the actual distribution functions of Equations (1.147) through (1.151) to the one obtained by using the simplified distribution function of Equation (1.37).

Problem 1.13

The paddle microcantilevers of Figure 1.90 are fabricated from the same material and have the same thickness t. Calculate and compare the bending and torsional resonant frequencies of the two configurations for the particular case where $w_1 = 2w_2$.

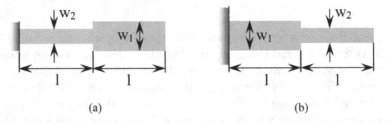

(a)   (b)

Figure 1.90 Two microcantilevers formed by serial connection of rectangular segments

Problem 1.14

A solid trapezoid microcantilever, as the one sketched in Figure 1.12, has specified values of the length l, minimum width w_1 and thickness t. Study the torsion-to-bending resonant frequency ratio as a function of the root width w_2.

Problem 1.15

By using the generic model of a homogeneous, two-segment constant rectangular cross-section microcantilever, find the precise distribution functions for bending and torsion for each segment of the circular corner-filleted design of Figure 1.91.

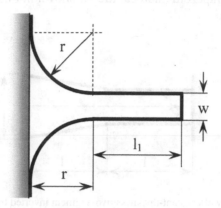

Figure 1.91 Top view of circular corner-filleted microcantilever with geometry

Problem 1.16

Calculate the resonant frequencies for the corner-filleted microcantilever of Figure 1.91 corresponding to out-of-the-plane bending and torsion.

Problem 1.17

A two-segment profile microcantilever with a rectangular cross-section of constant thickness t is shown in Figure 1.92. Determine its out-of-the-plane bending and torsion resonant frequencies: algebraically (symbolically)

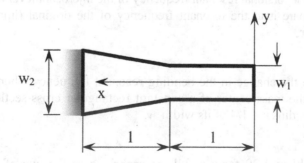

Figure 1.92 Top view of constant-thickness two-segment trapezoid microcantilever

and numerically for $l = 200$ μm, $w_1 = 15$ μm, $w_2 = 3\, w_1$, $t = 1$ μm, $E = 150$ GPa, $\mu = 0.28$, $\rho = 2400$ kg/m^3.

Problem 1.18

Solve the same problem as the one of the previous problem for the two-segment inverted trapezoid microcantilever shown in Figure 1.93.

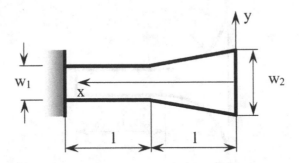

Figure 1.93 Top view of constant-thickness two-segment inverted trapezoid microcantilever

Problem 1.19

Holes need to be perforated in a constant rectangular cross-section microcantilever to reduce damping forces when the microcantilever operates in out-of-the-plane bending. Determine the change in the bending resonant frequency when four holes are perforated in a row along the longitudinal symmetry axis of the microcantilever. The diameter of a hole is $d = w/4$ (w is the microcantilever width), the spacing between two consecutive holes is w, and the spacing from the two edges to the extremity holes is $w/4$. Consider the numerical values: $w = 50$ μm, $t = 1$ μm, $E = 160$ GPa, $\rho = 2400$ kg/m^3.

Problem 1.20

Calculate the torsional resonant frequency of the microcantilever of Problem 1.19 and compare it to the resonant frequency of the original (imperforated) microcantilever.

Problem 1.21

Determine the change in the bending resonant frequency when a hole is perforated at the mass center of a constant rectangular cross-section micro-bridge with a radius $r = 1/4$ of its width w.

Problem 1.22

Solve Problem 1.20 for a paddle microbridge as the one of Fig. 1.63. Consider $l_1 = l_2$, $w_2 = 2w_1$, and $r = w_2/4$.

Problem 1.23

A paddle microcantilever has its paddle and root segments fabricated of different materials such that $E_1 = 0.6\,E_2$ and $\rho_1 = 0.8\,\rho_2$. Compare the bending resonant frequency of the inhomogeneous microcantilever to that of a homogeneous microcantilever, which is fabricated of the material with E_1 and ρ_1.

Problem 1.24

An unknown material attaches over the whole paddle of a microcantilever with a thickness equal to that of the original microcantilever. The resonant bending and torsion resonant frequencies are measured before and after attachment, and the mass of the deposited substance is also known. Evaluate the elastic modulus E and Poisson's ratio μ of the substance.

Problem 1.25

Deposition of an unknown substance is monitored by means of the change in the out-of-the-plane bending resonant frequency of a constant rectangular cross-section microcantilever. The options of functionalizing the microcantilever over half its length and whole length are considered. Which of the two variants will ensure the best frequency shift if the substance deposits in layers of equal thickness in both cases?

Problem 1.26

Use symbolic calculation to determine the out-of-plane bending and torsional resonant frequencies of the right circularly filleted microbridge whose top view is sketched in Figure 1.94 in terms of the geometric parameters r, w, and t (constant thickness) for a generic material defined by E and G. (Hint: Use the half-length model and the generic, compliance-based model).

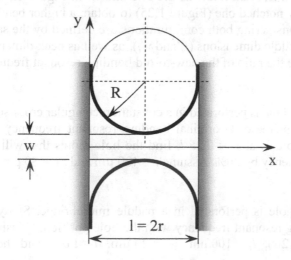

Figure 1.94   Top view of right circularly filleted microbridge

Problem 1.27

A hole is perforated in a constant rectangular cross-section microbridge at the midpoint. A cylinder having the dimensions of the perforated hole and the material properties of the original microbridge is deposited in the same position on an original microbridge. Calculate the bending resonant frequencies of the two modified configurations. Are the deviations from the original microbridge resonant frequency identical?

Problem 1.28

A circularly filleted microbridge and an elliptically filleted one (sketched in Figure 1.46) have the same length l, minimum width w, and thickness t. Compare the bending and torsion resonant frequencies of the two configurations.

Problem 1.29

Design a constant-thickness homogeneous paddle microbridge with an out-of-the-plane bending resonant frequency comprised in the ω_1 to ω_2 range. The microbridge has a maximum geometric envelope of $l \times w$ and the thickness is also specified by fabrication to t.

Problem 1.30

Compare the bending resonant frequencies of a circularly filleted microcantilever and of a constant rectangular cross-section one. Both configurations have the same thickness t, same length l, and the width of the constant cross-section design is equal to the tip width of the circularly filleted one.

Problem 1.31

A paddle microcantilever, as the one shown in Figure 1.21, is replaced by a circularly notched one (Figure 1.25) to obtain a higher bending resonant frequency. Considering both configurations are defined by the same constant thickness t, paddle dimensions (l_1 and w_1), as well as neck dimensions (l_2 and w_2), determine the ratio of the new-to-old bending resonant frequencies.

Problem 1.32

A circular hole is perforated in a constant rectangular cross-section microcantilever to increase its original bending resonant frequency. If the position of the hole is at $l_1 = 0.25\ l$, find the hole radius that will increase the resonant frequency by 20%. Assume $l = 400\ \mu m$ and $w = 80\ \mu m$.

Problem 1.33

A small hole is perforated in a paddle microbridge. Study the change in the bending resonant frequency with the hole position. Consider $r = w_1/4$, $l_2 = 2l_1$, $w_2 = 2w_1$, $l_1 = 100\ \mu m$, $w_1 = 20\ \mu m$, $t = 1\ \mu m$, and the material is polysilicon.

Problem 1.34

A constant rectangular cross-section microcantilever defined by a length l, width w, and thickness t is perforated in two variants: in the first variant, one hole is perforated at the symmetry center; in the second variant, eight holes are perforated at a distance $a = l/20$ from all the edges on two rows of four. Considering the pitch is the same about the x- and y-directions and equal to $p = 4r$ (r is the hole radius):

(a) Find a relationship between l and w.
(b) Compare the bending resonant frequencies of the two perforated micro-cantilevers.

Problem 1.35

Study the changes in the bending and torsional resonant frequencies of a bimorph cantilever with equal-length, equal-thickness layers in terms of the material properties. Consider $t = 1$ μm and $w = 40$ μm.

Problem 1.36

Solve Problem 1.35 by considering a bimorph microbridge instead of the microcantilever.

Problem 1.37

Find the density of an unknown substance that deposits in a layer-like manner at the tip of a constant rectangular cross-section microcantilever over a length of $l_1 = l/3$ (l is the cantilever length, its width is $w = l/6$, and the thickness is t_s) over the whole width and with a thickness $t_p = t_s/2$. Known is the shift in the bending frequency of $\Delta \omega_b$, E_s, E_t and ρ_s.

Problem 1.38

A symmetric trapezoid microbridge, as the one illustrated in Figure 1.48, is defined by $w_1 = 30$ μm, $w_2 = 70$ μm, $l = 300$ μm, and $t = 800$ nm. Find its material elastic properties E, G and μ, when the measured bending and torsion resonant frequencies are 1×10^6 Hz and 1×10^5 Hz, respectively, and the mass density is $\rho = 2200$ kg/m³. Assume the direct and shear elastic modulii are connected as: $G = E/[2(1+\mu)]$ where μ is Poisson's ratio.

Problem 1.39

A three-layer microbridge is studied, as the one analyzed in Example 1.26 and sketched in Figure 1.72, whose patches thickness is $t_p = 2t_s/3$ (t_s is the thickness of the supporting bridge) and length is $l_1 = l/2$ (l is the total length). Estimate its bending resonant frequency compared to the one of the microbridge alone (without the two patches).

Problem 1.40

Calculate the bending resonant frequency of the reversed trapezoid micro-bridge configuration shown in Figure 1.95 and compare it with the bending frequency of the trapezoid configuration of Figure 1.48.

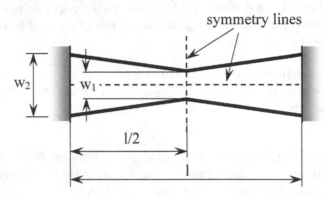

Figure 1.95 Top view of reversed trapezoid microbridge

Problem 1.41

Determine the torsion resonant frequency of the microbridge of Figure 1.95 and compare it to the similar frequency of the configuration shown in Figure 1.48.

Problem 1.42

The symmetric microbridge shown below in Figure 1.96 is designed to replace a constant rectangular cross-section microbridge having the same total length l, ($l = l_1 + 2r$) width w, and thickness t. Calculate the radius of the fillet, r, which will increase the bending resonant frequency by 25%. Known are: $r = 25$ μm, $l_1 = 150$ μm, $w = 10$ μm, and $t = 1$ μm.

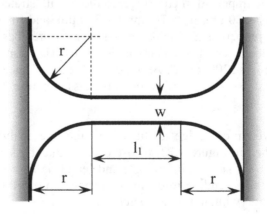

Figure 1.96 Top view of circular corner-filleted microbridge with geometry

Problem 1.43

A circular hole is perforated in a constant rectangular cross-section microcantilever and in an identical microbridge. If the dimensions of the two microresonators are l, w, and t, and knowing that the hole is positioned at the symmetry center, compare the two bending resonant frequencies as a function of the hole radius r.

Problem 1.44

Mass attaches in a point-like manner at the symmetry centers on a constant rectangular cross-section microcantilever and on an identical microbridge, as in Problem 1.43. Compare the two altered resonant frequencies.

References

1. S. Timoshenko, *Vibration Problems in Engineering*, D. van Nostrand Company, New York, 1928.
2. W.T. Thomson, *Theory of Vibrations with Applications*, Third Edition, Prentice Hall, Englewood Cliffs, 1988.
3. S.S. Rao, *Mechanical Vibrations*, Second Edition, Addison-Wesley Publishing Company, Reading, 1990.
4. N. Lobontiu, *Compliant Mechanisms: Design of Flexure Hinges*, CRC Press, Boca Raton, 2002.
5. N. Lobontiu, E. Garcia, *Mechanics of Microelectromechanical Systems*, Kluwer Academic Press, New York, 2004.
6. W.C. Young, R.G. Budynas, *Roark's Formulas for Stress and Strain*, Seventh Edition, McGraw-Hill, New York, 2002.
7. N. Lobontiu, *Mechanical Design of Microresonators: Modeling and Applications*, McGraw-Hill, New York, 2005.
8. N. Lobontiu, B. Ilic, E. Garcia, T. Reissman, H.G. Craighead, Modeling of nanofabricated paddle bridges for resonant mass sensing, *Review of Scientific Instruments*, 77, 2006, pp. 073301-2–073301-9.

Chapter 2

MICROMECHANICAL SYSTEMS: MODAL ANALYSIS

2.1 INTRODUCTION

This chapter studies the resonant/modal response of micromechanical systems by separating their components in either mass elements, which only contribute to the system's inertia, or spring elements, which only affect the overall elastic properties. The lumped-parameter method can thus conveniently be applied to model the free vibratory response of micromechanical systems. While some of them behave as single degree-of-freedom (DOF) systems, others undergo complex vibratory motion, which is defined by more than one DOF. For the latter category, it is possible at times to analyze each DOF individually, and such a motion is known as uncoupled. In other cases, two or more DOF combine in terms of stiffness and/or inertia, which make the respective motions to be coupled. Lagrange's equations are used to model the free response of multiple DOF micromechanical systems. Several example problems of mass-spring microsystems undergoing linear or/and rotary resonant vibrations are amply discussed and fully solved.

2.2 SINGLE DEGREE-OF-FREEDOM MASS-SPRING MICROMECHANICAL SYSTEMS

In many instances, microelectromechanical systems (MEMS) consist of masses (such as proof masses in accelerometers) and springs (such as microsuspensions), which perform either linear or rotary motion, depending on the characteristics of actuation and suspension. In such cases, it is advantageous to use lumped-parameter modeling, whereby the stiffness of the microsystem results solely from the spring suspension and the inertia fraction is given only by the proof mass. Some systems naturally behave as single DOF ones, with their motion being described by one physical parameter (commonly a displacement for translatory systems, and an angle for rotary one). For multiple DOF systems, where several physical coordinates define the motion, there are situations in which different motion types can be characterized individually, with no interference from other motions, and in such cases each individual motion can be regarded as a single DOF one. Chapter 1, which presented the resonant response of microcantilevers and microbridges, illustrated this category by studying the bending and torsion resonant frequencies separately. While Chapter 1 analyzed the cases in which inertia and stiffness fractions were produced by all segments composing a micromember, this subsection will assume that some segments of a microdevice are rigid, and therefore only contribute to the system's inertia, while others, which are compliant, behave as springs and therefore only alter the elastic properties of the microsystem. The concrete examples of paddle microcantilevers and paddle microbridges will be studied as belonging to this first single DOF category. The second category of single DOF micromechanical systems comprises designs that can behave as multiple DOF systems, but whose complex motion can be decomposed into several independent motions, each being equivalent to a single DOF system.

2.2.1 Paddle Microcantilever

The paddle microcantilever, whose sketch is shown in a three-dimensional (3D) view in Figure 2.1, is the simplest example in which the flexibility is provided by the slender root portion (the hinge) and the inertia results from the tip mass (the paddle). This example has been studied in Chapter 1 (see Figure 1.21 for the top-view dimensions) by assuming that both segments contribute flexibility and inertia to the overall system.

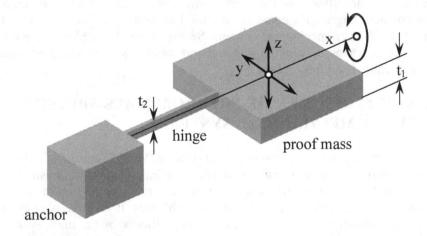

Figure 2.1 Paddle microcantilever with flexible, massless hinge and rigid proof mass

The in-plane bending (the hinge rotates about the z-axis) is defined by the following stiffness and mass:

$$\begin{cases} k_{b,z} = \dfrac{3EI_z}{l_2^3} = \dfrac{Ew_2^3 t_2}{4l_2^3} \\ m = \rho l_1 w_1 t_1 \end{cases} \qquad (2.1)$$

and the corresponding resonant frequency is:

$$\omega_{b,z} = \sqrt{\frac{k_{b,z}}{m}} = 0.5 \frac{w_2}{l_2} \sqrt{\frac{Et_2 w_2}{\rho l_1 l_2 t_1 w_1}} \qquad (2.2)$$

The out-of-the-plane bending resonant frequency (when the hinge rotates about the y-axis) is calculated similarly as:

$$\omega_{b,y} = \sqrt{\frac{k_{b,y}}{m}} = 0.5 \frac{t_2}{l_2} \sqrt{\frac{Et_2 w_2}{\rho l_1 l_2 t_1 w_1}} \qquad (2.3)$$

For a very thin rectangular cross-section, the lumped parameters defining the torsional resonant frequency are:

$$\begin{cases} k_t = \dfrac{GI_t}{l_2} = \dfrac{Gw_2 t_2^3}{3l_2} \\ J_t = \rho l_1 w_1 t_1 \dfrac{w_1^2 + t_1^2}{12} \end{cases} \qquad (2.4)$$

and therefore the corresponding resonant frequency is:

$$\omega_t = 2t_2 \sqrt{\frac{Gt_2 w_2}{\rho l_1 l_2 t_1 w_1 \left(t_1^2 + w_1^2\right)}} \qquad (2.5)$$

Example 2.1
Compare the torsion and out-of-the-plane resonant frequencies of a microcantilever in terms of the defining geometric parameters by using the lumped-parameter model previously derived.

Solution:
    By taking the ratio of the two frequencies (Equations (2.5) and (2.3)), the following equation is obtained:

$$\frac{\omega_t}{\omega_{b,y}} = 2\sqrt{2} \, \frac{l_2}{\sqrt{(1+\mu)(t_1^2 + w_1^2)}} \tag{2.6}$$

where μ is Poisson's ratio, which connects the E and G elastic modulii. When the two resonant frequencies are equal (the ratio of Equation (2.6) is equal to 1), the length of the hinge needs to be:

$$l_2^* = 0.35\sqrt{(1+\mu)(t_1^2 + w_1^2)} \tag{2.7}$$

The possible values of l_2^* are plotted in Figure 2.2 in terms of the thickness t_1 and width w_1 of the tip rigid segment and for polysilicon (with $\mu = 0.25$).

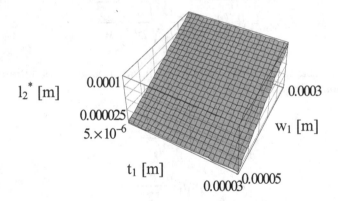

Figure 2.2 Limit length of hinge for equal bending and torsion resonant frequencies

Obviously, for values of the hinge length that are larger than the limit value of Equation (2.7), the torsional resonant frequency is larger than the out-of-the-plane bending one.

Example 2.2
    Compare the out-of-the-plane bending resonant frequencies as yielded by the lumped-parameter model (Equation (2.3)) and the fully compliant, full-inertia one (Equation (1.91)). Perform a similar comparison for the torsional resonant frequencies by using Equations (2.5) and (1.93). Assume the two segments have the same thickness (i.e., $t_1 = t_2 = t$).

Solution:

When the segments have identical thicknesses, the two resonant frequencies of Equations (2.3) and (2.5) simplify to:

$$\begin{cases} \omega_b^* = 0.5 \dfrac{t}{l_2} \sqrt{\dfrac{Ew_2}{\rho l_1 l_2 w_1}} \\[3mm] \omega_t^* = 2t \sqrt{\dfrac{Gw_2}{\rho l_1 l_2 w_1 \left(t^2 + w_1^2\right)}} \end{cases} \tag{2.8}$$

By combining Equation (2.8) with Equation (1.91), which formulates the bending resonant frequency of a paddle microcantilever where the equivalent stiffness and inertia fractions come from both segments, it is possible to express their ratio in terms of the width and length ratios as shown in the plot of Figure 2.3.

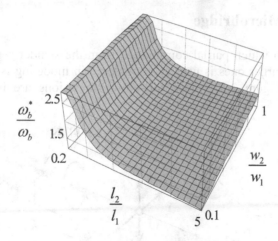

Figure 2.3 Out-of-the-plane bending resonant frequency ratio for a microcantilever: simplified model versus fully compliant, full-inertia model

A similar comparison is made between the two torsion resonant frequencies corresponding to the two models, Equation (2.58) and Equation (1.93). The ratio of the two frequencies is again plotted in terms of the length and width ratios, when $w_1 = 100$ μm and $t = 1$ μm.

As the two plots of Figures 2.3 and 2.4 indicate, the predictions by the simplified model are larger than the ones that consider compliance and inertia of both segments.

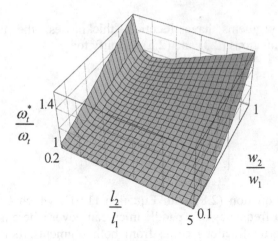

Figure 2.4 Torsional resonant frequency ratio for a microcantilever: simplified model versus fully compliant, full-inertia model

2.2.2 Paddle Microbridge

Another design in which partial compliance from the slender segments and inertia from the larger ones is usually considered in modeling is the paddle microbridge of Figure 2.5, whose top-view dimensions are indicated in Figure 1.55.

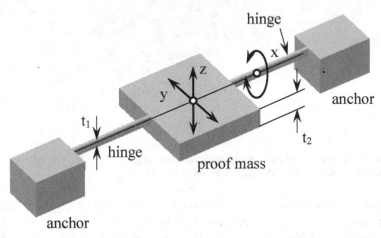

Figure 2.5 Paddle microbridge with flexible, massless hinges and rigid proof mass

Again, the main resonant motions are produced by in-plane and out-of-the-plane bending, as well as torsion of the end hinges. The mass is entirely provided by the proof mass and is considered lumped at the microbridge midpoint. The stiffness is produced by the two hinges, which act as two springs in parallel. In bending, the boundary conditions for a hinge are fixed and guided, because the middle mass undergoes pure translations during bending vibrations.

The following stiffness and mass correspond to in-plane bending (hinge rotation about the z-axis):

$$\begin{cases} k_{b,z} = 2\dfrac{12EI_z}{l_1^3} = \dfrac{2Ew_1^3 t_1}{l_1^3} \\ m = \rho l_2 w_2 t_2 \end{cases} \tag{2.9}$$

and the corresponding resonant frequency is:

$$\omega_{b,z} = \sqrt{\frac{k_{b,z}}{m}} = 1.41\frac{w_1}{l_1}\sqrt{\frac{Et_1 w_1}{\rho l_1 l_2 t_2 w_2}} \tag{2.10}$$

Similarly, the resonant frequency corresponding to out-of-the-plane vibrations (hinge rotation about the y-axis) is:

$$\omega_{b,y} = \sqrt{\frac{k_{b,y}}{m}} = 1.41\frac{t_1}{l_1}\sqrt{\frac{Et_1 w_1}{\rho l_1 l_2 t_2 w_2}} \tag{2.11}$$

The stiffness and inertia corresponding to torsion and very thin cross-sections are:

$$\begin{cases} k_t = 2\dfrac{GI_t}{l_1} = \dfrac{2Gw_1 t_1^3}{3l_1} \\ J_t = \dfrac{\rho l_2 w_2 t_2}{12}\left(t_2^2 + w_2^2\right) \end{cases} \tag{2.12}$$

Consequently, the torsion resonant frequency is:

$$\omega_t = \sqrt{\frac{k_t}{J_t}} = 2.82 t_1 \sqrt{\frac{Gt_1 w_1}{\rho l_1 l_2 t_2 w_2 \left(t_2^2 + w_2^2\right)}} \tag{2.13}$$

Example 2.3

Compare the torsion resonant frequency to the out-of-plane bending resonant frequency for a paddle microbridge by using the corresponding lumped-parameter models.

Solution:

By combining Equations (2.13) and (2.11), the ratio of the two resonant frequencies is expressed as:

$$\frac{\omega_t^*}{\omega_{b,y}} = \sqrt{2} \frac{l_1}{\sqrt{(1+\mu)(t_2^2 + w_2^2)}} \tag{2.14}$$

By enforcing that the out-of-the-plane bending resonant frequency be equal to the torsional one (the ratio of Equation (2.14) is one), results in the following condition:

$$l_1^* = 0.71\sqrt{(1+\mu)(t_2^2 + w_2^2)} \tag{2.15}$$

For $l_1 > l_1^*$, the torsional resonant frequency is larger than the out-of-the-plane bending one.

Example 2.4

Compare the out-of-the-plane bending resonant frequencies, as well as the torsional ones, by using the lumped-parameter models (Equations (2.11) and (2.13)) and the fully compliant, full-inertia ones (Equations (1.235) and (1.237)) for a paddle microbridge. Consider the three segments have the same thickness (i.e., $t_1 = t_2 = t$).

Solution:

The resonant frequencies of Equations (2.11) and (2.13) become:

$$\begin{cases} \omega_{b,y}^* = 0.71\dfrac{t}{l_1}\sqrt{\dfrac{Ew_1}{\rho l_1 l_2 w_2}} \\[4mm] \omega_t^* = 5.64t\sqrt{\dfrac{Gw_1}{\rho l_1 l_2 w_2 (t^2 + w_2^2)}} \end{cases} \tag{2.16}$$

A plot similar to the one of Figure 2.3 in which the paddle microcantilever has been studied is drawn for the paddle microbridge, as shown in Figure 2.6. A similar comparison is also made between the two torsion resonant frequencies corresponding to the two models. The ratio of the two frequencies is again plotted in terms of the length and width ratios, when $w_1 = 10$ μm and $t = 1$ μm.

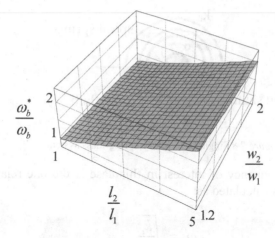

Figure 2.6 Out-of-the-plane bending resonant frequency ratio for a microbridge: simplified model versus fully compliant, full-inertia model

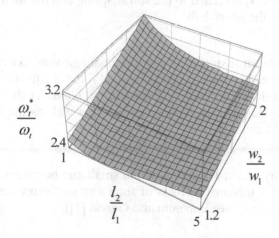

Figure 2.7 Torsional resonant frequency ratio for a microbridge: simplified model versus fully compliant, full-inertia model

As the two plots of Figures 2.6 and 2.7 suggest, the predictions by the simplified model are larger than the ones that consider compliance and inertia of both segments.

2.2.3 Rotary Motion Systems

A rigid body undergoing rotation about an axis that passes through its geometric center and perpendicular to its plane can be elastically supported by a torsion spring, as sketched in Figure 2.8.

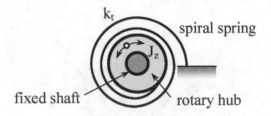

Figure 2.8 Single DOF rotary spring-mass system

The resonant frequency of interest in this case is the one related to the rotary motion and is calculated as:

$$\omega_z = \sqrt{\frac{k_t}{J_z}} \qquad (2.17)$$

where the stiffness k_t is related to the spiral spring and the moment of inertia J_z corresponds to the rotary hub.

Example 2.5

Find the resonant frequency of the torsional resonator sketched in Figure 2.8, whereby a central hub rotates about a fixed shaft and is elastically supported about the substrate by a spiral spring. Consider the case in which the spiral spring has a small number of turns as well as the design with a large number of turns.

Solution:

The geometry of a spiral spring with a small number of turns is shown in Figure 2.9, and the torsional stiffness of such a spring (expressed with respect to the free end) is (e.g., see Lobontiu and Garcia [1]):

$$k_t = \frac{2EI_z}{(r_1 + r_2)\alpha_{max}} \qquad (2.18)$$

Equation (2.18) is valid for a thin cross-section where bending effects are predominant. For relatively thick cross-sections, the torsional stiffness is (see also Lobontiu and Garcia [1]):

$$k_t = \frac{eAE}{\alpha_{max}} \qquad (2.19)$$

where e is the eccentricity between the cross-section centroidal axis and the neutral axis (the one where normal stresses are zero); the eccentricity (e.g., see Young and Budynas [2]) is calculated as:

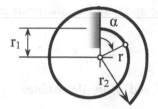

Figure 2.9 Geometry of a spiral spring with small number of turns

$$e = \begin{cases} r - \dfrac{w}{\ln\left[(2r+w)/(2r-w)\right]}, & for\ 0.6 < \dfrac{r}{w} < 8 \\ \dfrac{I_z}{rA}, & for\ 8 < \dfrac{r}{w} < 10 \end{cases} \tag{2.20}$$

The resonant frequency can easily be calculated by using the generic Equation (2.17) and either Equation (2.18) or (2.19) depending on the cross-section.

For a spiral spring with a large number of turns, the stiffness—as shown in Chironis [3], Wahl [4], or Lobontiu and Garcia [1]—is:

$$k_t = \frac{EI_z}{l} \tag{2.21}$$

where l is the length of the spiral. Equation (2.21) is identical to the one pertaining to a straight beam of length l under bending by a point moment.

Example 2.6

Compare the torsional resonant frequencies of two spiral-spring resonators predicted by the small and large number of turns models when considering the cross-section is thin and the spiral is logarithmic.

Solution:

When the accelerometers are identical, the large-to-small number of turns spiral spring resonant frequency ratio is:

$$r_\omega = \sqrt{\frac{(r_1 + r_2)\alpha_{\max}}{2l}} \tag{2.22}$$

The length of the spiral is calculated by integration of an element length dl:

$$l = \int_0^{\alpha_{\max}} dl = \int_0^{\alpha_{\max}} r\,d\alpha \tag{2.23}$$

In the case of a logarithmic spiral spring, the polar equation describing this curve is:

$$r = ae^{b\alpha} \tag{2.24}$$

The unknown constants a and b are determined by considering the limit conditions:

$$\begin{cases} r = r_1 \ for \ \alpha = 0 \\ r = r_2 \ for \ \alpha = \alpha_{max} \end{cases} \tag{2.25}$$

Therefore, Equation (2.24) becomes:

$$r = r_1 e^{\frac{\alpha}{\alpha_{max}} \ln \frac{r_2}{r_1}} \tag{2.26}$$

The length of the logarithmic spiral segment and the resonant frequency ratio can be calculated from Equations (2.23), (2.22), and (2.26) as:

$$\begin{cases} l = \alpha_{max} \dfrac{r_2 - r_1}{\ln \dfrac{r_2}{r_1}} \\[4em] r_\omega = \sqrt{\dfrac{r_2 + r_1}{2(r_2 - r)_1} \ln \dfrac{r_2}{r_1}} \end{cases} \tag{2.27}$$

By using the radii relationship $r_2 = cr_1$, the resonant frequency ratio of the second Equation (2.27) is plotted in Figure 2.10.

As Figure 2.10 illustrates, the resonant frequency prediction by the large number-of-turns model is slightly than the one by the small number of turns.

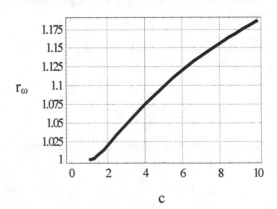

Figure 2.10 Torsional frequency ratio (Equation (2.27)) as a function of the radii ratio

Also suggested in the same Figure 2.10 is that the resonant frequency ratio increases with larger maximum-to-minimum radius ratios.

Example 2.7
    Determine the torsion-related resonant frequency of the rotary system of Figure 2.11, whereby the mobile hub is supported by n curved-beam springs. Study the influence of angle α on the resonant frequency.

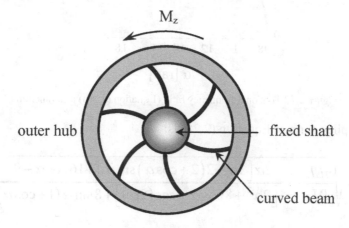

Figure 2.11 Set of curved beams acting as springs for the concentric hub-hollow shaft system

Solution:
    Equation (2.17) changes to:

$$\omega_z = \sqrt{\frac{nk_t}{J_z}} \qquad (2.28)$$

where k_t is the stiffness of one curved-beam spring and J_z is the mechanical moment of inertia of the mobile hub. The torsional stiffness of a single spring is given in Lobontiu and Garcia [1], based on Figure 2.12, and is not provided here, but in essence its equation is derived by expressing the relationship between the moment M_z and the corresponding angular deformation θ_z at the free end of the curved beam in Figure 2.12.

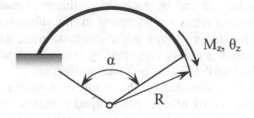

Figure 2.12 Curved-beam spring with geometry

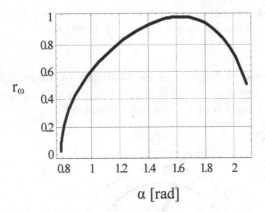

Figure 2.13 Torsional frequency ratio (Equation (2.30)) as a function of α

The explicit form of Equation (2.28) is:

$$\omega_z = \sqrt{\frac{nEI_z}{RJ_z} \times \frac{2\alpha\left[3\alpha - 2(2+\cos\alpha)\sin\alpha\right] + 16\cos\alpha - 9}{\alpha\left[2\alpha^2 + 8\cos\alpha + \cos(2\alpha)\right] + 8\sin\alpha(1+\cos\alpha)}} \quad (2.29)$$

One modality of studying the influence of angle α on the resonant frequency of Equation (2.29) is by means of the following ratio:

$$r_\omega = \frac{\omega_z(\alpha)}{\omega_z(\pi/2)} \quad (2.30)$$

and this function is plotted in Figure 2.13 in terms of α when α ranges from 1° to 120°. As Figure 2.13 indicates, the resonant frequency reaches a maximum at approximately α = 92°. In other words, the resonant frequency increases (nonlinearly) with the angle α up to 92° and then decreases (also nonlinearly) as the angle further increases past the 92° value.

2.2.4 Microaccelerometers

Microaccelerometers are employed to measure acceleration by using various means of transduction and by monitoring linear or rotary motion. Single DOF linear accelerometers are presented in this subsection in terms of their main resonant frequency. Rotary microaccelerometers, which are essentially single DOF systems, possibly supported by several dedicated springs, operate on the principle described in the previous subsection.

The bending free vibrations of the paddle microbridge have been analyzed in this chapter by means of a simple lumped-parameter model made up of a mass and two springs acting in parallel, as sketched in Figure 2.14.

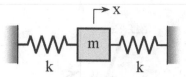

Figure 2.14 Single DOF system composed of a mass and two parallel springs

This system is a single DOF system, and its resonant frequency is calculated as:

$$\omega = \sqrt{\frac{2k}{m}} \qquad (2.31)$$

Microaccelerometers are often times designed based on the principle of Figure 2.14, and a few single-DOF accelerometers are analyzed next.

In what follows it will be assumed that the proof mass is defined by its mass *m* and that the hinge is defined by its in-plane dimensions *l* (the length) and w (width), whereas its thickness (the out-of-the-plane dimension) is *t*.

2.2.4.1 Inclined-Beam Microaccelerometer

The mass-spring microsystem of Figure 2.15 is used for linear motion about the indicated direction. Bending of the four beams ensures the alternative linear displacement of this system.

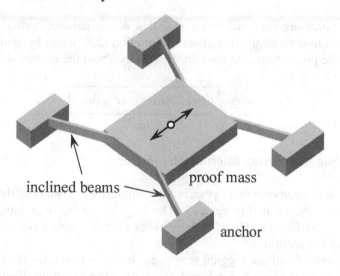

Figure 2.15 Single DOF accelerometer with four inclined beam-springs

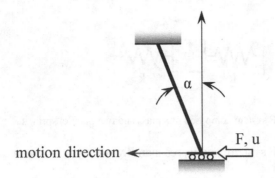

Figure 2.16 Load, displacement, and boundary conditions for an inclined beam-spring

The stiffness about the direction of motion for one inclined beam is derived by Lobontiu and Garcia [1] based on the geometry given in Figure 2.16 and by using a generic formulation where the hinge is not necessarily a constant cross-section one, as the case is here. The procedure of finding the stiffness about the motion direction implies applying the force F, determining the corresponding displacement u, and finding the ratio of the two, which is the stiffness. The formulation given by Lobontiu and Garcia [1] takes into consideration bending and axial deformations, and reduces for the constant rectangular cross-section hinge to:

$$k = Etw \frac{l^2 \sin \alpha + w^2 \cos \alpha}{l^3} \qquad (2.32)$$

Because there are four inclined beams that act as parallel springs, the total stiffness is four times the one given in Equation (2.32), and by also using the mass of the proof mass, the resonant frequency about the motion direction is:

$$\omega = \frac{2}{l} \sqrt{\frac{Etw \left(l^2 \sin \alpha + w^2 \cos \alpha \right)}{lm}} \qquad (2.33)$$

2.2.4.2 Saggital-Spring Microaccelerometer

Another accelerometer that operates as a single DOF system is the saggital-spring one, shown in Figure 2.17. The saggital spring was introduced by Lobontiu and Garcia [1], who derived its stiffness about the direction of motion of the proof mass.
The stiffness of half the saggital spring, as shown in Lobontiu and Garcia [1], is based on applying a force F and calculating the resulting displacement u (Figure 2.18) for half the saggital spring, namely:

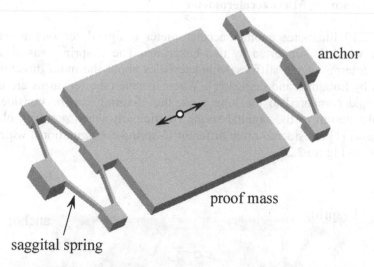

saggital spring

anchor

proof mass

Figure 2.17 Single DOF microaccelerometer with two end saggital springs

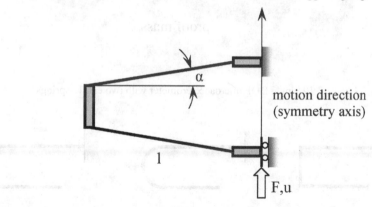

Figure 2.18 Load, displacement, and boundary conditions for half a saggital spring

$$k = \frac{6EI_z}{l^3 \cos^2 \alpha} \qquad (2.34)$$

A saggital spring is formed of two identical halves, which act as two springs in parallel, and therefore the total stiffness of a spring is twice the one given in Equation (2.34). By taking into account that two saggital springs are connected in parallel to the proof mass, as shown in Figure 2.17, the total stiffness is twice the one of a single spring; consequently, the resonant frequency of this accelerometer can be expressed as:

$$\omega = \frac{2\sqrt{6}}{l \cos \alpha} \sqrt{\frac{EI_z}{ml}} \qquad (2.35)$$

2.2.4.3 U-Spring Microaccelerometer

Figure 2.19 illustrates another accelerometer designed for one-directional motion, which is supported by two U-springs. The U-spring was also described in terms of the stiffness characteristics about the main directions of motion by Lobontiu and Garcia [1]. When its side (shorter) arms are considered rigid compared to the long ones, the U-spring can be considered a particular design of the saggital spring (particularly when $\alpha = 0$). Lobontiu and Garcia [1] considered three different U-spring configurations, which are sketched in Figure 2.20.

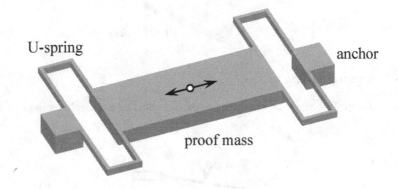

Figure 2.19 Single DOF microaccelerometer with two end U-springs

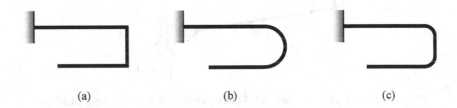

$$(a) \hspace{3cm} (b) \hspace{3cm} (c)$$

Figure 2.20 U-spring configurations with: (a) sharp corners; (b) semicircular corner; (c) rounded corners

The variant with sharp corners is only analyzed here, and the geometry of half a U-spring with sharp corners is given in Figure 2.21.

Lobontiu and Garcia [1] gave the stiffnesses about two in-plane and the out-of-the-plane directions for the three configurations shown in Figure 2.20 by taking the general case where the segments composing the three configurations are different. In case the three segments composing the U-spring of Figure 2.21 have the same constant cross-section, it can be shown(by taking the ratio of the force F and the corresponding displacement u at the same point about the motion direction) that the stiffness of half a U-spring is:

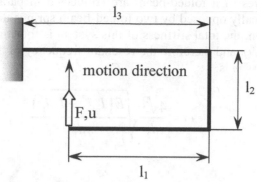

Figure 2.21 Geometry and boundary conditions of a half U-spring with sharp corners

$$k = \frac{3EI_z}{l_1^3 + 3l_1^2\left(l_2 + l_3\right) - 3l_1 l_3^2 + l_3^3} \tag{2.36}$$

The total stiffness corresponding to two springs is:

$$k_t = 4k \tag{2.37}$$

The resonant frequency of the U-spring accelerometer is:

$$\omega = 2\sqrt{3}\sqrt{\frac{EI_z}{\left[l_1^3 + 3l_1^2\left(l_2 + l_3\right) - 3l_1 l_3^2 + l_3^3\right]m}} \tag{2.38}$$

where m is the mass of the proof mass.

2.2.4.4 Folded-Beam Microaccelerometer

A proof mass that is elastically suspended and supported by two folded springs is shown in Figure 2.22. This configuration, too, can be considered as a single DOF system, when only the motion indicated in Figure 2.22 is of interest, and the two springs act in parallel. The stiffness about the motion direction has been derived by Lobontiu and Garcia [1] and is based on the static model of a half folded beam, illustrated in Figure 2.23. Assuming the two elastic members of the model sketched in Figure 2.22 have constant but different cross-sections, it can be shown (e.g., see Lobontiu and Garcia [1]) that the stiffness about the motion direction is:

$$k = 12E\left(\frac{I_{z1}}{l_1^3} + \frac{I_{z2}}{l_2^3}\right) \tag{2.39}$$

The two halves of a folded-beam are connected in parallel, and the proof mass is elastically opposed by two folded-beam springs that are connected in parallel. Again, the total stiffness of this system is four times the stiffness of Equation (2.39), and therefore the resonant frequency of the system having a point-like mass m is:

$$\omega = \frac{4\sqrt{3}}{l_1 l_2} \sqrt{\frac{E\left(I_{z1} l_2^3 + I_{z2} l_1^3\right)}{m l_1 l_2}} \tag{2.40}$$

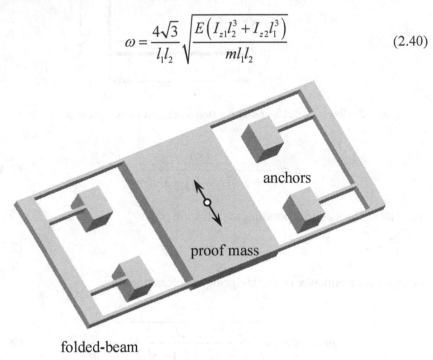

Figure 2.22 Single DOF microaccelerometer with two side folded-beam springs

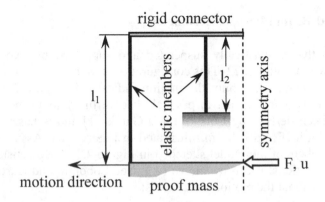

Figure 2.23 Geometry of a half folded-beam spring

2.2.4.5 Serpentine-Spring Microaccelerometer

The serpentine-spring microaccelerometer of Figure 2.24 is formed of a mass (which can be considered as a point-like mass) and two serpentine springs. The springs allow the point-like mass to undergo all six possible motions of a rigid solid, namely: in-plane translations about the x- and y-axes, out-of-the plane translation about the z-axis, and three rotations about the three axes mentioned above.

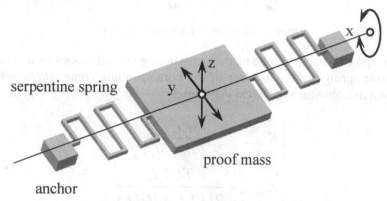

serpentine spring

proof mass

anchor

Figure 2.24 Serpentine-spring microaccelerometer

The top view of a microaccelerometer with two identical basic serpentine springs is sketched in Figure 2.25 and the dimensions of a basic serpentine spring are indicated in Figure 2.26.

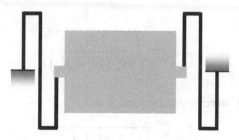

Figure 2.25 Top view of a microaccelerometer with basic serpentine springs

Two modal motions are studied here: the translation one about the x-axis and the rotation one about the same axis (Figure 2.24). The two motions can be regarded as independent and their resonant frequencies be found separately, although later in this chapter a problem of this type will be studied as a multiple DOF system, with the DOF being statically coupled.

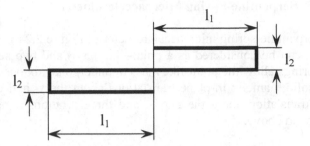

Figure 2.26 Geometry of a serpentine spring

An approximation can be made by assuming the stiffnesses of one basic serpentine spring are the inverses of the corresponding compliances, which are given in Lobontiu and Garcia [1], namely:

$$\begin{cases} K_{Fx-ux} = \dfrac{3EI_z}{2l_1^2\left(2l_1+3l_2\right)} \\[4mm] K_{Mx-\theta x} = \dfrac{EGI_yI_t}{2\left(EI_yl_2+2GI_tl_1\right)} \end{cases} \qquad (2.41)$$

where I_z and I_y are the spring cross-sectional moments of inertia related to bending, and I_t is the similar torsion-related moment of inertia. By taking into account that each modal motion of the serpentine-spring microaccelerometer has to include two springs that are in parallel, the translation frequency can be expressed as:

$$\omega_{tr} = \sqrt{\frac{3EI_z}{ml_1^2\left(2l_1+3l_2\right)}} \qquad (2.42)$$

whereas the rotation resonant frequency is:

$$\omega_{rot} = \sqrt{\frac{EGI_yI_t}{J_z\left(EI_yl_2+2GI_tl_1\right)}} \qquad (2.43)$$

Example 2.8
Design a serpentine-spring microaccelerometer whose translation resonant frequency is half the rotation resonant frequency. Consider the spring cross-section is rectangular with $w_s = 5\ t_s$ and $l_1 = 5\ l_2$. The proof mass is prismatic and its thickness is equal to the one of the serpentine spring. Consider the shear modulus is related to the longitudinal modulus as: $G = 0.4\ E$ and propose a design for the particular case where $t_s = 1\ \mu m$.

Solution:

By using Equations (2.42) and (2.43), together with the particular design conditions of this example, the following resonant frequency ratio is formulated:

$$\frac{\omega_{rot}}{\omega_{tr}} = 2.21 \frac{c_l}{\sqrt{1+c_w^2}} \tag{2.44}$$

with the non-dimensional parameters c_l and c_w defined as: $c_l = l_2/t_s$ and $c_w = w/t_s$. Figure 2.27 is the 3D plot of the frequency ratio.

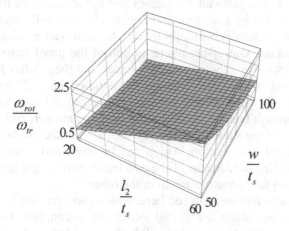

Figure 2.27 Rotation-to-translation resonant frequency ratio

By using the value of the 2 for this ratio, Equation (2.44) is reformulated as:

$$c_l = 0.9\sqrt{1+c_w^2} \tag{2.45}$$

and Figure 2.28 is the 2D plot of it.

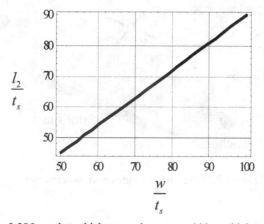

Figure 2.28 Length-to-thickness ratio versus width-to-thickness ratio

Numerically, when t_s = 1 µm, Equation (2.45) and the plot of Figure 2.28 enables expressing l_2 in terms of w. For a value of w = 100 µm it follows that l_2 = 90 µm. Given the other dimensional relationships of this problem, it follows that: l_1 = 450 µm and w_s = 5 µm. While the serpentine is fully designed, the length of the proof mass is not constrained mathematically, and therefore can be selected arbitrarily.

2.2.4.6 Bent-Beam Spring Microaccelerometer

Another accelerometer enabling planar motion is the bent-beam spring one shown in Figure 2.29. The relevant stiffnesses and compliances of the bent-beam spring are provided by Lobontiu and Garcia [1], as briefly mentioned in the following. This system is primarily used in translational motions about the planar x- and y- axes. Out-of-the-plane motion of the proof mass is also possible as well as rotations about the x- and y-axes, but these latter motions are mostly considered as parasitic and are not studied here. Figure 2.30 shows the geometry of a bent-beam spring. Motion about either x- or y-axis is chiefly realized through bending of the spring legs that are perpendicular to the motion direction. At the same time, the legs that are parallel to the motion direction undergo axial deformation, which can be neglected, compared to the bending. By adjusting the lengths of the two legs composing a bent-beam spring, it is possible to favor one motion over the other.

The simplest model will be analyzed here, which considers the two proof mass translation motions about the x- and y-axes are uncoupled. The (real) situation in which the motions are coupled will be analyzed later in this chapter, by considering the planar motion of the proof mass is a three DOF one.

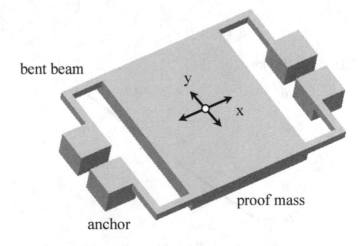

Figure 2.29 Bent-beam springs microaccelerometer

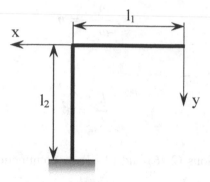

Figure 2.30 Top view of a bent-beam spring with geometric parameters

When the motion about the x-direction is only considered, it can simply be shown that the stiffness about that direction is (for one spring only):

$$k_x = \frac{3EI_z}{l_2^3} \tag{2.46}$$

Similarly, one spring's stiffness about the y-axis is:

$$k_y = \frac{3EI_z}{l_1^2 (l_1 + 3l_2)} \tag{2.47}$$

Example 2.9
Determine the legs proportion of a bent-beam spring in a four-spring microaccelerometer arrangement as the one of Figure 2.29, which will make the translation resonant frequency about the y-axis to be n times larger than the one about the x-axis.

Solution:
The translation resonant frequency of the microaccelerometer about the x-axis combines four springs in parallel and is:

$$\omega_{tr,x} = \sqrt{\frac{4k_x}{m}} = 2\sqrt{\frac{k_x}{m}} \tag{2.48}$$

where k_x is given in Equation (2.46) and m is the proof mass. An equation similar to Equation (2.48) can be formulated with k_y of Equation (2.47) instead of k_x. In requiring that:

$$\frac{\omega_{tr,y}}{\omega_{tr,x}} = n \qquad (2.49)$$

results in:

$$\frac{k_y}{k_x} = n^2 \qquad (2.50)$$

By using Equations (2.46) and (2.47) in conjunction with Equation (2.50) results in:

$$n = c\sqrt{\frac{c}{1+3c}} \qquad (2.51)$$

where $c = l_2/l_1$. Figure 2.31 is a 2D plot showing the variation of n (the resonant frequency ratio) in terms of c. The plot of Figure 2.31 indicates a quasi-linear increase of the frequency ratio of Equation (2.49) as a function of the leg length ratio. It can be seen, for instance, that for a frequency ratio of approximately 4, the length of the leg at the anchor needs to be seven times larger than the one of the other leg.

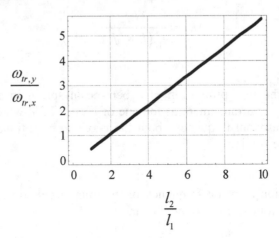

Figure 2.31 Resonant frequency ratio in terms of leg length ratio for a bent-beam spring microaccelerometer modeled as a single DOF system

2.3 MULTIPLE DEGREE-OF-FREEDOM MASS-SPRING MICROMECHANICAL SYSTEMS

The motion of multiple DOF mechanical microsystems is defined by several physical parameters (either linear displacement for a translatory DOF or angle for a rotary DOF). The simplest situation is that of a single member that is rigid (and therefore behaves as a mass element) and connects to several spring elements. The same MEMS device can perform two different types of motion, in terms of its actuation. One motion is taking place in a plane parallel to the substrate whereby the mass involved in the motion lends the whole microsystem the characteristics of a three DOF system, as three coordinates (two translation displacements about the x- and y-directions, and one rotation about the z-axis) define the complex motion. Another motion that can be emulated through actuation is out-of-the-plane and is generally a combination of the remaining three elementary motions: the z-translation, and the x- and y-rotations. The two motion categories and their DOF are sketched in Figure 2.32. More complex multi DOF MEMS contain more than one mass element, each mass element being capable of moving spatially about six DOF.

Formulating the dynamic equations that describe the vibratory free response of multiple DOF micromechanical systems is achieved by lumped-parameter modeling and utilization of Lagrange's equations method, which is introduced next for classical spring systems, as well as for MEMS-type spring ones. Single-mass and multiple-mass mechanical microsystem examples, with both in-plane and out-of-the-plane motions, are then studied.

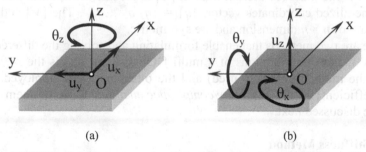

(a)   (b)

Figure 2.32 DOF of single-mass mechanical microsystem: (a) planar motion; (b) out-of-the-plane motion

2.3.1 Lagrange's Equations

For a multi DOF vibrating system where energy conserves, Lagrange's equations are:

$$\frac{d}{dt}\left(\frac{\partial T}{\partial \dot{q}_i}\right) - \frac{\partial T}{\partial q_i} + \frac{\partial U}{\partial q_i} = 0 \qquad (2.52)$$

where q_i are generalized coordinates and the dotted q_i are generalized velocities. T represents the kinetic energy and U is the potential energy of the multi DOF system. The generalized coordinates set comprises the minimum number of parameters that completely define the state of a (vibrating) system. More details on this theorem can be found in specialized texts (such as those of Thomson [5] or Timoshenko [6]). It is considered in the examples analyzed here that the generalized coordinates are actually physical displacements (linear and rotary), which define the positions of rigid bodies that are interconnected with various springs.

For a conservative system (where no forcing is exerted on the system), Equation (2.17) results in a n equations system, which governs the free vibrations of the system. Solving that system produces the resonant frequencies and the associated mode shapes.

The first step in Lagrange's equations method consists of identifying the independent motions of the system, which are identical to the number of DOF. Once the generalized coordinates q_i are determined, the kinetic and potential energy terms need to be formulated, followed by derivation of the dynamic equations of motion corresponding to the system's free response. The number of equations is equal to the number of DOF and they can all be collected in a matrix equation of the form:

$$[M]\{\ddot{q}\}+[K]\{q\} = \{0\} \qquad (2.53)$$

where $[M]$ is the mass (or inertia) matrix, $[K]$ is the stiffness matrix and $\{q\}$ is the generalized coordinates vector, $\{q\} = \{q_1\ q_2\ ...q_n\}^t$. The $[M]$ and $[K]$ matrices are of $n \times n$ dimension and are symmetric.

There are two methods that enable formulating and solving the differential equations of the free vibrations of a multi DOF system: one is the *stiffness method* (the method generally used) and the other one (less employed, but equally efficient) is the *flexibility/compliance method*. Each of them will briefly be discussed next.

2.3.1.1 Stiffness Method

Left multiplication of Equation (2.53) by the inverse of the mass matrix leads to:

$$\{\ddot{q}\}+[M]^{-1}[K]\{q\} = \{0\} \qquad (2.54)$$

The solution to the homogeneous system (2.54) is of harmonic form, namely:

$$\{q\} = \{Q\}\sin(\omega t) \qquad (2.55)$$

where $\{Q\}$ is a vector containing the amplitudes of the $\{q\}$ vector. Substituting Equation (2.55) into Equation (2.54) results in the following equation:

$$\left(-\omega^2[I]+[M]^{-1}[K]\right)\{Q\} = \{0\} \qquad (2.56)$$

which is a homogeneous system of n linear algebraic equations with n unknowns. In order for the system to have nontrivial solutions, it is necessary that the determinant of the matrix in the left-hand side of Equation (2.56) be 0, namely:

$$\det\left(-\lambda[I]+[D_s]\right)=0 \tag{2.57}$$

where:

$$\lambda=\omega^2 \tag{2.58}$$

is the *eigenvalue* corresponding to the resonant frequency ω, and:

$$[D_s]=[M]^{-1}[K] \tag{2.59}$$

is the *stiffness-based dynamic matrix* (the subscript s indicates it is based on *stiffness*). Equation (2.57)—the *characteristic equation*—is an n-degree algebraic equation that yields the eigenvalues λ. The theory of vibration shows that the first resonant frequency (or the natural frequency) and therefore its corresponding eigenvalue is limited as follows (e.g., see Thomson [5]):

$$\frac{1}{\sum_{i=1}^{n}(C_{ii}M_i)}<\lambda_1<\sum_{i=1}^{n}D_{s,ii} \tag{2.60}$$

where $D_{s,ii}$ are the diagonal terms of the dynamic matrix $[D_s]$ (the sum of the diagonal terms is the *trace* of that matrix), M_i are the diagonal terms of the mass matrix $[M]$ (which is usually diagonal), and C_{ii} are the diagonal terms of the *compliance matrix*, which is defined as the inverse of the stiffness matrix $[K]$.

For every eigenvalue, Equation (2.56) allows solving for an *eigenvector* $\{Q\}_i$. Through normalization, one component of $\{Q\}_i$ is equal to 1 and all other $n-1$ components are less than 1 in their absolute value. Knowing all the components of an eigenvector gives a visual representation of the corresponding *eigenmode*, which collects the relative amplitudes of all DOF at a resonant frequency (or eigenvalue).

2.3.1.2 Flexibility/Compliance Method

Another method that models the free-vibratory response of a multi DOF system is the flexibility/compliance method, because it uses the compliance matrix $[C]$ instead of the stiffness matrix $[K]$. It is known that $[C]$ is the inverse of $[K]$ and therefore, Equation (2.53) can be written as:

$$[M]\{\ddot{q}\}+[C]^{-1}\{q\} = \{0\} \tag{2.61}$$

By left multiplying Equation (2.61) by $[C]$ and by taking into account Equation (2.55), results in:

$$\left(-\frac{1}{\omega^2}[I]+[C][M]\right)\{Q\} = \{0\} \tag{2.62}$$

Again, nontrivial solution for $\{Q\}$ requires that:

$$\det\left(-\frac{1}{\lambda}[I]+[D_c]\right)=0 \tag{2.63}$$

where:

$$[D_c]=[C][M] \tag{2.64}$$

is the *compliance-based dynamic matrix* (the subscript c indicates the matrix is formulated based on *compliance*). Equation (2.63) is the characteristic equation corresponding to the compliance-based approach and the remaining steps in characterizing the modal response through this procedure are identical to those of the stiffness-based approach.

2.3.1.3 Classical Spring-Mass Systems

For systems that are formed of regular springs (designed for linear or rotary motion) and masses, applying Lagrange's equations of Equation (2.52) is straightforward. An example is fully solved next detailing all the steps involved in finding the eigenvalues, eigenvectors, and describing the eigenmodes for a two DOF system.

Example 2.10
   Analyze the modal response of the mechanical system of Figure 2.33 consisting of two rigid bodies undergoing translatory motions and two linear springs.

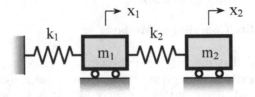

Figure 2.33 Two DOF mass-spring system

Solution:

The two coordinates x_1 and x_2 fully define the state of the system (namely, the positions of the two masses) and therefore this is a two DOF system. The kinetic energy is related to the masses and their velocities as:

$$T = \frac{1}{2}m_1\dot{x}_1^2 + \frac{1}{2}m_2\dot{x}_2^2 \qquad (2.65)$$

The potential energy is stored in the two springs, which deform when two non-zero displacements x_1 and x_2 are set. The potential energy is:

$$U = \frac{1}{2}k_1 x_1^2 + \frac{1}{2}k_2\left(x_2 - x_1\right)^2 \qquad (2.66)$$

Lagrange's equations applied to the kinetic and potential energy, Equations (2.65) and (2.66), result in:

$$\begin{cases} m_1\ddot{x}_1 + \left(k_1 + k_2\right)x_1 - k_2 x_2 = 0 \\ m_2\ddot{x}_2 - k_2 x_1 + k_2 x_2 = 0 \end{cases} \qquad (2.67)$$

In matrix form, Equation (2.67) is written as:

$$[M]\{\ddot{x}\} + [K]\{x\} = \{0\} \qquad (2.68)$$

where:

$$[M] = \begin{bmatrix} m_1 & 0 \\ 0 & m_2 \end{bmatrix} \qquad (2.69)$$

is the mass matrix and:

$$[K] = \begin{bmatrix} k_1 + k_2 & -k_2 \\ -k_2 & k_2 \end{bmatrix} \qquad (2.70)$$

is the stiffness matrix. The coordinate vector of Equation (2.68) is:

$$\{x\} = \begin{Bmatrix} x_1 \\ x_2 \end{Bmatrix} \qquad (2.71)$$

The mass matrix of Equation (2.69) is diagonal and because there are no non-zero off-diagonal terms, the system is *dynamically uncoupled*. On the other hand, this system is *statically coupled* because there are non-zero off-diagonal terms in the stiffness matrix of Equation (2.70), which indicates elastic connectivity between the two DOF.

As the case was with single DOF systems, the solution to Equation (2.68) is sought in harmonic form:

$$\{x\} = \begin{Bmatrix} X_1 \\ X_2 \end{Bmatrix} \sin(\omega t) = \{X\}\sin(\omega t) \tag{2.72}$$

where X_1 and X_2 are unknown amplitudes and ω is an arbitrary (also unknown) resonant frequency. By substituting Equation (2.72) into Equation (2.68), the latter changes to:

$$\left(-\omega^2[M]+[K]\right)\{X\}\sin(\omega t) = \{0\} \tag{2.73}$$

The characteristic equation corresponding to nontrivial solutions $\{X\}$ of Equation (2.73) is:

$$\omega^4 - \left(\frac{k_1+k_2}{m_1} - \frac{k_2}{m_2}\right)\omega^2 + \frac{k_1 k_2}{m_1 m_2} = 0 \tag{2.74}$$

The two roots of Equation (2.74) are:

$$\omega_{1,2}^2 = \frac{k_1 m_2 + k_2(m_1+m_2) \pm \sqrt{\left[k_1 m_2 + k_2(m_1+m_2)\right]^2 - 4k_1 k_2 m_1 m_2}}{2m_1 m_2} \tag{2.75}$$

In the situation in which the two bodies have identical masses, $m_1 = m_2 = m$, and the springs are also identical, $k_1 = k_2 = k$, the resonant frequencies become:

$$\omega_{1,2}^2 = \frac{3 \pm \sqrt{5}}{2}\frac{k}{m} \tag{2.76}$$

or, approximately:

$$\begin{cases} \omega_1 \approx 0.62\sqrt{\dfrac{k}{m}} \\[3mm] \omega_2 \approx 1.62\sqrt{\dfrac{k}{m}} \end{cases} \tag{2.77}$$

A direct consequence of the Equation system (2.73) being homogeneous is that absolute values of the amplitudes composing the vector $\{X\}$ cannot be determined because one component has to be arbitrary. As a result, the equation system can only provide amplitude ratios, in our case X_2/X_1 (or X_1/X_2) for either of the two resonant frequencies. Explicitly, Equation (2.73) can be written as:

$$\begin{cases} \left(\dfrac{2k}{m} - \omega_i^2 \right) X_1^{(i)} - \dfrac{k}{m} X_2^{(i)} = 0 \\[2mm] -\dfrac{k}{m} X_1^{(i)} + \left(\dfrac{k}{m} - \omega_i^2 \right) X_2^{(i)} = 0 \end{cases} \tag{2.78}$$

where $i = 1, 2$ (for the two resonant frequencies of Equation (2.77). For ω_1 of the first Equation (2.77), the following ratio is obtained from either of the two parts of Equation (2.78):

$$\frac{X_2^{(1)}}{X_1^{(1)}} = \frac{1+\sqrt{5}}{2} \tag{2.79}$$

Equation (2.79) suggests two things in relation to the vibratory state of the system corresponding to the resonant frequency ω_1: the first indication is that the two bodies move (vibrate) either about their positive directions x_1 and x_2 of Figure 2.33 or about the negative directions of the same axes; the second indication is that the amplitude of the second body (denoted by m_2 in the same Figure 2.33) is always larger than the one of the first body (because the amplitude ratio is larger than 1). This system motion is named *mode* and can be represented graphically as shown in Figure 2.34 (a).

Similarly, for ω_2 of Equation (2.77), both parts of Equation (2.78) produce the following amplitude ratio:

$$\frac{X_2^{(2)}}{X_1^{(2)}} = \frac{1-\sqrt{5}}{2} \tag{2.80}$$

The minus sign of this ratio shows that the two bodies move in opposite directions during the resonant motion ω_2 and that the amplitude of the second body is smaller than the one of the first body. The second mode is sketched in Figure 2.34 (b). In both parts of Figure 2.34, an amplitude of 1 was selected for the first body's vibratory motion, which results in the respective modes being *normalized*. A more explicit representation of the two modal motions is shown in Figure 2.35, in which the actual free vibrations of the two bodies are sketched for the two resonant frequencies of Equation (2.77).

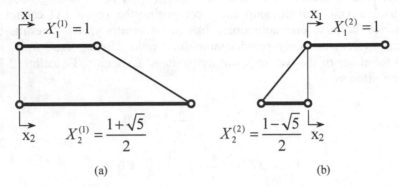

Figure 2.34 Modes for the two DOF mass-spring system: (a) first mode; (b) second mode

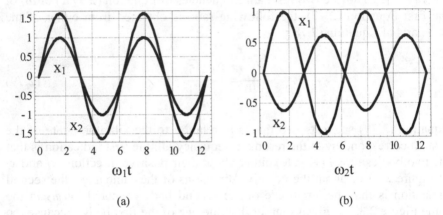

Figure 2.35 Resonant vibrations of the two DOF mass-spring system: (a) first mode; (b) second mode

2.3.1.4 Micromechanical Spring-Mass Systems

In micromechanical systems, the springs, masses, and their specific functions are not always clearly defined as in classical macro-domain systems, where springs (for instance) are designed for either linear or rotary motion and their inertia is usually disregarded. In MEMS, often times a vibrating flexible component contributes with both inertia and stiffness—illustrative examples are the microcantilevers and microbridges studied in Chapter 1. Moreover, a beam vibrating in a plane cumulates the functions of three different springs as its elastic properties are defined by three stiffnesses, namely: the direct linear K_l (connecting a force to the resulting deflection), direct rotary K_r (setting up the relationship between a moment and its corresponding rotation angle), and coupled (or cross) K_c (which relates a force to a rotation angle or a moment to a deflection).

Example 2.11

Demonstrate that the strain (potential energy) corresponding to the planar bending vibrations of a cantilever is expressed as a quadratic form in the free endpoint degrees of freedom u_z (deflection) and θ_y (slope/rotation).

Solution:

For a two DOF system, the quadratic-equation strain energy should be of the generic form:

$$U = au_z^2 + bu_z\theta_y + c\theta_y^2 \tag{2.81}$$

and therefore if constants a, b, and c can be found to satisfy Equation (2.81) and to be physically meaningful and adequate to this problem, it follows the problem's assertion was proven. By applying *Castigliano's first theorem*, the tip force and moment that correspond to the deflection and slope at the same point are derived from the bending strain energy U as:

$$\begin{cases} F_z = \dfrac{\partial U}{\partial u_z} = 2au_z + b\theta_y \\[2mm] M_y = \dfrac{\partial U}{\partial \theta_y} = bu_z + 2c\theta_y \end{cases} \tag{2.82}$$

In matrix form, Equation (2.82) is written as:

$$\begin{Bmatrix} F_z \\ M_y \end{Bmatrix} = \begin{bmatrix} 2a & b \\ b & 2c \end{bmatrix} \begin{Bmatrix} u_z \\ \theta_y \end{Bmatrix} \tag{2.83}$$

At the same time, as known from mechanics of materials, the load vector and the deformation/displacement one are related by means of the stiffness matrix as:

$$\begin{Bmatrix} F_z \\ M_y \end{Bmatrix} = \begin{bmatrix} K_l & K_c \\ K_c & K_r \end{bmatrix} \begin{Bmatrix} u_z \\ \theta_y \end{Bmatrix} \tag{2.84}$$

where K_l, K_c, and K_r are the direct linear, cross (coupled), and direct rotary stiffnesses. By comparing Equations (2.83) and (2.84), it follows that:

$$\begin{cases} a = \dfrac{1}{2}K_l \\[2mm] b = K_c \\[2mm] c = \dfrac{1}{2}K_r \end{cases} \tag{2.85}$$

and consequently, the quadratic form of Equation (2.81) becomes:

$$U = \frac{1}{2}K_l u_z^2 + K_c u_z \theta_y + \frac{1}{2}K_r \theta_y^2 \tag{2.86}$$

Equation (2.86) can also be written in matrix form as:

$$U = \frac{1}{2}\left\{\begin{matrix} u_z \\ \theta_y \end{matrix}\right\}^t \left[\begin{matrix} K_l & K_c \\ K_c & K_r \end{matrix}\right] \left\{\begin{matrix} u_z \\ \theta_y \end{matrix}\right\} \tag{2.87}$$

which is the second order formulation of the generic (*n*-th order for *n* DOF) equation of the strain (potential) energy for a linear elastic system.

Example 2.12
   Using the compliance approach, find the resonant frequencies and the related eigenvectors and eigenmodes for the cantilever beam with a tip mass of Figure 2.36 during free vibrations in the *xz* plane.

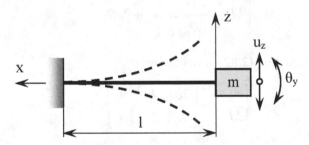

Figure 2.36 Cantilever with tip mass vibrating in the *xz* plane

Solution:
   Considering the stiffness results only from the flexible beam and the inertia from the tip mass, the system of Figure 2.36 vibrating in the *xz* plane behaves as a two DOF system, its generalized coordinates being the *z*-displacement of the mass (which is identical to the cantilever tip deflection u_z) and its rotation (which is the same as the slope at the cantilever's free end, θ_y).
   The elastic strain energy, as shown in the Example 2.9, is of quadratic form in u_z and θ_y, as shown in Equation (2.86), and the corresponding stiffness matrix is the one given in Equation (2.84). The flexibility/compliance approach needs the compliance matrix, which is the inverse of the stiffness matrix. For a constant cross-section microcantilever, and when considering the displacement vector at the free end is $\{u\} = \{u_z\ \theta_y\}^t$, the compliance matrix is:

$$[C] = \frac{1}{EI_y}\begin{bmatrix} \dfrac{l^3}{3} & \dfrac{l^2}{2} \\ \dfrac{l^2}{2} & l \end{bmatrix} \tag{2.88}$$

The kinetic energy of the two DOF system results from the z-translation and y-rotation of the tip mass, namely:

$$T = \frac{1}{2}m\dot{u}_z^2 + \frac{1}{2}J_y\dot{\theta}_y^2 = \frac{1}{2}\{\dot{u}_z \quad \dot{\theta}_y\}\begin{bmatrix} m & 0 \\ 0 & J_y \end{bmatrix}\begin{Bmatrix} \dot{u}_z \\ \dot{\theta}_y \end{Bmatrix} \tag{2.89}$$

which indicates the inertia matrix is:

$$[M] = \begin{bmatrix} m & 0 \\ 0 & J_y \end{bmatrix} \tag{2.90}$$

By solving the characteristic equation (Equation (2.63)), the following resonant frequencies are obtained:

$$\omega_{1,2}^2 = \frac{2EI_y}{J_y l^3 m}\left[3J_y + l^2 m \pm \sqrt{\left(3J_y + l^2 m\right)^2 - 3J_y l^2 m} \right] \tag{2.91}$$

Two eigenvectors, $\{U^{(1)}\}$ and $\{U^{(2)}\}$ (the letter U is used for modal amplitudes to make correspondence to the cantilever's free end displacement vector $\{u\}$), each having two components, $U_1^{(1)}$ and $U_2^{(1)}$ for the first eigenvector and first eigenfrequency of Equation (2.91) and $U_1^{(2)}$ and $U_2^{(2)}$ for the second eigenvector and second eigenfrequency of Equation (2.91). Because the amplitude of the harmonic motion of the first DOF (the translation u_z) has the dimension of length and the amplitude of the second DOF is an angle (measured in radians), it follows that the ratio of the two DOF modal amplitudes has either the dimension of a length or length to the power of -1, depending on the order in the ratio. To make those ratios non-dimensional, either multiplication or division by a length factor is necessary. For instance, the following non-dimensional ratios can be formulated:

$$\begin{cases} r_1 = l\dfrac{Q_2^{(1)}}{Q_1^{(1)}} = 1 + \dfrac{\sqrt{\left(3+r_i\right)^2 + 3r_i} - r_i}{3} \\[4mm] r_2 = l\dfrac{Q_2^{(2)}}{Q_1^{(2)}} = -1 + \dfrac{\sqrt{\left(3+r_i\right)^2 + 3r_i} + r_i}{3} \end{cases} \tag{2.92}$$

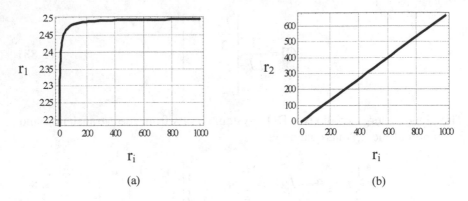

Figure 2.37 Modal amplitude ratios: (a) first mode; (b) second mode

with:

$$r_i = \frac{l^2 m}{J_y} \tag{2.93}$$

Figure 2.37 shows the variation of the two modal ratios of Equation (2.92) in terms of the inertia ratio of Equation (2.93). The two ratios are positive, which indicates that the two modal motions take place either both about the positive directions of the z- and y-axes or both about the negative directions of these axes (the motions are *synchronous*). The following limits are taken to the first non-dimensional ratio of Equation (2.92):

$$\begin{cases} \lim_{r_i \to 0} r_1 = 2 \\ \lim_{r_i \to \infty} r_1 = 2.5 \end{cases} \tag{2.94}$$

Equations (2.94) indicate that the transformed motion of the second DOF (θ_y, whose amplitude is U_2 and which by multiplication through l becomes a displacement) is always larger than the motion of the first DOF (the amplitude U_1 of u_z). Similar limits taken to the second ratio of Equation (2.92) results in:

$$\begin{cases} \lim_{r_i \to 0} r_2 = 0 \\ \lim_{r_i \to \infty} r_2 = \infty \end{cases} \tag{2.95}$$

Equations (2.95) also indicate that the two DOF motions for the second mode are synchronous. However, the magnitude relationship between the two motions is not immediately clear. By solving the equation $r_2 = 1$, the solution is $r_i = 9/7$, which, coupled with Equation (2.93), results in:

$$J_y^* = \frac{7}{9}l^2 m \tag{2.96}$$

For values of J_y that are smaller than the threshold value J_y^* of Equation (2.96), $r_i < 9/7$ and, as shown in the plot of Figure 2.37 (b), $r_2 < 1$, therefore the transformed motion of the second DOF is smaller than the one of the first DOF. The reverse relationship is in place for $J_y > J_y^*$ when $r_i > 9/7$ and $r_2 > 1$.

2.3.1.5 Rayleigh's Quotient in Matrix Form

Chapter 1 introduced and applied Rayleigh's quotient method to determine the resonant frequency in either bending or torsion for microcantilevers and microbridges. Essentially, because bending and torsion were decoupled, the two free vibratory motions have been treated separately, each as a single DOF system. Micromechanical systems, however, may possess several DOF, and for such cases, Rayleigh's quotient is formulated in matrix form as shown in the following.

A mechanical system defined by n DOF and which undergoes free undamped vibrations has the following kinetic and potential energies:

$$\begin{cases} T = \dfrac{1}{2}\{\dot{q}\}'[M]\{\dot{q}\} \\[2mm] U = \dfrac{1}{2}\{q\}'[K]\{q\} \end{cases} \tag{2.97}$$

By considering harmonic response of the form given in Equation (2.55), the maximum kinetic and potential energies are:

$$\begin{cases} T_{max} = \dfrac{1}{2}\omega^2 \{Q\}'[M]\{Q\} \\[2mm] U_{max} = \dfrac{1}{2}\{Q\}'[K]\{Q\} \end{cases} \tag{2.98}$$

where $\{Q\}$ is the amplitude vector of $\{q\}$, according to the same Equation (2.55). By equating the maximum kinetic energy to the maximum potential one (because the system is conservative), the resulting resonant frequency is expressed as:

$$\omega^2 = \frac{\{Q\}'[K]\{Q\}}{\{Q\}'[M]\{Q\}} = \sum_{i=1}^{n} D_{s,ii} \tag{2.99}$$

The second equality of Equation (2.99) is a consequence of Equation (2.60), which showed that the upper limit to the resonant frequency series is the trace of the stiffness-based dynamic matrix, and this limit is shown to be identical to the value calculated by means of Rayleigh's quotient approach (see Thomson [5] for more details).

2.3.2 Single-Mass Mechanical Microsystems

This section analyzes multiple DOF mechanical microsystems that are formed of a single rigid body (mass) and several springs that can be either simple (such as straight-line beams) or more complex, as the one studied in the Section 2.2 dedicated to microresonators as single DOF systems. A body in space has six DOF (three translations and three rotations), and therefore the systems analyzed in this section are six DOF ones. To simplify the corresponding analysis, it is generally possible (as mentioned previously) to split the problem into two subproblems, namely: one *planar* (which considers two translations in one plane and a rotation about an axis perpendicular to the plane), and the other *out-of-the-plane* (which focuses on the remaining three DOF: two rotations about axes contained in the plane of the first motion category and one translation about the axis perpendicular to the plane). This division is also suggested by the operation of MEMS, many of which vibrate in a plane parallel to the substrate (and therefore belong to the first category), such as microgyroscopes, microactuators, and microsensors, as well as many others that vibrate out-of-the-plane (and are therefore representative of the second category), such as torsional micromirrors or switches.

For each design category, two modeling methods are proposed: the first one uses the *geometry of deformation* and addresses systems of relatively simple geometries with springs placed symmetrically; the second methodology is *generic* and can model any single-mass multiple-spring mechanical microsystem with springs that can be dissimilar and located at different distances from the rigid body's center in a non-symmetric manner.

2.3.2.1 In-Plane Motion

The two modeling procedures mentioned previously will be discussed in this subsection that addresses the planar motion of single-mass, multiple-spring mechanical microsystems undergoing planar motion.

2.3.2.1.1 Formulation based on deformation geometry

A simple method, which is based on the geometry of deformation of the compliant members (the springs), can be developed to analyze the planar motion of single-mass, multiple-spring mechanical microsystems. The method, as shown in the following, can be applied to devices of relatively uncomplicated geometry with symmetric disposition of the springs.

Stiffnesses of Multi-Segment Springs

While for a single-segment (straight or curved line) spring, which deforms trough bending, there are two tip deformations corresponding to a point moment and force acting at that tip, as shown in Figure 2.38 (a), for a spring formed of at least two lines (straight or curved) that are not collinear, there are three tip deformations in relation to three agents: two point forces and one moment, as illustrated in Figure 2.38 (b).

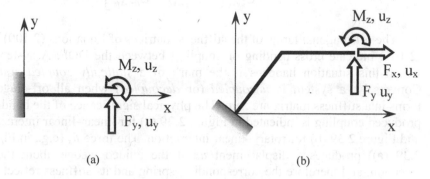

Figure 2.38 Loads and degrees of freedom at the free tip of a: (a) single segment beam; (b) two-segment beam

The load-deformation at the free tip of the single-segment spring is of the form:

$$\begin{Bmatrix} F_y \\ M_z \end{Bmatrix} = \begin{bmatrix} K_{Fy-uy} & K_{Fy-\theta z} \\ K_{Fy-\theta z} & K_{Mz-\theta z} \end{bmatrix} \begin{Bmatrix} u_y \\ \theta_z \end{Bmatrix} \qquad (2.100)$$

The stiffness subscripts indicate the corresponding load-deformation relationships. For the two-segment beam of Figure 2.38 (b), the counterpart of Equation (2.100) is:

$$\begin{Bmatrix} F_x \\ F_y \\ M_z \end{Bmatrix} = \begin{bmatrix} K_{Fx-ux} & K_{Fx-uy} & K_{Fx-\theta z} \\ K_{Fx-uy} & K_{Fy-uy} & K_{Fy-\theta z} \\ K_{Fx-\theta z} & K_{Fy-\theta z} & K_{Mz-\theta z} \end{bmatrix} \begin{Bmatrix} u_x \\ u_y \\ \theta_z \end{Bmatrix} \qquad (2.101)$$

As mentioned in Lobontiu and Garcia [1], for instance, there are two families of stiffnesses for these types of problems in which bending produces cross effects (a force can produce rotation and/or a moment results in a deflection). In cases in which compliances can easily be determined and compounded in matrices similar to the stiffness matrices of either Equation (2.100) or (2.101), the stiffness matrix is calculated by inverting the corresponding compliance matrix. The compliance matrix corresponding to the stiffness matrix of Equation (2.101) is of the generic form:

$$[C] = \begin{bmatrix} C_{ux-Fx} & C_{ux-Fy} & C_{ux-Mz} \\ C_{ux-Fy} & C_{uy-Fy} & C_{uy-Mz} \\ C_{ux-Mz} & C_{uy-Mz} & C_{\theta z-Mz} \end{bmatrix} \tag{2.102}$$

The off-diagonal terms of the stiffness matrices of Equations (2.100) and (2.101) indicate cross-bending or coupling between the DOF. A system in which this situation happens is the mark of a *statically coupled system*. Conversely, a system is *uncoupled* (or *decoupled*) when all off-diagonal terms in a stiffness matrix are zero. The physical significance of the bending-produced coupling is indicated in Figure 2.39 (a) for linear–linear interaction and Figure 2.39 (b) for rotary–linear interaction. The force F_x (e.g., in Figure 2.39 (a)) produces a displacement u_y of the guided wedge about the y-direction, and therefore the corresponding spring and its stiffness reflect this relationship in the subscript, K_{Fx-uy}. Similarly, rotation of the eccentric disk, as shown in Figure 2.39 (b) and that is caused by a moment M_z, determines the linear motion of the guided body about the y-direction. Consequently, the spring capturing this interaction has a stiffness denoted by K_{Mz-uy}.

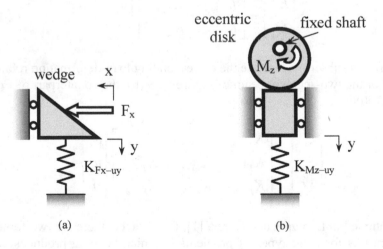

(a)   (b)

Figure 2.39 Physical representation of coupled stiffness: (a) linear–linear case; (b) rotary–linear case

Two-Beam Microaccelerometer

When the planar free vibratory motion of a generic accelerometer with two end beam springs (as the one whose top view is sketched in Figure 2.40) is considered, the system behaves as a three DOF system, its generalized coordinates being u_x, u_y, and θ_z (the motions of the plate's mass center), as indicated in Figure 2.41.

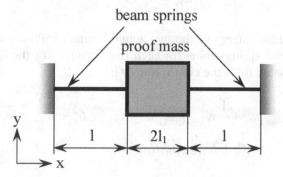

beam springs

proof mass

1        $2l_1$ 1

y

x

Figure 2.40 Top view of two-beam microaccelerometer

The coordinates of points 1' and 2' where the original points 1 and 2 displace after the proof mass moves by u_x, u_y, and θ_z are determined as:

$$\begin{cases} u_{1x} = u_x + l\left(1 - \cos\theta_z\right) \\ u_{1y} = u_y - l\sin\theta_z \end{cases}$$

(2.103)

and:

$$\begin{cases} u_{2x} = u_x - l\left(1 - \cos\theta_z\right) \\ u_{2y} = u_y + l\sin\theta_z \end{cases}$$

(2.104)

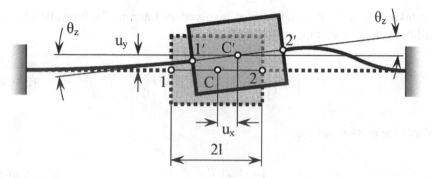

θ_z

u_y

1'   C'

2'

θ_z

1

C

2

u_x

2l

Figure 2.41 Top view of microaccelerometer with deformed beam-springs

By considering small deformations, it follows that $\sin\theta_z \approx \theta_z$ and $\cos\theta_z \approx 1$, and, therefore, Equations (2.103) and (2.104) simplify to:

$$
\begin{cases}
u_{1x} = u_{2x} \approx u_x \\
u_{1y} = u_y - l_1\theta_z \\
u_{2y} = u_y + l_1\theta_z
\end{cases}
\tag{2.105}
$$

The elastic potential energy of the two actual beams combines contributions from 18 individual springs (nine for each real spring). For the left spring of Figure 2.40, denoted by 1, the elastic energy is:

$$
U_1 = \frac{1}{2}K_{Fx-ux,1}u_{1x}^2 + \frac{1}{2}K_{Fy-uy,1}u_{1y}^2 + \frac{1}{2}K_{Mz-\theta z,1}\theta_z^2 + K_{Fx-uy,1}u_{1x}u_{1y}
$$
$$
+ K_{Fx-\theta z,1}u_{1x}\theta_z + K_{Fy-\theta z,1}u_{1y}\theta_z
\tag{2.106}
$$

The potential energy of the other spring can be expressed similarly to the equation above. By adding up the elastic energies of the two beams, and by taking into account the two springs behave as pure beams, it follows that: $K_{Fx-ux} = 0$, $K_{Fx-uy} = 0$ and $K_{Fx-\theta z} = 0$, and the total potential elastic energy becomes:

$$
U = K_{Fy-uy,1}u_y^2 + 2K_{Fy-\theta z,1}u_y\theta_z + \left(K_{Mz-\theta z,1} + l_1^2 K_{Fy-uy,1}\right)\theta_z^2
\tag{2.107}
$$

The kinetic energy corresponding to the inertia properties of the proof mass is:

$$
T = \frac{1}{2}m\dot{u}_y^2 + \frac{1}{2}J_z\dot{\theta}_z^2
\tag{2.108}
$$

By taking the U and T derivatives involved in Lagrange's formulation, the following matrix-form equation is obtained:

$$
[M]\begin{Bmatrix} \ddot{u}_y \\ \ddot{\theta}_z \end{Bmatrix} + [K]\begin{Bmatrix} u_y \\ \theta_z \end{Bmatrix} = \begin{Bmatrix} 0 \\ 0 \end{Bmatrix}
\tag{2.109}
$$

where the mass matrix is:

$$
[M] = \begin{bmatrix} m & 0 \\ 0 & J_z \end{bmatrix}
\tag{2.110}
$$

and the stiffness matrix is:

$$[K] = \frac{4EI_z}{l}\begin{bmatrix} \dfrac{6}{l^2} & \dfrac{3}{l} \\ \dfrac{3}{l} & 2\left(1+3\dfrac{l_1^2}{l^2}\right) \end{bmatrix} \qquad (2.111)$$

The stiffness matrix has been obtained after considering the following stiffness terms of a fixed-free beam (the beam-spring): $K_{Fy-uy,1} = K_{Fy-uy,2} = 12EI_z/l^3$, $K_{Fy-\theta z,1} = K_{Fy-\theta z,2} = 6EI_z/l^2$, $K_{Mz-\theta z,1} = K_{Mz-\theta z,2} = 4EI_z/l$. By solving the resulting characteristic equation, the following resonant frequencies are obtained:

$$\omega^2 = \frac{4EI_z}{J_z m l^3}\left\{3J_z + m\left(l^2 + 3l_1^2\right) \pm \sqrt{\left[3J_z + m\left(l^2 + 3l_1^2\right)\right]^2 - 3J_z m\left(l^2 + 12l_1^2\right)}\right\} (2.112)$$

Two-Spring Microaccelerometer

A proof mass that undergoes planar motion can be suspended over the substrate by means of two identical springs, instead of two beams, as the case was in the previous subsection. This design is sketched in Figure 2.42. Each spring will move at its tip, which is connected to the proof mass by two linear displacements (about its local x- and y-axes), and one rotation about the z-axis, which is perpendicular to the motion plane. The modal response for this microsystem will be analyzed where all six stiffnesses of a spring are taken into consideration (and the system is statically coupled) by using the Lagrange's equations method. The six stiffnesses are gathered in the stiffness matrix denoted by $[K_1]$ in Figure 2.42.

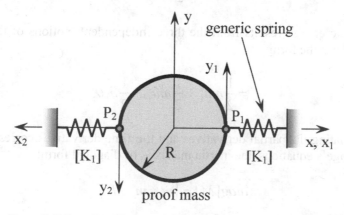

Figure 2.42 Top view of microaccelerometer with two generic springs

The displacements of the two points can be expressed in terms of the disk center's displacements u_x and u_y and its rotation θ_z under the assumption of small displacements. For point P_1, these displacements are:

$$\begin{cases} u_{1x} = u_x \\ u_{1y} = u_y + R\theta_z \end{cases} \tag{2.113}$$

and for P_2 they are:

$$\begin{cases} u_{2x} = -u_x \\ u_{2y} = -u_y + R\theta_z \end{cases} \tag{2.114}$$

The displacements of Equations (2.113) and (2.114) serve at expressing the strain energy that is stored in the two generic springs of the microaccelerometer sketched in Figure 2.42 when the disk is displaced by the quantities u_x and u_y and rotated by an angle θ_z, namely:

$$U = K_{Fx-ux,1}u_x^2 + K_{Fy-uy,1}u_y^2 + \left(K_{Mz-\theta z,1} + 2RK_{Fy-\theta z,1} + R^2 K_{Fy-uy,1} \right)\theta_z^2 \tag{2.115}$$
$$+ 2K_{Fx-uy,1}u_x u_y$$

By calculating the partial derivatives of U in terms of the three DOF, u_x, u_y, and θ_z, the resulting stiffness matrix is of the form given in Equation (2.101) with:

$$\begin{vmatrix} K_{Fx-ux} = 2K_{Fx-ux,1}; \ K_{Fx-uy} = 2K_{Fx-uy,1}; \ K_{Fx-\theta z} = 0; \\ K_{Fy-uy} = 2K_{Fy-uy,1}; \ K_{Fy-\theta z} = 0; \ K_{Mz-\theta z} = 2\left[K_{Mz-\theta z,1} + R\left(2K_{Fy-\theta z,1} + RK_{Fy-uy,1}\right) \right] \end{vmatrix} \tag{2.116}$$

The kinetic energy results from the three independent motions of the proof mass and has the form:

$$T = \frac{1}{2}m\dot{u}_x^2 + \frac{1}{2}m\dot{u}_y^2 + \frac{1}{2}J_z\dot{\theta}_z^2 \tag{2.117}$$

By calculating the partial derivatives and the time derivatives corresponding to Lagrange's equations, the inertia matrix is of diagonal form:

$$diag[M] = \{m \quad m \quad J_z\} \tag{2.118}$$

Having the stiffness and inertia matrices defined in Equations (2.116) and (2.118), the three resonant frequencies of the microsystem shown in Figure 2.42 can be determined in terms of the specific springs.

Bent-Beam Microaccelerometer

An example of the generic procedure derived previously, the bent-beam microaccelerometer of Figure 2.43, is analyzed next. Figure 2.44 is a sketch with the top view of a bent-beam spring showing the tip loading, which serves at determining the six compliances and related stiffnesses.

More details on finding the six compliances of a bent-beam spring are given in Lobontiu and Garcia [1]. In essence, when only bending is taken into account, the tip deflections (about the x- and y-axes) and tip rotation can be

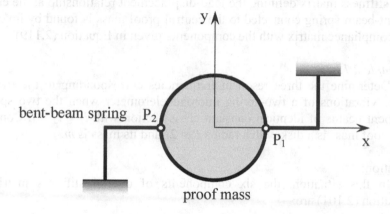

Figure 2.43 Top view of microaccelerometer with two bent-beam springs

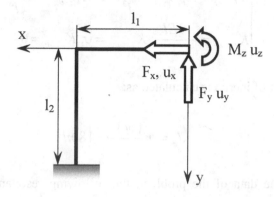

Figure 2.44 Bent-beam spring with geometry and tip loading

determined by using *Castigliano's second* (or *displacement*) *theorem*, which is based on the strain energy stored in the bent beam. When the two portions of the bent beam have the same (constant) cross-section and the lengths of Figure 2.44, the six stiffnesses in the matrix of Equation (2.101) are calculated by inverting the compliance matrix of Equation (2.102). Its terms are calculated from the model proposed by Lobontiu and Garcia [1], which are:

$$\left| \begin{matrix} C_{ux-Fx,1} = \dfrac{l_2^3}{3EI_z}; C_{ux-Fy,1} = \dfrac{l_1 l_2^2}{2EI_z}; C_{ux-Mz,1} = -\dfrac{l_2^2}{2EI_z} \\ C_{uy-Fy,1} = \dfrac{l_1^2\left(l_1+3l_2\right)}{3EI_z}; C_{uy-Mz,1} = -\dfrac{l_1\left(l_1+2l_2\right)}{2EI_z}; C_{\theta z-Mz,1} = \dfrac{l_1+l_2}{EI_z} \end{matrix} \right. \tag{2.119}$$

The stiffness matrix defining the load-displacement relationship at the end of a bent-beam spring connected to the central proof mass is found by inverting the compliance matrix with the components given in Equation (2.119).

Example 2.13
    Determine the three resonant frequencies corresponding to the free in-plane vibrations of a two-spring microaccelerometer when the two springs are bent beams of identical constant cross-section with $l_1 = l_2 = l$. Consider the proof mass is a disk with a radius $R = 2l$ and its mass is m.

Solution:
    In this situation, the six components of overall stiffness matrix of Equation (2.101) are:

$$\left| \begin{matrix} K_{Fx-ux} = K_{Fy-uy} = \dfrac{15EI_z}{l^3}; K_{Mz-\theta z} = \dfrac{655EI_z}{l^3} \\ K_{Fx-uy} = -\dfrac{9EI_z}{l^3}; K_{Fx-\theta z} = 0; K_{Fy-\theta z} = 0 \end{matrix} \right. \tag{2.120}$$

The moment of inertia is calculated as:

$$J_z = m\frac{\left(6l\right)^2}{2} = 18ml^2 \tag{2.121}$$

By using the data of the problem, the following resonant frequencies are obtained:

$$\begin{cases} \omega_1 = \dfrac{2.45}{l} \sqrt{\dfrac{EI_z}{ml}} \\[2ex] \omega_2 = \dfrac{4.9}{l} \sqrt{\dfrac{EI_z}{ml}} \\[2ex] \omega_3 = \dfrac{6}{l} \sqrt{\dfrac{EI_z}{ml}} \end{cases} \qquad (2.122)$$

Four-Spring Microaccelerometer

A similar problem is the one of a four-spring microaccelerometer. The in-plane free vibrations are first studied, followed by the out-of-plane motion. Figure 2.45 shows the configuration of a proof mass elastically supported by four generic and identical springs.

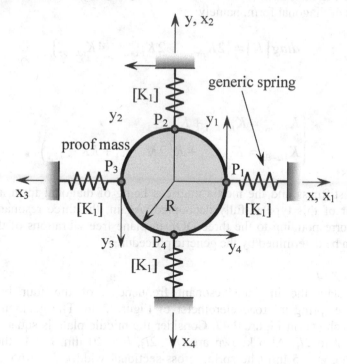

Figure 2.45 Top view of a four-spring accelerometer

When the plate center moves by the quantities u_x, u_y and rotates by an angle θ_z, the displaced positions of points P_1, P_2, P_3, and P_4 (indicated in Figure 2.45) are defined by the following quantities:

$$\begin{cases} u_{ix} = u_x \cos\alpha_i + u_y \sin\alpha_i \\ u_{iy} = -u_x \sin\alpha_i + u_y \cos\alpha_i + R\theta_z \end{cases} \quad (2.123)$$

where i = 1, 2, 3, and 4. In the case the four points are placed on the circumference, such that: $\alpha_1 = 0$, $\alpha_2 = \pi/2$, $\alpha_3 = \pi$, and $\alpha_4 = 3\pi/2$, the total strain energy can be expressed as:

$$U = \left(K_{Fx-ux,1} + K_{Fy-uy,1}\right)u_x^2 + \left(K_{Fx-ux,1} + K_{Fy-uy,1}\right)u_y^2$$
$$+2\left[K_{Mz-\theta z,1} + R\left(2K_{Fy-\theta z,1} + RK_{Fy-uy,1}\right)\right]\theta_z^2 \quad (2.124)$$

The kinetic energy corresponding to the proof mass is the one given in Equation (2.117) for the three DOF planar motion. By applying Lagrange's equations, the inertia matrix is the one of Equation (2.118), whereas the stiffness matrix is of diagonal form, namely:

$$diag[K] = \{2K_{Fx-ux} \quad 2K_{Fx-ux} \quad 4K_{Mz-\theta z}\} \quad (2.125)$$

with:

$$\begin{cases} K_{Fx-ux} = K_{Fx-ux,1} + K_{Fy-uy,1} \\ K_{Mz-\theta z} = K_{Mz-\theta z,1} + R\left(2K_{Fy-\theta z,1} + RK_{Fy-uy,1}\right) \end{cases} \quad (2.126)$$

The stiffness and the inertia matrices being of diagonal form, a microresonator of this type is fully decoupled. Again, the three resonant frequencies corresponding to the three DOF in-plane free vibrations of the proof mass can be determined by the generic procedure.

Example 2.14
Calculate the in-plane resonant frequencies of the four bent-beam serpentine spring microaccelerometer of Figure 2.46. The geometry of the spring is shown in Figure 2.47. Consider the middle plate is square with its side equal to $2l_1$. Also known are $l_1 = 2l_2$, $l_2 = 20$ μm, $t = 1$ (the device thickness), $w = 5$ μm (the spring cross-sectional width), $E = 165$ GPa, $\rho = 2500$ kg/m^3.

Solution:
The stiffnesses characterizing the elastic behavior of the bent-beam serpentine spring of Figure 2.47 are determined by calculating the inverse of the compliance matrix given in Lobontiu and Garcia [1]. For the particular data of this example, the relevant stiffnesses are:

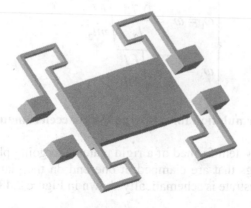

Figure 2.46 Microaccelerometer with four bent-beam serpentine springs

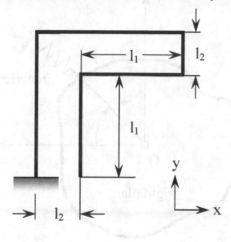

Figure 2.47 Geometry of bent-beam spiral spring

$$\begin{cases} K_{Fx-ux} = K_{Fy-uy} = \dfrac{3986EI_y}{7113l_2^3} \\ \\ K_{Mz-\theta z} = \dfrac{71776EI_z}{7113l_2} \end{cases} \qquad (2.127)$$

In this case, the radius of Figure 2.46 is calculated in terms the proof mass side as: $R = 2^{1/2} l_1$, whereas the area moment of the spring cross-section is $I_z = w^3 t/12$. The proof mass moment of inertia is $J_z = 8ml_2^2$, where m is the plate's mass. After carrying out the calculations, the following resonant frequencies are obtained:

$$\begin{cases} \omega_1 = \omega_2 = \dfrac{0.75}{l_2}\sqrt{\dfrac{EI_z}{ml_2}} \\[3mm] \omega_3 = \dfrac{1.12}{l_2}\sqrt{\dfrac{EI_z}{ml_2}} \end{cases} \qquad (1.128)$$

2.3.2.1.2 Generic formulation for *n*-spring microaccelerometer

A micromechanical system formed of a rigid plate undergoing planar motion and *n* identical springs that are clamped at one end on the plate and at the other on the fixed substrate is schematically shown in Figure 2.48.

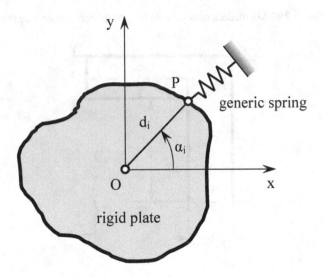

Figure 2.48 Solid plate with a generic spring in planar motion (the microsystem contains *n* springs)

This system is defined by three DOF, expressed by two translations and one rotation of a point belonging to the rigid plate (usually its center of mass/-symmetry). The lumped-parameter mass matrix is the one of Equation (2.118). The stiffness matrix will be formulated next, which will enable calculation of the three resonant frequencies associated with the system's planar motion and the three DOF.

   As mentioned previously, the two translation motions are coupled; more-over, the proof mass in its planar motion behaves as a three DOF system, the motions associated with its center of mass being the x and y translations and a rotation θ_z. Figure 2.49 indicates the displaced position of a proof mass, whose center moved by the quantities u_x and u_y, whereas the whole body rotates by an angle θ_z. This situation is characteristic of planar motions, and proof masses are a good illustration of this particular vibratory system (such as a microgyroscope).

It is of interest to determine the position of several points on the periphery of the proof mass after the planar motion of it. These points are locations where various springs can attach to the proof mass and therefore determining the displaced position of these points will enable evaluating the elastic deformations of the corresponding springs.

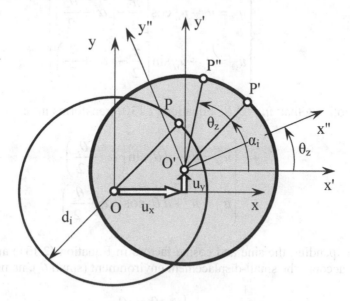

Figure 2.49 Three DOF of a solid in planar motion

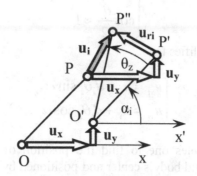

Figure 2.50 Displacement of a generic point P on a solid undergoing planar motion

Figures 2.49 and 2.50 show the planar motion of the solid in planar motion: initially, the point P is on the circumference of radius d_i and its angular position is defined by an angle α_i. The two translations of the solid, u_x and u_y, displace this point at P', whereas the third motion, the planar rotation θ_z, further moves the point at P" by means of the vector u_{ri}. Consequently, the total displacement u can be expressed as:

$$\bar{u}_i = \bar{u}_x + \bar{u}_y + \bar{u}_{ri} \qquad (2.129)$$

The Cartesian projections of this vector are:

$$\begin{cases} u_{xi} = u_x - u_{ri} \cos\left[\dfrac{\pi}{2} - \left(\alpha_i + \dfrac{\theta_z}{2}\right)\right] \\[4mm] u_{yi} = u_y + u_{ri} \sin\left[\dfrac{\pi}{2} - \left(\alpha_i + \dfrac{\theta_z}{2}\right)\right] \end{cases} \qquad (2.130)$$

By noticing that: $u_{ri} = d_i\,\theta_z$, Equation (2.130) transforms into:

$$\begin{cases} u_{xi} = u_x - d_i\theta_z \sin\left(\alpha_i + \dfrac{\theta_z}{2}\right) \\[4mm] u_{yi} = u_y + d_i\theta_z \cos\left(\alpha_i + \dfrac{\theta_z}{2}\right) \end{cases} \qquad (2.131)$$

By expanding the sine and cosine factors in Equation (2.131) and by taking into account the small-displacement environment (small θ_z), namely:

$$\begin{cases} \sin\dfrac{\theta_z}{2} \approx \dfrac{\theta_z}{2} \\[4mm] \cos\dfrac{\theta_z}{2} \approx 1 \end{cases} \qquad (2.132)$$

Equation (2.131) simplifies to:

$$\begin{cases} u_{xi} = u_x - d_i\theta_z \sin\alpha_i \\[2mm] u_{yi} = u_y + d_i\theta_z \cos\alpha_i \end{cases} \qquad (2.133)$$

Equation (2.133) enables one to find the position of a point laying at a distance d_i from the rigid body's center and positioned by angle α_i with respect to the x- (horizontal) axis when the center of mass of a solid moves in a plane by the quantities u_x and u_y and when the solid rotates by a small angle θ_z.

Equation (2.133) needs to be used in expressing the potential energy of the microoscillator formed of the central mass and n identical springs. As shown previously, each spring can be defined by six individual stiffness parameters, namely three direct-bending ones: K_{Fx-ux}, K_{Fy-uy}, $K_{Mz-\theta z}$, and three cross-bending ones: K_{Fx-uy}, $K_{Fx-\theta z}$, and $K_{Fy-\theta z}$. All these stiffnesses are expressed in the local frames of every individual spring, and these axes are, in general, not parallel to the global reference frame axes x and y. It is therefore necessary to express the relevant deformations of a spring (which give the potential energy of that spring in combination to the corresponding stiffnesses) in terms of the displacements at the spring end, which are determined by means of Equation (2.133). To achieve this goal, a coordinate transformation between the local frame of a generic spring and the global reference frame is performed, based on Figure 2.51, which allows formulating the following relationship:

$$\begin{Bmatrix} \Delta u_{xi} \\ \Delta u_{yi} \\ \theta_{zi} \end{Bmatrix} = \begin{bmatrix} \cos\alpha_i & \sin\alpha_i & 0 \\ -\sin\alpha_i & \cos\alpha_i & 0 \\ 0 & 0 & 1 \end{bmatrix} \begin{Bmatrix} u_{xi} \\ u_{yi} \\ \theta_z \end{Bmatrix} \qquad (2.134)$$

where the vector in the left-hand side collects the spring deformations about the local x_i and y_i axes, and the vector in the right-hand side was calculated in Equation (2.133). The rotation matrix of Equation (2.134) took into account that the rotation at point P equals the rotation of the rigid plate.

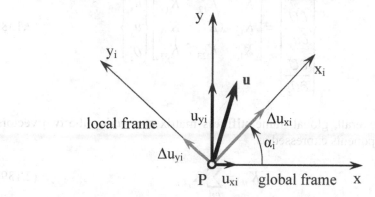

Figure 2.51 Global and local frames at the endpoint of a generic spring

By combining Equations (2.133) and (2.134), the following relationship is obtained connecting the spring deformations to the displacements of the plate's center:

$$\begin{Bmatrix} \Delta u_{xi} \\ \Delta u_{yi} \\ \theta_{zi} \end{Bmatrix} = \begin{bmatrix} \cos\alpha_i & \sin\alpha_i & 0 \\ -\sin\alpha_i & \cos\alpha_i & d_i \\ 0 & 0 & 1 \end{bmatrix} \begin{Bmatrix} u_x \\ u_y \\ \theta_z \end{Bmatrix} \qquad (2.135)$$

At the same time, the potential energy of the generic spring i expressed in terms of its stiffnesses and local deformations is:

$$U_i = \frac{1}{2}\left(K_{Fx-ux,i}\,\Delta u_{xi}^2 + K_{Fy-uy,i}\,\Delta u_{yi}^2 + K_{Mz-\theta z,i}\,\theta_{zi}^2\right) + K_{Fx-uy,i}\,\Delta u_{xi}\Delta u_{yi}$$
$$+ K_{Fx-\theta z,i}\,\Delta u_{xi}\theta_{zi} + K_{Fy-\theta z,i}\,\Delta u_{yi}\theta_{zi} \tag{2.136}$$

If Equation (2.135) is substituted in Equation (2.136), the potential energy of a generic spring will be a function of the system's three DOF, u_x, u_y, and θ_z, as well as of its stiffnesses. It is thus possible to calculate the total potential energy owing to the n springs as:

$$U = \sum_{i=1}^{n} U_i \tag{2.137}$$

By taking the partial derivatives of the total potential energy in terms of the three coordinates, u_x, u_y, and θ_z, the following equation is obtained:

$$\left\{ \begin{array}{c} \dfrac{\partial U}{\partial u_x} \\[2mm] \dfrac{\partial U}{\partial u_y} \\[2mm] \dfrac{\partial U}{\partial \theta_z} \end{array} \right\} = \begin{bmatrix} K_{11} & K_{12} & K_{13} \\ K_{12} & K_{22} & K_{23} \\ K_{13} & K_{23} & K_{33} \end{bmatrix} \left\{ \begin{array}{c} u_x \\ u_y \\ \theta_z \end{array} \right\} \tag{2.138}$$

where the overall, global-frame stiffness matrix connecting the two vectors has its components expressed as:

$$K_{jk} = \sum_{i=1}^{n} K_{jk,i} \tag{2.139}$$

with j, $k = 1, 2, 3$ for $j \le k$ and :

$$\begin{cases} K_{11,i} = K_{Fx-ux,i}\cos^2\alpha_i - K_{Fx-uy,i}\sin(2\alpha_i) + K_{Fy-uy}\sin^2\alpha_i \\[2mm] K_{12,i} = K_{Fx-uy,i}\cos(2\alpha_i) + \dfrac{K_{Fx-ux,i} - K_{Fy-uy,i}}{2}\sin(2\alpha_i) \\[2mm] K_{13,i} = \left(K_{Fx-\theta z,i} + d_i K_{Fx-uy,i}\right)\cos\alpha_i - \left(K_{Fy-\theta z,i} + d_i K_{Fy-uy,i}\right)\sin\alpha_i \\[2mm] K_{22,i} = K_{Fy-uy,i}\cos^2\alpha_i + K_{Fx-uy,i}\sin(2\alpha_i) + K_{Fx-ux,i}\sin^2\alpha_i \\[2mm] K_{23,i} = \left(K_{Fy-\theta z,i} + d_i K_{Fy-uy,i}\right)\cos\alpha_i + \left(K_{Fx-\theta z,i} + d_i K_{Fx-uy,i}\right)\sin\alpha_i \\[2mm] K_{33,i} = d_i^2 K_{Fy-uy,i} + 2 d_i K_{Fy-\theta z,i} + K_{Mz-\theta z,i} \end{cases} \tag{2.140}$$

To conclude, determining the total stiffness matrix in the global frame, the following steps have been followed:

1. Express the tip displacements of an individual spring in the global reference frame in terms of the system's three DOF.
2. Express the deformations of an individual spring (which are in the local reference frame) in terms of the system's three DOF.
3. Express the potential energy corresponding to one spring in quadratic form in terms of its deformations (in the local frame).
4. Express the potential energy of one spring in terms of the system's three DOF.
5. Sum all individual potential energies to obtain the total strain energy.
6. Calculate the partial derivatives of the total potential energy in terms of the system's DOF and determine the overall, global-frame stiffness matrix.

It should be noted that phases 5 and 6 are interchangeable, and therefore, if more convenient, partial derivatives of individual spring's potential energy should be calculated first, followed by summing of all the corresponding stiffness terms.

2.3.2.1.3 Fishhook spring microaccelerometer

A microaccelerometer with two fishhook springs, as sketched in Figure 2.52, is analyzed in terms of its in-plane free vibratory motion. A fishhook spring

Figure 2.52 Top view of a microaccelerometer with two fishhook springs

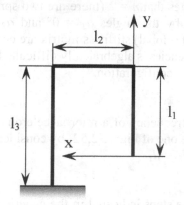

Figure 2.53 Geometry, boundary conditions, and planar coordinates of a fishhook spring as a half U-spring

is actually a half U-spring, as the one pictured in Figure 2.21. Figure 2.53 retakes this configuration and indicates the x- and y-axes as directions of the planar motion.

The generic inertia matrix is given in Equation (2.101), whereas the stiffness components of a fishhook spring are calculated by inverting the compliance matrix, which, for constant cross-section, is derived from Lobontiu and Garcia [1], and whose components are:

$$C_{ux-Fx} = \frac{l_1^3 + 3l_1^2 l_2 + l_3\left(3l_1^2 - 3l_1 l_3 + l_3^2\right)}{3EI_z} \tag{2.141}$$

$$C_{ux-Fy} = -\frac{l_2\left[l_1 l_2 + l_3\left(2l_1 - l_3\right)\right]}{2EI_z} \tag{2.142}$$

$$C_{ux-Mz} = \frac{l_1^2 + 2l_1 l_2 + l_3\left(2l_1 - l_3\right)}{2EI_z} \tag{2.143}$$

$$C_{uy-Fy} = \frac{l_2^2\left(l_2 + 3l_3\right)}{3EI_z} \tag{2.144}$$

$$C_{uy-Mz} = -\frac{l_2\left(l_2 + 2l_3\right)}{2EI_z} \tag{2.145}$$

$$C_{\theta z-Mz} = \frac{l_1 + l_2 + l_3}{EI_z} \tag{2.146}$$

Figure 2.52 also indicates that $n = 2$ (there are two springs) and that the two springs are positioned by the angles $\alpha_1 = 0°$ and $\alpha_2 = 180°$. Closed-form equations of the overall (global) stiffness matrix are complicated, and finding the three resonant frequencies is algebraically difficult. The following example will analyze a simplified configuration.

Example 2.15

   Find the resonant frequency of a microaccelerometer suspended by two fishhook springs, as the one of Figure 2.52, by considering that $l_1 = l_2 = l$, $l_3 = 2l_1$, $J_z = ml_1^2$, and $d_1 = d_2 = 4l_1$.

Solution:

   By following all the steps indicated in the generic procedure, the follow-ing overall, global-frame stiffness matrix is obtained:

$$[K^*] = \frac{EI_z}{59l} \begin{bmatrix} \dfrac{37}{l^2} & -\dfrac{21}{l^2} & 0 \\ -\dfrac{21}{l^2} & \dfrac{165}{l^2} & 0 \\ 0 & 0 & 3637 \end{bmatrix} \tag{2.147}$$

Equation (2.147) shows that the rotational DOF θ_z is decoupled from the translational ones, u_x and u_y, in terms of stiffness (statically). The mass matrix is simply:

$$[M] = m \begin{bmatrix} 1 & 0 & 0 \\ 0 & 1 & 0 \\ 0 & 0 & l^2 \end{bmatrix} \tag{2.148}$$

The three resonant frequencies can now be determined from the stiffness matrix (Equation (2.147)) and mass matrix (Equation (2.148)) as:

$$\begin{cases} \omega_1 = \dfrac{0.75}{l} \sqrt{\dfrac{EI_z}{ml}} \\[3mm] \omega_2 = \dfrac{1.69}{l} \sqrt{\dfrac{EI_z}{ml}} \\[3mm] \omega_1 = \dfrac{7.85}{l} \sqrt{\dfrac{EI_z}{ml}} \end{cases} \tag{2.149}$$

2.3.2.2 Out-of-the-Plane Motion

The two modeling procedures previously exposed (the geometry of deformation method and the generic method) for the planar motion problem can also be implemented to study out-of-the-plane free vibrations of single-mass, multiple-spring mechanical microsystems, as discussed in this subsection.

2.3.2.2.1 Geometry of deformation formulation

For relatively simple geometry configurations that are designed based on spring symmetry, the geometry of deformation method can be used, as exemplified next through two- and four-spring microaccelerometers.

Two-Spring Microaccelerometer

Consider the out-of-the plane motion of a two-spring microaccelerometer, whose side view of its undeformed state is shown in Figure 2.54. The particular case is studied here with the motion taking place in the *xz* plane.

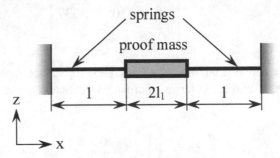

Figure 2.54 Side view of two-spring microaccelerometer

An arbitrary out-of-the-plane state of the proof mass and the deformed springs are uniquely defined by two geometric parameters connected to the proof mass center (as shown in Figure 2.55), namely: the *z*-translation, u_z, and *y*-rotation, θ_y.

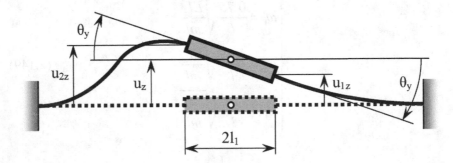

Figure 2.55 Side view of microaccelerometer with deformed springs

The relationships of the displacements of points 1 and 2 to the center plate motion are:

$$\begin{cases} u_{1z} = u_z - l_1\theta_y \\ u_{2z} = u_z + l_1\theta_y \end{cases} \tag{2.150}$$

The elastic potential energy of the two actual springs is expressed by adding up the elastic potential energies of eight individual springs (four for each actual spring) in the form:

$$U = \sum_{i=1}^{2} \left(\frac{1}{2} K_{Fz-uz,1} u_{iz}^2 + \frac{1}{2} K_{My-\theta y,1} \theta_y^2 + K_{Fz-\theta y,1} u_{iz} \theta_y \right) \quad (2.151)$$

where the subscript portion 1 shows the two springs are identical. By taking into account the two springs are identical, and by considering Equation (2.150), Equation (2.151) changes to:

$$U = K_{Fz-uz,1} u_z^2 + \left(K_{My-\theta y,1} + l_1^2 K_{Fz-uz,1} \right) \theta_y^2 + 2 K_{Fz-\theta y,1} u_z \theta_y \quad (2.152)$$

The kinetic energy of this system contains the parts contributed by the central plate in its translation about the z-axis and rotation about the y-axis, namely:

$$T = \frac{1}{2} m \dot{u}_z^2 + \frac{1}{2} J_y \dot{\theta}_y^2 \quad (2.153)$$

By applying now Lagrange's equations method, corresponding to the u_z and θ_y DOF, the following matrix-form equation is obtained:

$$[M]\begin{Bmatrix} \ddot{u}_z \\ \ddot{\theta}_y \end{Bmatrix} + 2[K]\begin{Bmatrix} u_z \\ \theta_y \end{Bmatrix} = \begin{Bmatrix} 0 \\ 0 \end{Bmatrix} \quad (2.154)$$

where the mass matrix is:

$$[M] = \begin{bmatrix} m & 0 \\ 0 & J_y \end{bmatrix} \quad (2.155)$$

and the stiffness matrix is:

$$[K] = \begin{bmatrix} K_{Fz-uz,1} & K_{Fz-\theta y,1} \\ K_{Fz-\theta y,1} & K_{My-\theta y,1} + l_1^2 K_{Fz-uz,1} \end{bmatrix} \quad (2.156)$$

The solutions to the characteristic equation are:

$$\omega_{1,2}^2 = \frac{B \pm \sqrt{4 m J_y \left[K_{Fz-\theta y,1}^2 - K_{Fz-uz,1} \left(K_{My-\theta y,1} + l_1^2 K_{Fz-uz,1} \right) \right] + B^2}}{m J_y} \quad (2.157)$$

with:

$$B = m\left(K_{My-\theta y,1} + l_1^2 K_{Fz-uz,1}\right) + J_y K_{Fz-uz,1} \qquad (2.158)$$

Example 2.16

Consider the two springs are simple constant cross-section beams of length $l = l_1$. Study whether it is possible to have a rotational resonant frequency that is n times the translational one, $\omega_1 = n\,\omega_2$.

<u>Solution:</u>

By using the following substitution:

$$m = xJ_y \qquad (2.159)$$

into the problem's condition that relates the two resonant frequencies, and by using the expressions of ω_1 and ω_2 (Equation (2.157)), the following equation is obtained:

$$\left(n^2 - 1\right)\left[K_{Fz-uz,1} + \left(K_{My-\theta y,1} + l_1^2 K_{Fz-uz,1}\right)x\right] = $$
$$\left(n^2 + 1\right)\sqrt{4K_{Fz-\theta y,1}^2 x + \left[K_{My-\theta y,1}x + K_{Fz-uz,1}\left(l_1^2 x - 1\right)\right]^2} \qquad (2.160)$$

The following stiffnesses are used in Equation (2.160):

$$\begin{cases} K_{Fz-uz,1} = \dfrac{12EI_y}{l_b^3} \\[3mm] K_{Fz-\theta y,1} = \dfrac{6EI_y}{l_b^2} \\[3mm] K_{My-\theta y,1} = \dfrac{4EI_y}{l_b} \end{cases} \qquad (2.161)$$

There are two solutions of x resulting from Equation (2.160), namely:

$$x_{1,2} = \dfrac{3\left[13n^4 - 6n^2 + 13 \pm \sqrt{13}\left(n^2 + 1\right)\sqrt{13n^4 - 38n^2 + 13}\right]}{128n^2 l_1^2} \qquad (2.162)$$

In order for the two roots x_1 and x_2 to be real, the quantity under the square root sign in Equation (2.162) needs to be positive, and this leads to the following conditions:

$$\begin{cases} 0 \le n \le 0.63 \\ n \ge 1.59 \end{cases} \tag{2.163}$$

Another condition relates to the solution x_2 as that one needs also to be positive, but it can simply be shown that the numerator in Equation (2.162) is always a positive number.

Four-Spring Microaccelerometer

The out-of-the-plane free vibrations of the four-spring microaccelerometer shown in Figure 2.45 are defined by three DOF of the plate center, namely: u_z, θ_x, and θ_y. If a rotation of the proof mass about the x-axis is considered together with another rotation about the y-axis, both superimposed to an out-of-the-plane translation about the z-axis, the resulting coordinates of the points 1, 2, 3, and 4 are for the linear case:

$$\begin{cases} u_{1z} = u_z - R\theta_y \\ u_{2z} = u_z + R\theta_x \\ u_{3z} = u_z + R\theta_y \\ u_{4z} = u_z - R\theta_x \end{cases} \tag{2.164}$$

These vibratory motions involve bending and torsion of each pair of springs. Plate center translation by a quantity u_z, for instance, involves bending of all springs, while plate rotation about the x-axis means the springs that are aligned with this axis will be subjected to torsion, whereas the other two springs will bend. Consequently, the strain energy stored by the four real springs due to bending is:

$$\begin{aligned} U_b = & \frac{1}{2} K_{Fz-uz,1} u_{1z}^2 + \frac{1}{2} K_{My-\theta y,1} \theta_y^2 + K_{Fz-\theta y,1} u_{1z} \theta_y \\ & + \frac{1}{2} K_{Fz-uz,1} u_{3z}^2 + \frac{1}{2} K_{My-\theta y,1} \theta_y^2 + K_{Fz-\theta y,1} u_{3z} \theta_y \\ & + \frac{1}{2} K_{Fz-uz,1} u_{2z}^2 + \frac{1}{2} K_{My-\theta y,1} \theta_x^2 + K_{Fz-\theta y,1} u_{2z} \theta_x \\ & + \frac{1}{2} K_{Fz-uz,1} u_{4z}^2 + \frac{1}{2} K_{My-\theta y,1} \theta_x^2 + K_{Fz-\theta y,1} u_{4z} \theta_x \end{aligned} \tag{2.165}$$

The strain energy corresponding to torsion of the four springs is:

$$U_t = 2\left(\frac{1}{2}K_{Mx-\theta x,1}\theta_y^2\right) + 2\left(\frac{1}{2}K_{Mx-\theta x,1}\theta_x^2\right) \tag{2.166}$$

The total elastic strain energy is the sum of the two energy components of Equations (2.165) and (2.166), namely:

$$U = U_b + U_t = K_{Fz-uz,1}\left[2u_z^2 + R^2\left(\theta_x^2 + \theta_y^2\right)\right]$$
$$+2K_{Fz-\theta y,1}u_z\left(\theta_x + \theta_y\right) + \left(K_{My-\theta y,1} + K_{Mx-\theta x,1}\right)\left(\theta_x^2 + \theta_y^2\right) \tag{2.167}$$

The kinetic energy involves the three motions of the plate, and is of the form:

$$T = \frac{1}{2}m\dot{u}_z^2 + \frac{1}{2}J_x\dot{\theta}_x^2 + \frac{1}{2}J_y\dot{\theta}_y^2 \tag{2.168}$$

By applying Lagrange's equations procedure, the following matrix equation results:

$$[M]\begin{Bmatrix}\ddot{u}_z\\\ddot{\theta}_x\\\ddot{\theta}_y\end{Bmatrix} + [K]\begin{Bmatrix}u_z\\\theta_x\\\theta_y\end{Bmatrix} = \begin{Bmatrix}0\\0\\0\end{Bmatrix} \tag{2.169}$$

where the mass matrix is:

$$[M] = \begin{bmatrix}m & 0 & 0\\0 & J_x & 0\\0 & 0 & J_y\end{bmatrix} \tag{2.170}$$

and the stiffness matrix is:

$$[K] = \begin{bmatrix}K_{11} & K_{12} & K_{13}\\K_{12} & K_{22} & 0\\K_{13} & 0 & K_{33}\end{bmatrix} \tag{2.171}$$

with:

$$K_{11} = 4K_{Fz-uz,1}; \; K_{12} = K_{13} = 2K_{Fz-\theta y,1};$$
$$K_{22} = K_{33} = 2\left(R^2 K_{Fz-uz,1} + K_{My-\theta y,1} + K_{Mx-\theta x,1}\right); \tag{2.172}$$

Example 2.17

Consider a circular plate of radius R and thickness t is supported by four identical beams of rectangular cross-section (w and t), as shown in Figure 2.56. Calculate the three resonant frequencies corresponding to the free out-of-the-plane vibrations of the proof mass and beams system. The beams are defined by a length of $l = 120$ μm and cross-sectional dimensions of $w = 8$ μm and $t = 0.5$ μm. The material is polysilicon with $E = 160$ GPa and $\rho = 2500$ kg/m^3. Consider three different values of the disk radius, namely: $R = 100$ μm, 150 μm, and 200 μm.

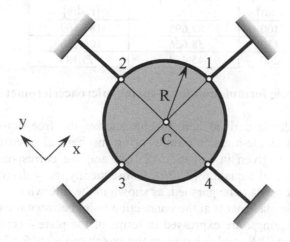

Figure 2.56 Accelerometer with four simple-beam springs

Solution:

In this case, the stiffnesses defining each of the four springs (expressed in local coordinate frames for each) are:

$$\left| \begin{matrix} K_{Fz-uz} = \dfrac{12EI_y}{l^3}; \; K_{Fz-\theta y} = \dfrac{6EI_y}{l^2}; \\ \\ K_{My-\theta y} = \dfrac{4EI_y}{l}; \; K_{Mx-\theta x} = \dfrac{GI_t}{l} \end{matrix} \right. \tag{2.173}$$

In Equations (2.173), I_y is the spring's cross-sectional moment of area calculated as: $I_y = wt^3/12$ and I_t is the spring's cross-sectional torsion moment of area (which for very thin cross-sections is $I_t = wt^3/3$, where w and t are the spring's constant cross-sectional width and thickness). The mass of the central disk is simply $m = \rho \pi R^2 t$. For a solid disk, the moments of inertia are: $J_x = J_y = mR^2/4$.

It is expected that this accelerometer produce a piston-type modal motion (out-of-the-plane z-translation) and two identical rotation modes of the disk about the x- and y- axes, the last two modal frequencies needing to be equal. However, due to round-off errors, the rotation resonant frequencies were slightly different for each of the three different radius values, and therefore the results for them have been averaged. These values are shown in Table 2.1.

Table 2.1 Resonant frequencies of the three DOF system

R [μm]	ω_1 [rad/s]	$\omega_2 = \omega_3$ [rad/s]
100	82,695	181,169
150	58,626	106,120
200	45,464	75,183

2.3.2.2.2 Generic formulation for n-spring microaccelerometer

A generic model is derived here to characterize the free out-of-the-plane vibrations of a single-mass, multiple-spring mechanical microsystem. The kinetic energy is given in Equation (2.168) and the corresponding inertia matrix is the one of Equation (2.170). Consequently, only derivation of the stiffness matrix needs to be pursued, as shown in the following.

The local displacements at the connection point between the central plate and a generic spring i are expressed in terms of the plate's center displacements, which are θ_x, θ_y, and u_z based on the sketch of Figure 2.57.

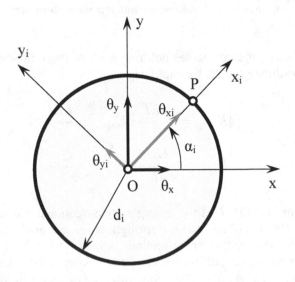

Figure 2.57 Angular DOF of a solid in out-of-the-plane motion

By following a procedure similar to the one applied for the planar motion, it can be shown that:

$$\begin{cases} \Delta u_{zi} = u_z + d_i \theta_x \sin \alpha_i - d_i \theta_y \cos \alpha_i \\ \theta_{xi} = \theta_x \cos \alpha_i + \theta_y \sin \alpha_i \\ \theta_{yi} = -\theta_x \sin \alpha_i + \theta_y \cos \alpha_i \end{cases} \qquad (2.174)$$

Equation (2.174) provides the deformations of a generic spring whose end is connected to a central plate when the center of mass of a solid moves out-of-the-plane plane by the quantities u_z, θ_x, and θ_y.

Each spring connecting to the central plate can be defined by six individual stiffness parameters, namely three direct ones: K_{Fz-uz}, $K_{Mx-\theta x}$, $K_{My-\theta y}$ and three cross-bending ones: $K_{Fz-\theta x}$, $K_{Fz-\theta y}$, and $K_{Mx-\theta y}$. These stiffnesses are expressed in the local frames and it is therefore necessary to express the local-frame spring deformations as a function of the system's three DOF.

The potential energy of a generic spring expressed in terms of local deformations is:

$$\begin{aligned} U_i = \frac{1}{2}\Big(& K_{Fz-uz,i} \Delta u_{zi}^2 + K_{Mx-\theta x,i} \theta_{xi}^2 + K_{My-\theta y,i} \theta_{yi}^2 + \Big) + K_{Fz-\theta x,i} \Delta u_{zi} \theta_{xi} \\ & + K_{Fz-\theta y,i} \Delta u_{zi} \theta_{yi} + K_{Mx-\theta y,i} \theta_{xi} \theta_{yi} \end{aligned} \qquad (2.175)$$

As the case was with the planar motion of the n-spring single-mass microsystem, the potential energy of Equation (2.175) is expressed in terms of the system's three DOF, u_z, θ_x, and θ_y by means of the transformation Equation (2.174). Then, again, the total strain energy is found by adding up all individual strain energies and, by calculating partial derivatives of the total potential energy, the following matrix equation is derived:

$$\begin{Bmatrix} \dfrac{\partial U}{\partial u_z} \\[2mm] \dfrac{\partial U}{\partial \theta_x} \\[2mm] \dfrac{\partial U}{\partial \theta_y} \end{Bmatrix} = \begin{bmatrix} K_{11} & K_{12} & K_{13} \\ K_{12} & K_{22} & K_{23} \\ K_{13} & K_{23} & K_{33} \end{bmatrix} \begin{Bmatrix} u_z \\ \theta_x \\ \theta_y \end{Bmatrix} \qquad (2.176)$$

where the terms K_{jk} of Equation (2.176) are calculated by means of the summation of Equation (2.139) with:

$$
\left\{
\begin{aligned}
K_{11,i} &= K_{Fz-\iota z,i} \\
K_{12,i} &= K_{Fz-\theta x,i}\cos\alpha_i - \left(K_{Fz-\theta y,i} - d_i K_{Fz-\iota z,i}\right)\sin\alpha_i \\
K_{13,i} &= K_{Fz-\theta x,i}\sin\alpha_i + \left(K_{Fz-\theta y,i} - d_i K_{Fz-\iota z,i}\right)\cos\alpha_i \\
K_{22,i} &= K_{Mx-\theta x,i}\cos^2\alpha_i + \left(d_i^2 K_{Fz-\iota z,i} - 2d_i K_{Fz-\theta y,i} + K_{My-\theta y,i}\right)\sin^2\alpha_i \\
&\quad - \left(K_{Mx-\theta y,i} - d_i K_{Fz-\theta x,i}\right)\sin(2\alpha_i) \\
K_{23,i} &= -\frac{d_i^2 K_{Fz-\iota z,i} - 2d_i K_{Fz-\theta y,i} - K_{Mx-\theta x,i} + K_{My-\theta y,i}}{2}\sin(2\alpha_i) \\
&\quad + \left(K_{Mx-\theta y,i} - d_i K_{Fz-\theta x,i}\right)\cos(2\alpha_i) \\
K_{33,i} &= \left(d_i^2 K_{Fz-\iota z,i} - 2d_i K_{Fz-\theta y,i} + K_{My-\theta y,i}\right)\cos^2\alpha_i + K_{Mx-\theta x,i}\sin^2\alpha_i \\
&\quad + \left(K_{Mx-\theta y,i} - d_i K_{Fz-\theta x,i}\right)\sin(2\alpha_i)
\end{aligned}
\right.
\qquad (2.177)
$$

Example 2.18

Determine the out-of-the-plane resonant frequency of the microresonator shown in Figure 2.58. The proof mass is supported by three identical beams of constant circular cross-section, which are placed at a 120° relative center angle. Consider that $d = l$ (l is the length of the beam) and $J_x = J_y = ml^2$.

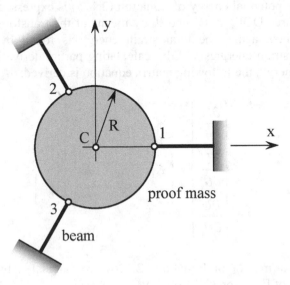

Figure 2.58 Three-beam microaccelerometer

Solution:

The compliance matrix of a constant cross-section beam expressing deformation-load relationships based on the three DOF of the plate u_z, θ_x, and θ_y is:

$$[C_i] = \frac{l}{EI_z} \begin{bmatrix} \dfrac{l^2}{3} & 0 & \dfrac{l}{2} \\ 0 & 1+\mu & 0 \\ \dfrac{l}{2} & 0 & 1 \end{bmatrix} \tag{2.178}$$

where it has been taken into account that $I_p = 2I_z = 2I_y$ and $G = E/[2(1+\mu)]$. By following up the procedure exposed previously, the overall global-frame stiffness matrix (which combines contributions from the three identical beam springs) is:

$$[K] = \frac{EI_z}{l^3} \begin{bmatrix} 36 & 0 & 0 \\ 0 & 43.2l^2 & 0 \\ 0 & 0 & 43.2l^2 \end{bmatrix} \tag{2.179}$$

Because the stiffness matrix of Equation (2.179) is diagonal, it follows that the microresonator of Figure 2.58 is statically decoupled, and because the inertia matrix for a lumped-parameter system of this type is also of diagonal form (which indicates dynamic decoupling) it follows that the analyzed mechanical microsystem is fully decoupled.

By using the numerical data of this example, the following resonant frequencies are obtained:

$$\begin{cases} \omega_1 = \dfrac{6}{l}\sqrt{\dfrac{EI_z}{ml}} \\[3mm] \omega_2 = \omega_3 = \dfrac{6.57}{l}\sqrt{\dfrac{EI_z}{ml}} \end{cases} \tag{2.180}$$

The first resonant frequency of Equation (2.180) is connected to the translation DOF u_z and therefore the resulting mode is a "piston"-type one, whereby the central plate translates about the z-axis and the beams are subjected to bending only. The other two modes have identical resonant frequencies and are connected to the rotations θ_x and θ_y.

2.3.3 Two-Mass Mechanical Microsystems

A system composed of two masses and two sets of springs (one set connecting the two bodies one to another, and the second spring set connecting the outer body to the substrate, as sketched in Figure 2.59) is analyzed here. Again, the in-plane and out-of-the-plane motions will be studied separately, and stiffness and inertia matrices will be derived for both design categories by means of generic modeling.

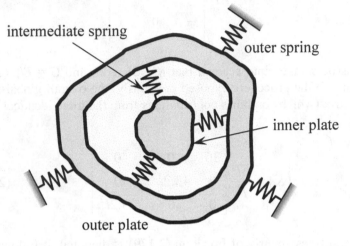

intermediate spring

outer spring

inner plate

outer plate

Figure 2.59 Two-mass multi-spring mechanical microsystem

The springs are just symbolically represented as linear springs—in actuality they can be any spring capable of elastically reacting spatially when acted upon by any of the six possible point loads. The assumption that the longitudinal spring axis passes through the two plates center (the plates are also assumed concentric) is used here. These two hypotheses do not reduce the generality of the problem. In addition (as shown in the solved examples), this situation is encountered in practical MEMS design.

2.3.3.1 In-the-Plane Motion

Figure 2.60 shows the main features of the microsystem's planar motion. This system has six DOF, as shown in Figure 2.61 (a), namely two translations and one rotation for each of the bodies denoted by *a* (the inner one) and *b* (the outer one). A generic inner spring denoted by *i* (there are *n* such springs) is fixed at points P (on the inner plate) and Q (on the outer one). A generic outer spring, *j*, is also shown (there can be *m* such springs), which is connected to the outer body and to the substrate.

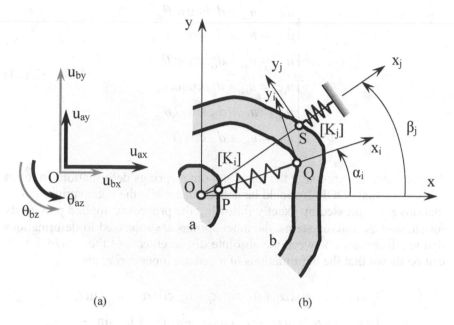

(a)   (b)

Figure 2.60 Two-mass, multi-spring mechanical microsystem in planar motion: (a) degrees of freedom; (b) model geometry

2.3.3.1.1 Inertia matrix

The six DOF can be arranged into the following vector:

$$\{u\} = \{u_{ax} \quad u_{ay} \quad \theta_{az} \quad u_{bx} \quad u_{by} \quad \theta_{bz}\}^t \qquad (2.181)$$

Assuming the two plates are defined by the masses m_a and m_b as well as by the mechanical moments of inertia J_{az} and J_{bz}, it is simple to demonstrate that the lumped-parameter inertia matrix corresponding to the DOF of Equation (2.181) is diagonal and of the form:

$$diag\,[M] = \{m_a \quad m_a \quad J_{az} \quad m_b \quad m_b \quad J_{bz}\} \qquad (2.182)$$

2.3.3.1.2 Stiffness matrix

The procedure here follows the steps covered in the similar problems of a single-mass, multi-spring system in planar motion. The first step, therefore, is expressing the global displacement components at points P, Q, and S (Figure 2.60 (b)) in terms of the system six DOF, namely:

$$\begin{cases} u_{Pxi} = u_{ax} - d_{Pi} \sin \alpha_i \theta_{az} \\ u_{Pyi} = u_{ay} + d_{Pi} \cos \alpha_i \theta_{az} \\ u_{Qxi} = u_{bx} - d_{Qi} \sin \alpha_i \theta_{bz} \\ u_{Qyi} = u_{by} + d_{Qi} \cos \alpha_i \theta_{bz} \\ u_{Sxj} = u_{bx} - d_{Sj} \sin \beta_j \theta_{bz} \\ u_{Syj} = u_{by} + d_{Sj} \cos \beta_j \theta_{bz} \end{cases} \qquad (2.183)$$

The next step is expressing the inner and outer springs deformations in terms of the system six DOF. It should be noticed that while the outer springs deformations are expressed by exactly following the procedure applied previously for single-mass microsystems, the inner springs are subjected to deformations that are differences between the absolute displacements of their endpoints. It can be shown that the deformations of a generic inner spring are:

$$\begin{cases} \Delta u_{xi} = u_{ax} \cos \alpha_i + u_{ay} \sin \alpha_i - u_{bx} \cos \alpha_i - u_{by} \sin \alpha_i \\ \Delta u_{yi} = -u_{ax} \sin \alpha_i + u_{ay} \cos \alpha_i + d_{Pi} \theta_{az} + u_{bx} \sin \alpha_i \\ \qquad\qquad - u_{by} \cos \alpha_i - d_{Qi} \theta_{bz} \\ \theta_{zi} = \theta_{az} - \theta_{bz} \end{cases} \qquad (2.184)$$

Similarly, the deformations of a generic outer spring j are:

$$\begin{cases} \Delta u_{xj} = u_{bx} \cos \beta_j + u_{by} \sin \beta_j \\ \Delta u_{yj} = -u_{bx} \sin \beta_j + u_{by} \cos \beta_j + d_{Sj} \theta_{bz} \\ \theta_{zi} = \theta_{bz} \end{cases} \qquad (2.185)$$

The potential energy stored by a generic spring is:

$$U_i = \frac{1}{2} \left(K_{Fx-ux,k} \Delta u_{xk}^2 + K_{Fy-uy,k} \Delta u_{yk}^2 + K_{Mz-\theta z,k} \theta_{zk}^2 \right) + K_{Fx-uy,k} \Delta u_{xk} \Delta u_{yk}$$
$$+ K_{Fx-\theta z,k} \Delta u_{xk} \theta_{zk} + K_{Fy-\theta z,k} \Delta u_{yk} \theta_{zk} \qquad (2.186)$$

where $k = i$ for the inner spring ($i = 1, 2, \ldots, n$) and $k = j$ for the outer spring ($j = 1, 2, \ldots, m$).

By substituting Equations (2.184) and (2.185) into the corresponding Equation (2.186), the individual potential energies of the inner and outer springs become functions of the system's six DOF. The total potential energy is determined by adding all individual spring contributions, and then the partial derivatives connected to Lagrange's equations (as shown for single-mass systems) are calculated, which enables formulating the 6 × 6 lumped-parameter stiffness matrix according to the DOF of Equation (2.181). The stiffness matrix components are calculated as:

$$K_{kl} = \sum_{i=1}^{n} K_{kl,i} \tag{2.187}$$

with $k, l = 1, 2, 3$ and $k \le l$, and as:

$$K_{kl} = \sum_{i=1}^{n} K_{kl,i} + \sum_{j=1}^{m} K_{kl,j} \tag{2.188}$$

with $k, l = 4, 5, 6$ and $k \le l$.

The individual stiffnesses for the inner spring entering Equations (2.187) and (2.188) are expressed per row. The first row contains the terms:

$$\begin{cases} K_{11,i} = -K_{14,i} = K_{44,i} = K_{Fx-ux,i}\cos^2\alpha_i - K_{Fx-uy,i}\sin(2\alpha_i) + K_{Fy-uy,i}\sin^2\alpha_i \\[2mm] K_{12,i} = -K_{15,i} = -K_{24,i} = K_{45,i} = K_{Fx-uy,i}\cos(2\alpha_i) + \dfrac{K_{Fx-ux,i}-K_{Fy-uy,i}}{2}\sin(2\alpha_i) \\[2mm] K_{13,i} = -K_{34,i} = \left(K_{Fx-\theta z,i}+d_{Pi}K_{Fx-uy,i}\right)\cos\alpha_i - \left(K_{Fy-\theta z,i}+d_{Pi}K_{Fy-uy,i}\right)\sin\alpha_i \\[2mm] K_{16,i} = -K_{46,i} = -\left(K_{Fx-\theta z,i}+d_{Qi}K_{Fx-uy,i}\right)\cos\alpha_i + \left(K_{Fy-\theta z,i}+d_{Qi}K_{Fy-uy,i}\right)\sin\alpha_i \end{cases} \tag{2.189}$$

The second line is formed of:

$$\begin{cases} K_{22,i} = -K_{25,i} = K_{55,i} = K_{Fy-uy,i}\cos^2\alpha_i + K_{Fx-uy,i}\sin(2\alpha_i) + K_{Fx-ux,i}\sin^2\alpha_i \\[2mm] K_{23,i} = -K_{35,i} = \left(K_{Fy-\theta z,i}+d_{Pi}K_{Fy-uy,i}\right)\cos\alpha_i \\[2mm] \qquad\quad +\left(K_{Fx-\theta z,i}+d_{Pi}K_{Fx-uy,i}\right)\sin\alpha_i \\[2mm] K_{26,i} = -K_{56,i} = -\left(K_{Fy-\theta z,i}+d_{Qi}K_{Fy-uy,i}\right)\cos\alpha_i \\[2mm] \qquad\quad -\left(K_{Fx-\theta z,i}+d_{Qi}K_{Fx-uy,i}\right)\sin\alpha_i \end{cases} \tag{2.190}$$

The elements of the third line are:

$$\begin{cases} K_{33,i} = d_{Pi}^2 K_{Fy-uy,i} + 2d_{Pi}K_{Fy-\theta z,i} + K_{Mz-\theta z,i} \\[2mm] K_{36,i} = -d_{Pi}d_{Qi}K_{Fy-uy,i} - \left(d_{Pi}+d_{Qi}\right)K_{Fy-\theta z,i} - K_{Mz-\theta z,i} \end{cases} \tag{2.191}$$

The element of the sixth line is:

$$K_{66,i} = d_{Qi}^2 K_{Fy-uy,i} + 2d_{Qi} K_{Fy-\theta z,i} + K_{Mz-\theta z,i} \qquad (2.192)$$

The elements corresponding to the outer spring are:

$$\begin{cases} K_{44,j} = K_{Fx-ux,j} \cos^2 \beta_j - K_{Fx-uy,j} \sin(2\beta_j) + K_{Fy-uy,j} \sin^2 \beta_j \\[2mm] K_{45,j} = K_{Fx-uy,j} \cos(2\beta_j) + \dfrac{K_{Fx-ux,j} - K_{Fy-uy,j}}{2} \sin(2\beta_j) \\[2mm] K_{46,j} = \left(K_{Fx-\theta z,j} + d_{Sj} K_{Fx-uy,j} \right) \cos \beta_j - \left(K_{Fy-\theta z,j} + d_{Sj} K_{Fy-uy,j} \right) \sin \beta_j \\[2mm] K_{55,j} = K_{Fy-uy,j} \cos^2 \beta_j + K_{Fx-uy,j} \sin(2\beta_j) + K_{Fx-ux,j} \sin^2 \beta_j \\[2mm] K_{56,j} = \left(K_{Fy-\theta z,j} + d_{Sj} K_{Fy-uy,j} \right) \cos \beta_j + \left(K_{Fx-\theta z,j} + d_{Sj} K_{Fx-uy,j} \right) \sin \beta_j \\[2mm] K_{66,j} = d_{Sj}^2 K_{Fy-uy,j} + 2d_{Sj} K_{Fy-\theta z,j} + K_{Mz-\theta z,j} \end{cases} \qquad (2.193)$$

Example 2.19

Determine the resonant frequencies of the microgyroscope sketched in Figure 2.61 by considering that $J_{az} = J_{bz} = 4ml^2$. The four line springs are identical and their length is l. Known are also $Q_1Q_2 = 4l$ and $S_1S_2 = 6l$. Consider only the bending deformations of the springs.

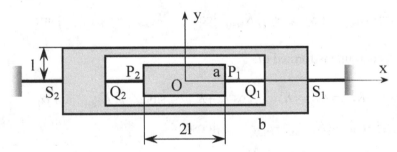

Figure 2.61 Two-mass, four-beam microgyroscope in planar motion

Solution:

The distances of interest are $d_{P1} = d_{P2} = l$, $d_{Q1} = d_{Q2} = 2l$, $d_{S1} = d_{S2} = 3l$ and the position angles are $\alpha_1 = \beta_1 = 0°$, $\alpha_2 = \beta_2 = 180°$. The system has only four DOF because the axial deformation of the beams not being accounted for, the displacements u_{ax} and u_{bx} are zero. This also caries over the beam allowed deformations, which are only deflections and slopes. In this situation, the following element stiffnesses are also zero: K_{Fx-ux}, K_{Fx-uy}, and $K_{Fx-\theta z}$. The remaining stiffnesses are the usual bending ones, namely: $K_{Fy-uy} = 12EI_z/l^3$, $K_{Fy-\theta z} = 6EI_z/l^2$ and $K_{Mz-\theta z} = 4EI_z/l$. With those values, the stiffness matrix is calculated as:

$$[K] = \frac{4EI_z}{l} \begin{bmatrix} \dfrac{6}{l^2} & 0 & -\dfrac{6}{l^2} & 0 \\ 0 & 14 & 0 & -23 \\ -\dfrac{6}{l^2} & 0 & \dfrac{12}{l^2} & 0 \\ 0 & -23 & 0 & 112 \end{bmatrix}$$

(2.194)

The inertia matrix is:

$$[M] = m \begin{bmatrix} 1 & 0 & 0 & 0 \\ 0 & 4l^2 & 0 & 0 \\ 0 & 0 & 1 & 0 \\ 0 & 0 & 0 & 4l^2 \end{bmatrix}$$

(2.195)

The two matrices have been calculated for the following DOF vector: $\{u\} = \{u_{ay} \; \theta_{az} \; u_{by} \; \theta_{bz}\}^t$. The following resonant frequencies have been obtained:

$$\omega_i = \frac{c_i}{l} \sqrt{\frac{EI_z}{ml}}$$

(2.196)

where $c_1 = 3.02$, $c_2 = 4.21$, $c_3 = 7.92$, and $c_4 = 15.3$.

2.3.3.2 Out-of-the-Plane Motion

The out-of-the-plane motion of the generic microsystem of Figure 2.60 (b) is studied now. In this case, the other six DOF of the two bodies, namely out-of-the-plane (z-axis) translation and rotations about the x- and y-axes of the plates common centers, are of importance (Figure 2.62 indicates these DOF).

2.3.3.2.1 Inertia matrix

The six DOF can be arranged into the following vector:

$$\{u\} = \{u_{az} \quad \theta_{ax} \quad \theta_{ay} \quad u_{bz} \quad \theta_{bx} \quad \theta_{by}\}^t$$

(2.197)

By expressing the kinetic energy corresponding to these DOF and by performing the partial and time derivatives implied by the kinetic energy term in the Lagrange's equations, the inertia matrix is of diagonal form, namely:

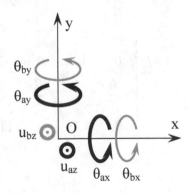

Figure 2.62 Degrees of freedom for a two-mass microsystem undergoing free out-of-the-plane vibrations

$$diag\,[M] = \left\{ m_a \quad J_{ax} \quad J_{ay} \quad m_b \quad J_{bx} \quad J_{by} \right\} \qquad (2.198)$$

2.3.3.2.2 Stiffness matrix

The displacements of the three points of interest, P and Q for the inner spring, and S for the outer spring (see Figure 2.60 (b)) are first expressed by the aid of Figure 2.62 in terms of the system's six DOF shown in Equation (2.197), namely:

$$\begin{cases} u_{Pzi} = u_{az} + d_{Pi} \sin \alpha_i \theta_{ax} - d_{Pi} \cos \alpha_i \theta_{ay} \\ \theta_{Pxi} = \theta_{ax} \cos \alpha_i + \theta_{ay} \sin \alpha_i \\ \theta_{Pyi} = -\theta_{ax} \sin \alpha_i + \theta_{ay} \cos \alpha_i \end{cases} \qquad (2.199)$$

$$\begin{cases} u_{Qzi} = u_{bz} + d_{Qi} \sin \alpha_i \theta_{bx} - d_{Qi} \cos \alpha_i \theta_{by} \\ \theta_{Qxi} = \theta_{bx} \cos \alpha_i + \theta_{by} \sin \alpha_i \\ \theta_{Qyi} = -\theta_{bx} \sin \alpha_i + \theta_{by} \cos \alpha_i \end{cases} \qquad (2.200)$$

$$\begin{cases} u_{Szj} = u_{bz} + d_{Sj} \sin \beta_j \theta_{bx} - d_{Sj} \cos \beta_j \theta_{by} \\ \theta_{Sxj} = \theta_{bx} \cos \beta_j + \theta_{by} \sin \beta_j \\ \theta_{Syj} = -\theta_{bx} \sin \beta_j + \theta_{by} \cos \beta_j \end{cases} \qquad (2.201)$$

The next step, as mentioned previously, is to express the deformations of the generic inner and outer springs at their end closest to the plates center in terms of the system's six DOF. Again, the deformations of the inner spring are to be expressed as differences between its ends displacements, the end-points being anchored at the plates a and b. The connection equations for the generic inner spring i are:

$$
\begin{cases}
\Delta u_{zi} = \left(u_{az} + d_{Pi} \sin \alpha_i \theta_{ax} - d_{Pi} \cos \alpha_i \theta_{ay} \right) \\
\qquad - \left(u_{bz} + d_{Qi} \sin \alpha_i \theta_{bx} - d_{Qi} \cos \alpha_i \theta_{by} \right) \\
\theta_{xi} = \left(\theta_{ax} \cos \alpha_i + \theta_{ay} \sin \alpha_i \right) - \left(\theta_{bx} \cos \alpha_i + \theta_{by} \sin \alpha_i \right) \\
\theta_{yi} = \left(-\theta_{ax} \sin \alpha_i + \theta_{ay} \cos \alpha_i \right) - \left(-\theta_{bx} \sin \alpha_i + \theta_{by} \cos \alpha_i \right)
\end{cases}
\tag{2.202}
$$

The equations for the generic spring j are:

$$
\begin{cases}
\Delta u_{zj} = u_{bz} + d_{Sj} \sin \beta_j \theta_{bx} - d_{Sj} \cos \beta_j \theta_{by} \\
\theta_{xj} = \theta_{bx} \cos \beta_j + \theta_{by} \sin \beta_j \\
\theta_{yj} = -\theta_{bx} \sin \beta_j + \theta_{by} \cos \beta_j
\end{cases}
\tag{2.203}
$$

The potential energy stored by either the inner spring or the outer is:

$$
\begin{aligned}
U_k = \frac{1}{2} \Big(& K_{Fz-uz,k} \Delta u_{zk}^2 + K_{Mx-\theta x,k} \theta_{xk}^2 + K_{My-\theta y,k} \theta_{yk}^2 \Big) \\
& + K_{Fz-\theta x,k} \Delta u_{zk} \theta_{xk} + K_{Fz-\theta y,k} \Delta u_{zk} \theta_{yk} + K_{Mx-\theta y,k} \theta_{xk} \theta_{yk}
\end{aligned}
\tag{2.204}
$$

where $k = i$ for the inner spring and $k = j$ for the outer spring.

By substituting the connection Equations (2.202) and (2.203) into Equation (2.204) with the appropriate k, the potential energies of inner and outer springs are expressed in terms of the six DOF. The total potential energy again is the sum of all individual energies, and by taking the partial derivatives of the total potential energy in terms of the six DOF, as required by Lagrange's equations, the overall 6 × 6 stiffness matrix is obtained. The sub-matrix corresponding to the inner springs is also 6 × 6, as all six DOF are involved, whereas the submatrix of the outer springs is 3 × 3 because only u_{bz}, θ_{bx}, and θ_{bx} are involved. The terms of the overall stiffness matrix are computed by means of Equations (2.187) and (2.188), as given next. For the generic inner spring, the stiffness elements of the first row are:

$$\begin{cases} K_{11,i} = -K_{14,i} = K_{44,i} = K_{Fz-uz,i} \\ K_{12,i} = K_{Fz-\theta x,i}\cos\alpha_i - \left(K_{Fz-\theta y,i} - d_{Pi}K_{Fz-uz,i}\right)\sin\alpha_i \\ K_{13,i} = K_{Fz-\theta x,i}\sin\alpha_i + \left(K_{Fz-\theta y,i} - d_{Pi}K_{Fz-uz,i}\right)\cos\alpha_i \\ K_{15,i} = -K_{45,i} = -K_{Fz-\theta x,i}\cos\alpha_i + \left(K_{Fz-\theta y,i} - d_{Qi}K_{Fz-uz,i}\right)\sin\alpha_i \\ K_{16,i} = -K_{46,i} = -K_{Fz-\theta x,i}\sin\alpha_i - \left(K_{Fz-\theta y,i} - d_{Qi}K_{Fz-uz,i}\right)\cos\alpha_i \end{cases} \qquad (2.205)$$

The elements on the second row are:

$$\begin{cases} K_{22,i} = K_{Mx-\theta x,i}\cos^2\alpha_i + \left(d_{Pi}^2 K_{Fz-uz,i} - 2d_{Pi}K_{Fz-\theta y,i} + K_{My-\theta y,i}\right)\sin^2\alpha_i \\ \qquad -\left(K_{Mx-\theta y,i} - d_{Pi}K_{Fz-\theta X,i}\right)\sin(2\alpha_i) \\ K_{23,i} = -\dfrac{\left(d_{Pi}^2 K_{Fz-uz,i} - 2d_{Pi}K_{Fz-\theta y,i} + K_{My-\theta y,i} - K_{Mx-\theta x,i}\right)\sin(2\alpha_i)}{2} \\ \qquad +\left(K_{Mx-\theta y,i} - d_{Pi}K_{Fz-\theta x,i}\right)\cos(2\alpha_i) \\ K_{24,i} = -K_{Fz-\theta x,i}\cos\alpha_i + \left(K_{Fz-\theta y,i} - d_{Pi}K_{Fz-uz,i}\right)\sin\alpha_i \\ K_{25,i} = -K_{Mx-\theta x,i}\cos^2\alpha_i + \dfrac{2K_{Mx-\theta y,i} - \left(d_{Pi} + d_{Qi}\right)K_{Fz-\theta x,i}}{2}\sin(2\alpha_i) \\ \qquad -\left[d_{Pi}d_{Qi}K_{Fz-uz,i} - \left(d_{Pi} + d_{Qi}\right)K_{Fz-\theta y,i} + K_{My-\theta y,i}\right]\sin^2\alpha_i \\ K_{26,i} = -\left(K_{Mx-\theta y,i} - d_{Qi}K_{Fz-\theta x,i}\right)\cos^2\alpha_i + \left(K_{Mx-\theta y,i} - d_{Pi}K_{Fz-\theta x,i}\right)\sin^2\alpha_i \\ \qquad +\dfrac{d_{Pi}d_{Qi}K_{Fz-uz,i} - \left(d_{Pi} + d_{Qi}\right)K_{Fz-\theta y,i} + K_{My-\theta y,i} - K_{Mx-\theta x,i}}{2}\sin(2\alpha_i) \end{cases}$$

$$(2.206)$$

The elements on the third row are:

$$
\begin{cases}
K_{33,i} = K_{Mx-\theta x,i}\sin^2\alpha_i + \left(d_{Pi}^2 K_{Fz-uz,i} - 2d_{Pi}K_{Fz-\theta y,i} + K_{My-\theta y,i}\right)\cos^2\alpha_i \\
\qquad + \left(K_{Mx-\theta y,i} - d_{Pi}K_{Fz-\theta x,i}\right)\sin(2\alpha_i) \\[4pt]
K_{34,i} = -K_{Fz-\theta x,i}\sin\alpha_i - \left(K_{Fz-\theta y,i} - d_{Pi}K_{Fz-uz,i}\right)\cos\alpha_i \\[4pt]
K_{35,i} = -\left(K_{Mx-\theta y,i} - d_{Pi}K_{Fz-\theta x,i}\right)\cos^2\alpha_i + \left(K_{Mx-\theta y,i} - d_{Qi}K_{Fz-\theta x,i}\right)\sin^2\alpha_i \qquad (2.207) \\
\qquad + \dfrac{d_{Pi}d_{Qi}K_{Fz-uz,i} - \left(d_{Pi} + d_{Qi}\right)K_{Fz-\theta y,i} + K_{My-\theta y,i} - K_{Mx-\theta x,i}}{2}\sin(2\alpha_i) \\[4pt]
K_{36,i} = -K_{Mx-\theta x,i}\sin^2\alpha_i - \dfrac{2K_{Mx-\theta y,i} - \left(d_{Pi} + d_{Qi}\right)K_{Fz-\theta x,i}}{2}\sin(2\alpha_i) \\
\qquad - \left[d_{Pi}d_{Qi}K_{Fz-uz,i} - \left(d_{Pi} + d_{Qi}\right)K_{Fz-\theta y,i} + K_{My-\theta y,i}\right]\cos^2\alpha_i
\end{cases}
$$

The elements on the fifth row are:

$$
\begin{cases}
K_{55,i} = K_{Mx-\theta x,i}\cos^2\alpha_i + \left(d_{Qi}^2 K_{Fz-uz,i} - 2d_{Qi}K_{Fz-\theta y,i} + K_{My-\theta y,i}\right)\sin^2\alpha_i \\
\qquad - \left(K_{Mx-\theta y,i} - d_{Qi}K_{Fz-\theta X,i}\right)\sin(2\alpha_i) \\[4pt]
K_{56,i} = -\dfrac{\left(d_{Qi}^2 K_{Fz-uz,i} - 2d_{Qi}K_{Fz-\theta y,i} + K_{My-\theta y,i} - K_{Mx-\theta x,i}\right)\sin(2\alpha_i)}{2} \qquad (2.208) \\
\qquad + \left(K_{Mx-\theta y,i} - d_{Qi}K_{Fz-uz,i}\right)\cos(2\alpha_i)
\end{cases}
$$

The relevant element on the sixth row is:

$$
\begin{aligned}
K_{66,i} &= K_{Mx-\theta x,i}\sin^2\alpha_i + \left(d_{Qi}^2 K_{Fz-uz,i} - 2d_{Qi}K_{Fz-\theta y,i} + K_{My-\theta y,i}\right)\cos^2\alpha_i \\
&\quad + \left(K_{Mx-\theta y,i} - d_{Qi}K_{Fz-\theta x,i}\right)\sin(2\alpha_i)
\end{aligned} \qquad (2.209)
$$

The stiffness elements corresponding to the outer spring j are:

$$
\begin{cases}
K_{44,j} = K_{Fz-uz,j} \\[4pt]
K_{45,j} = K_{Fz-\theta x,j} \cos\beta_j - \left(K_{Fz-\theta y,j} - d_{Sj} K_{Fz-uz,j}\right)\sin\beta_j \\[4pt]
K_{46,j} = K_{Fz-\theta x,j} \sin\beta_j + \left(K_{Fz-\theta y,j} - d_{Sj} K_{Fz-uz,j}\right)\cos\beta_j \\[4pt]
K_{55,j} = K_{Mx-\theta x,j} \cos^2\beta_j + \left(d_{Sj}^2 K_{Fz-uz,j} - 2d_{Sj} K_{Fz-\theta y,j} + K_{My-\theta y,j}\right)\sin^2\beta_j \\[4pt]
\qquad\quad - \left(K_{Mx-\theta y,j} - d_{Sj} K_{Fz-\theta x,j}\right)\sin(2\beta_j) \\[10pt]
K_{56,j} = -\dfrac{d_{Sj}^2 K_{Fz-uz,j} - 2d_{Sj} K_{Fz-\theta y,j} - K_{Mx-\theta x,j} + K_{My-\theta y,j}}{2}\sin(2\beta_j) \\[8pt]
\qquad\quad + \left(K_{Mx-\theta y,j} - d_{Sj} K_{Fz-\theta x,j}\right)\cos(2\beta_j) \\[6pt]
K_{66,j} = \left(d_{Sj}^2 K_{Fz-uz,j} - 2d_{Sj} K_{Fz-\theta y,j} + K_{My-\theta y,j}\right)\cos^2\beta_j + K_{Mx-\theta x,j}\sin^2\beta_j \\[4pt]
\qquad\quad + \left(K_{Mx-\theta y,j} - d_{Sj} K_{Fz-\theta x,j}\right)\sin(2\beta_j)
\end{cases}
\qquad (2.210)
$$

Example 2.20

Determine the resonant frequencies corresponding to the out-of-the-plane motion of the two-mass, four-spring microgyroscope sketched in Figure 2.63. Consider that all beams are identical with their length being $l = 200$ μm and of circular cross-section with a diameter $d = 2$ μm. The material is polysilicon with $E = 150$ GPa and $\mu = 0.25$. The two plates have the same mass $m = 0.002$ kg and their relevant mechanical moments of inertia are: $J_{ax} = J_{bx} = ml^2/4$, $J_{ay} = J_{by} = ml^2$.

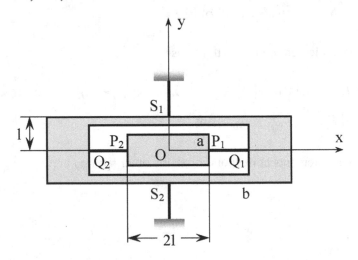

Figure 2.63 Two-mass, four-beam microgyroscope in out-of-the-plane free vibration

Solution:
        The configuration of Figure 2.63 brings a slight alteration to the micro-gyroscope of the previous example through changing the position of the outer springs by 90°. It can be checked that all six DOF are thus enabled. Also, the following particular amounts apply in this example: $d_{P1} = d_{P2} = l$, $d_{Q1} = d_{Q2} = 2l$, $d_{S1} = d_{S2} = l$, and also: $\alpha_1 = 0°$, $\alpha_2 = 180°$, $\beta_1 = 90°$, $\beta_2 = 270°$.
        The vector defining the relevant DOF of a beam element in the local frame is $\{u_k\} = \{u_{zk}, \theta_{xk}, \theta_{yk}\}^t$ with $k = i, j$. Consequently, the following local stiffnesses are applying here: $K_{Fz\text{-}uz} = 12EI_z/l^3$, $K_{Fz\text{-}\theta y} = 6EI_z/l^2$, $K_{My\text{-}\theta y} = 4EI_z/l$, and $K_{Mx\text{-}\theta x} = EI_z/[(1+\mu)l]$. By carrying out the necessary calculations, the overall stiffness matrix is found to be:

$$[K] = \frac{4EI_z}{l} \begin{bmatrix} \dfrac{6}{l^2} & 0 & 0 & -\dfrac{6}{l^2} & 0 & 0 \\ 0 & 0.4 & 0 & 0 & -0.4 & 0 \\ 0 & 0 & 14 & 0 & 0 & -23 \\ -\dfrac{6}{l^2} & 0 & 0 & \dfrac{12}{l^2} & 0 & 0 \\ 0 & -0.4 & 0 & 0 & 14.4 & 0 \\ 0 & 0 & -23 & 0 & 0 & 38.4 \end{bmatrix} \tag{2.211}$$

The inertia matrix is diagonal, its diagonal terms being:

$$diag\,[M] = \left\{ m \quad \frac{ml^2}{4} \quad ml^2 \quad m \quad \frac{ml^2}{4} \quad ml^2 \right\} \tag{2.212}$$

The resonant frequencies have the following numerical values: $\omega_1 = 485$ rad/s, $\omega_2 = 4{,}578$ rad/s, $\omega_3 = 6{,}750$ rad/s, $\omega_4 = 46{,}264$ rad/s, $\omega_5 = 153{,}846$ rad/s, $\omega_6 = 169{,}781$ rad/s.

Problems

Problem 2.1
        Design a paddle microcantilever with equal torsion and bending resonant frequencies. Its thickness is t and the dimensions of the geometric rectangular envelope are l and w. Known are also the material parameters.

Problem 2.2
        Mass attaches uniformly over the whole paddle area of a paddle micro-cantilever. By using a lumped-parameter model with the paddle considered

rigid and the root considered massless and compliant calculate the change in the bending resonant frequency. Consider $l_1 = l_2$ and $w_1 = 2w_2$.

Problem 2.3
    Solve Problem 2.2 by considering a paddle microbridge instead of the paddle microcantilever. Consider $w_2 = 2w_1$.

Problem 2.4
    Mass attaches in a point-like manner to the symmetry center of a paddle microcantilever. Calculate the bending frequency shift by using a model with the paddle considered rigid and the root segment considered massless and compliant. Also calculate the frequency shift by considering inertia and compliance are contributed by both the paddle and root segments.

Problem 2.5
    Solve Problem 2.4 by considering a paddle microbridge instead of the paddle microcantilever.

Problem 2.6
    Study the torsion-to-bending resonant frequency ratio of a paddle micro-cantilever when considering inertia is resulting only from the paddle and compliance is being produced by the root segment.

Problem 2.7
    Rework Problem 2.6 for a paddle microbridge.

Problem 2. 8
    A paddle microcantilever as the one sketched in Figure 2.1 has a bending frequency that is too low. Calculate the increase in the resonant frequency gained through converting the cantilever into a paddle bridge by addition of another flexure, identical to the original one, to its free end. Use the lumped-parameter model.

Problem 2.9
    A torsional microresonator as the one pictured in Figure 2.8 is supported by a spiral spring with $n = 1$ turn, a minimum radius of $r_1 = 20$ μm and maximum radius $r_2 = 30$ μm. The microfabrication technology results in a uniform thickness of $t = 2$ μm. Determine the spring's cross-sectional width w and inertia of the resonator which will produce a resonant frequency of 30 kHz. Consider the cross-section is thin, and $E = 155$ GPa and $\rho = 2300$ kg/m^3.

Problem 2.10

A rotary microresonator as the one pictured in Figure 2.11 has a specified resonant frequency ω_t. Given are also the shaft diameter d_s and the hub inner diameter d_i, as well as the material properties. Determine the radius of curvature of the curved springs as well as the number of springs.

Problem 2.11

A single DOF rotary spring-mass microsystem as the one of Figure 2.8 has a spring with $n = 1.5$ turns and its cross-section is square with a known side t, in the situation where the $r/t < 8$ (r is the variable curvature radius). Study the design variants that will generate a torsion resonant frequency ranging anywhere in the interval $\omega_{t1} \rightarrow \omega_{t2}$. Consider J is also known.

Problem 2.12

A microcantilever of constant circular cross-section (of known length l and diameter d) has a rigid body (of known mass m and moment of inertia J_x about the microcantilever's longitudinal axis) at its free end. Identify the material modulus of elasticity E and Poisson's ratio μ by using the bending and torsional resonant response of this paddle microcantilever.

Problem 2.13

An inclined-beam microaccelerometer, as the one shown in Figure 2.15, to obtain a resonant frequency of 50 kHz when the proof mass is defined by $m = 2 \times 10^{-15}$ kg, $E = 150$ GPa and $w = t = 2$ μm. What is the expected resonant frequency when l spans the [50; 300] μm range and α spans the [0°; 20°] range?

Problem 2.14

Find the in-plane resonant frequency of the single DOF microresonator shown in Figure 2.64. Known are the mass of the central plate, as well the geometry and material properties of the two identical constant cross-section beams. Also derive the result by using Problem 2.13.

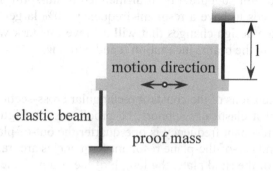

Figure 2.64 Single DOF microaccelerometer with in-plane motion

Problem 2.15

Compare the resonant frequencies of a saggital-spring microaccelero-meter (Figure 2.17) and a U-spring one (Figure 2.19). Consider the proof masses are identical and also the long beams composing the springs are identical. Consider the short segments of the U-springs are rigid and study the influence of the inclination angle of the legs in the saggital springs.

Problem 2.16

A microaccelerometer is composed of a rigid plate (of mass m and central moment of inertia J_x) and two identical beams, as sketched in Figure 2.65. Considering the system is a single DOF one, calculate:

a) the resonant frequency corresponding to the in-plane free vibrations about the direction perpendicular to the beam axis;
b) the resonant frequency corresponding to the free rotations about the beams axes.

Known are the length l, the cross-sectional dimensions w and t (t is very small), and the material properties.

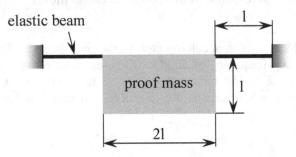

Figure 2.65 Microaccelerometer with offset beams

Problem 2.17

A linear microresonator supported by four inclined beam-springs (as the one of Figure 2.15) needs to have a resonant frequency 50% larger than its current one. Propose the design changes that will achieve this task when the only modifications are in the beams inclination α and width w.

Problem 2.18

Determine the dimensions of the constant rectangular cross-section of the four identical beams that elastically support the proof mass in Figure 2.66, such that the in-plane resonant frequency is one-quarter the out-of-plane one. Assume both the in- and out-of-the-plane resonant frequencies are translatory. Known is the mass m of the rigid plate, the length of the beams l, the beams' cross-sectional dimensions w and t, as well as the material's linear modulus of elasticity E.

elastic beam

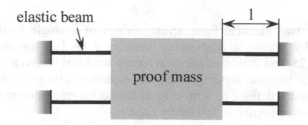

Figure 2.66 Microaccelerometer with in-plane and out-of-the-plane motions

Problem 2.19
Compare the resonant frequencies of an inclined beam-spring micro-accelerometer and of a saggital spring one (both are assumed to be single DOF systems). Assume the two configurations have beams that are identical in both dimensions and inclination and identical proof masses, too.

Problem 2.20
A single DOF microaccelerometer (as the one of Figure 2.67) is supported by U-springs at its ends. Find the percentage change in the resonant frequency after the length $l_3 = 2l_1 = 2l_2$ (see Figure 2.21) is increased by 50%. The mass of the rigid plate is m.

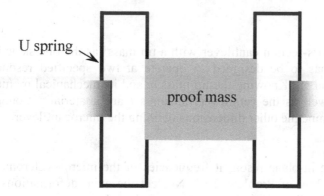

Figure 2.67   Microaccelerometer with U springs

Problem 2.21
Analyze the resonant frequency of the single DOF microaccelerometer with two folded beam suspensions shown in Figure 2.22 by considering that the length l_2 (see Figure 2.23) ranges from $0.2 \times l_1$ to $0.8 \times l_1$, as well as when the moment of inertia I_{z2} varies from $0.1 \times I_{z1}$ to $2 \times I_{z1}$.

Problem 2.22

Design the two identical basic spiral springs of a single DOF micro-accelerometer by finding a relationship between l_1 and l_2 (as pictured in Figures 2.24, 2.25, and 2.26) that needs to have a torsional resonant frequency that is twice the in-plane bending resonant frequency. The proof mass is a parallelepiped and is defined by a mass m and a mechanical moment of inertia J. The springs' cross-section is defined by t and w.

Problem 2.23

Compare the planar-motion resonant frequencies of two single DOF microaccelerometers that have identical proof masses. One design has its proof mass suspended by means of two end folded-beam springs, and the other one by means of two basic spiral springs. Consider the dimensions l_1 and l_2, which define both springs, are identical; the springs cross-sections and material properties are also identical.

Problem 2.24

A bent-beam spring microaccelerometer, as the one sketched in Figure 2.29 with the spring being shown in Figure 2.30, has to ensure the maximum separation between the two resonant translations about the x- and the y-axes. Knowing that a leg's length is limited by l_{min} and l_{max}, propose a design that will produce the maximum separation between the two in-plane resonant frequencies.

Problem 2.25

A constant cross-section cantilever with a tip mass (as the one sketched in Figure 2.36) has to be designed to operate at two specified resonant frequencies, ω_1 and ω_2. Knowing the tip mass m and its mechanical moment of inertia J_y, as well as the cantilever thickness t and material's Young's modulus E, determine the other dimensions defining the microcantilever.

Problem 2.26

Determine the in-plane resonant frequencies of the microaccelerometer shown in Figure 2.64 of Problem 2.14. Neglect the axial deformations of the two beams. The distance from the plate center to the beams' connection points to the plate is l. The mass of the plate is m, the beams' cross-sectional dimensions are w and t, and Young's modulus is E.

Problem 2.27

A two-mass system, as the one schematically shown in Figure 2.68, can function as a mechanical or electrical microfilter, altering the input to the system (which can be a displacement or a voltage, for instance). Calculate the resonant frequencies of this system by using Lagrange's equations and describe the corresponding modes.

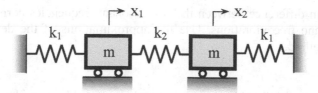

Figure 2.68 Lumped-parameter model of a two-mass microfilter

Problem 2.28

By applying Lagrange's equations, calculate the resonant frequencies of a mechanical microsystem whose lumped-parameter schematic representation is given in Figure 2.69. Describe the corresponding modes.

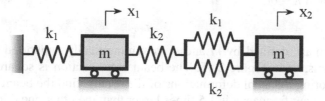

Figure 2.69 Lumped-parameter model of a two-mass microresonator

Problem 2.29

Determine the resonant frequencies corresponding to out-of-the-plane bending of a constant cross-section microbridge with a point-like body at its midpoint whose mass m and mechanical moment of inertia J_y are known. Known are also the length l, cross-sectional moment of inertia I_y and elasticity modulus E for the microbridge.

Problem 2.30

For a constant-thickness paddle microcantilever, as the one shown in Figure 1.21, use the lumped-parameter modeling method to derive the equivalent tip inertia and compliance properties corresponding to out-of-the-plane bending. Use this model to determine the resonant frequencies by using the compliance approach.

Problem 2.31

Repeat Problem 2.30 for the microcantilever sketched in Fig. 1.92 of Problem 1.17 when $w_2 = 2w_1$.

Problem 2.32

A two-beam spring accelerometer as the one of Figure 2.40 is confined within a rectangular envelope of $2L \times L$. Knowing that the thickness of the microdevice is constant and equal to t, determine the design that will realize the maximum difference between the two resonant frequencies corresponding to the in-plane free vibrations. Use the approximations of the deformation geometry method.

Problem 2.33

A proof mass defined by m and J_z is end-supported by two identical basic spiral springs. Knowing the compliances at one spring end (Lobontiu and Garcia [1]), namely: $C_{ux-Fx} = 2l_1^2 (2l_1 + 3l_2)/(3EI_z)$, $C_{ux-Fy} = - l_1l_2 (l_1 + 2l_2)/(EI_z)$, $C_{ux-Mz} = l_1 (l_1 + 2l_2)/(EI_z)$, $C_{uy-Fy} = 2l_2^2 (9l_1 + 4l_2)/(3EI_z)$, $C_{uy-Mz} = - 2l_2 (2l_1 + l_2)/(EI_z)$, $C_{\theta z-Mz} = 2(2l_1 + l_2)/(EI_z)$, calculate the three resonant frequencies corresponding to the in-plane free vibrations. The distance from the plate's center to the spring connection points is l_1. It is also known that $l_1 = 2l$, $l_2 = l$ and $R = 6l$.

Problem 2.34

A proof mass defined by m and J_z is symmetrically supported by four identical beams, as in Figure 2.45. The beam cross-section is square with a side of t. Ignoring the axial deformations of the beams, find the beam length l when one resonant frequency is 1.5 times larger than the other one during free in-plane vibrations. It is also known that the radius of the disk plate is $R = l$.

Problem 2.35

A circular disk of mass m and mechanical moment of inertia J_z is symmetrically and radially supported by three fishhook springs (as the one of Figure 2.53). Knowing that $l_1 = l_2 = l_3 = R$ (R is the disk radius), also knowing Young's modulus E, determine the resonant frequencies of the in-plane motion.

Problem 2.36

A two-beam accelerometer (as the one shown in Figure 2.54) has the ratio of its two out-of-the-plane resonant frequencies equal to 5. Determine a relationship between the plate's mechanical moment of inertia J_y and mass m when $l_1 = 2l$ (l is the length of the beam and l_1 is the plate half-length).

Problem 2.37

Determine the percentage change in the piston-type resonant frequency (the one corresponding to translatory out-of-the-plane vibrations about the z-axis) between a two-beam accelerometer and a four-beam one when all beams are identical and the proof mass is the same.

Problem 2.38

Find the out-of-the-plane resonant frequencies of the microgyroscope sketched in Figure 2.61. All four beams are identical and have circular constant cross-section. Known is also that $m_b = 2m_a$, $J_{ax} = J_{bx}$, $J_{ay} = 4J_{by}$.

Problem 2.39

Compare the minimum resonant frequency of the in-plane free vibrations for the microgyroscope of Figure 2.61, to the minimum one of the microgyroscope of Figure 2.63, considering similar designs with identical constant cross-section beams.

Problem 2.40

Calculate the in-plane resonant frequencies for the microgyroscope sketched in Figure 2.70. All beams are identical and $m_a = m_b$, $J_{bz} = 8J_{az}$. The inner disc's radius is equal to R/4.

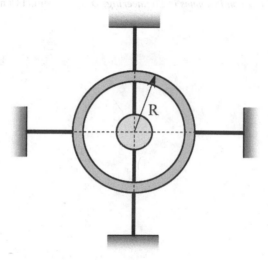

Figure 2.70 Microgyroscope with beam springs

Problem 2.41

A mechanical microfilter can be modeled by means of the lumped-parameter model sketched in Figure 2.71. For this particular system it is known that $m_{i+1} = q_m \times m_i$ (i = 1, 2 and $q_m > 1$), also that $k_{j+1} = q_k k_j$ (j = 1, 2, 3 and $0 > q_k < 1$). Determine the minimum and the maximum resonant frequencies of this system.

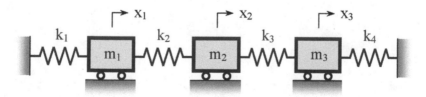

Figure 2.71 Lumped-parameter model of a two-mass microresonator

References

1. N. Lobontiu, E. Garcia, *Mechanics of Microelectromechanical Systems*, Kluwer Academic Press, New York, 2004.
2. W.C. Young, R.G. Budynas, *Roark's Formulas for Stress and Strain*, Seventh Edition, McGraw-Hill, New York, 2002.
3. N.P. Chironis (Editor), *Spring Design and Application*, McGraw-Hill Book Company, New York, 1961.
4. A.M. Wahl, *Mechanical Springs*, McGraw-Hill Book Company, Second Edition, New York, 1963.
5. W.T. Thomson, *Theory of Vibrations with Applications*, Third Edition, Prentice Hall, Englewood Cliffs, 1988.
6. S. Timoshenko, *Vibration Problems in Engineering*, D. van Nostrand Company, New York, 1928.

Chapter 3

ENERGY LOSSES IN MEMS AND EQUIVALENT VISCOUS DAMPING

3.1 INTRODUCTION

Energy losses change the behavior of mechanical microsystems and limit their performance. The response of a single degree-of-freedom (DOF) mechanical system, for instance, is conditioned by a damping term (force in translatory motion and torque in rotary motion), which can be formulated as a viscous damping agent whose magnitude is proportional to velocity. The damping coefficient is the proportionality constant and various forms of energy losses can be expressed as viscous damping ones, either naturally or by equivalence so that a unitary formulation is obtained. For oscillatory micro/nanoelectro-mechanical systems (MEMS/NEMS), losses can be quantified by means of the quality factor (Q-factor), which is the ratio of the energy stored to the energy lost during one cycle of vibration, and the damping coefficient can be expressed in terms of the Q-factor. Energy losses in MEMS/NEMS are the result of the interaction between external and internal mechanisms. Fluid–structure interaction (manifested as squeeze- or slide-film damping), anchor (connection to substrate) losses, thermoelastic damping (TED), surface/volume losses and phonon-mediated damping are the most common energy loss mechanisms discussed in this chapter.

3.2 LUMPED-PARAMETER VISCOUS DAMPING

3.2.1 Viscous Damping Coefficient and Damping Ratio

Viscous damping in a lumped-parameter system that performs linear motion is expressed by a resistance force, which is proportional to velocity, namely:

$$F_d = c\dot{x} \tag{3.1}$$

where c is the damping coefficient (a similar relationship is obtained for rotary motion where a torque is set through damping resistance and that is proportional to the angular velocity).

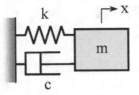

Figure 3.1 Mass-dashpot single DOF system

The damped free vibrations of the single DOF system of Figure 3.1 are described by the equation:

$$m\ddot{x} + c\dot{x} + kx = 0 \tag{3.2}$$

For *linear systems*, the damping coefficient c is constant, as well as the mass m and stiffness k coefficients. However, as shown in the following, situations may appear in damped MEMS/NEMS system where c depends on the vibration frequency (particularly in driven systems). Equation (3.2) in such instances becomes *nonlinear*, and its integration is not pursued in this chapter.

By using the following notations:

$$\begin{cases} \omega_r^2 = \dfrac{k}{m} \\[2mm] \zeta = \dfrac{c}{2m\omega_r} \end{cases} \tag{3.3}$$

where ω_r is the resonant frequency and ζ is the *damping ratio*, Equation (3.2) can be rewritten as:

$$\ddot{x} + 2\zeta\omega_r\dot{x} + \omega_r^2 x = 0 \tag{3.4}$$

which is the standard form known from vibrations. There are three different cases and their corresponding solutions depend on the value of the damping ratio ζ. When $0 < \zeta < 1$, which leads to *underdamped* free vibrations, the solution to Equation (3.4) (e.g., see Thomson [1]) is:

$$x(t) = e^{-\zeta\omega_r t}\left[\left(\frac{\zeta}{1-\zeta^2}x(0) + \frac{\dot{x}(0)}{\omega_d}\right)\sin(\omega_d t) + x(0)\cos(\omega_d t)\right] \tag{3.5}$$

where ω_d is the damped resonant frequency and is defined as:

$$\omega_d = \sqrt{1-\zeta^2}\,\omega_r \qquad (3.6)$$

Equation (3.5) can be rewritten as:

$$x(t) = Xe^{-\zeta\omega_r t}\sin(\omega_d t + \varphi) \qquad (3.7)$$

where:

$$
\begin{cases}
X = \dfrac{\dfrac{\zeta}{1-\zeta^2}x(0) + \dfrac{\dot{x}(0)}{\omega_d}}{\cos\varphi} \\[3ex]
\varphi = \tan^{-1}\dfrac{1}{\dfrac{\zeta}{1-\zeta^2} + \dfrac{\dot{x}(0)}{\omega_d x(0)}}
\end{cases}
\qquad (3.8)
$$

For $\zeta = 1$, the vibrations of the system shown in Figure 3.1 are *critically damped* and the solution to Equation (3.4) is:

$$x(t) = e^{-\zeta\omega_r t}\left[x(0)(1+\omega_r t) + \dot{x}(0)t \right] \qquad (3.9)$$

The *overdamped* vibrations occur when $\zeta > 1$, and the solution to Equation (3.4) is in that case:

$$x(t) = ae^{-\left(\zeta+\sqrt{\zeta^2-1}\right)\omega_r t} + be^{-\left(\zeta-\sqrt{\zeta^2-1}\right)\omega_r t} \qquad (3.10)$$

with:

$$
\begin{cases}
a = \dfrac{-\zeta+\sqrt{\zeta^2-1}}{2\sqrt{\zeta^2-1}}x(0) - \dfrac{\dot{x}(0)}{2\omega_r\sqrt{\zeta^2-1}} \\[3ex]
b = \dfrac{\zeta+\sqrt{\zeta^2-1}}{2\sqrt{\zeta^2-1}}x(0) + \dfrac{\dot{x}(0)}{2\omega_r\sqrt{\zeta^2-1}}
\end{cases}
\qquad (3.11)
$$

Figure 3.2 plots the three damped vibration cases under zero initial displacement and non-zero initial velocity. An exponentially decaying envelope is the asymptote curve to the underdamped response curve. The overdamped response curve is rapidly decaying. The critically damped response shows no harmonicity, as well as the overdamped one, and they both rapidly converge towards zero.

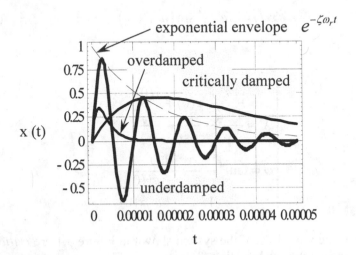

Figure 3.2 Free damped response of underdamped, overdamped, and critically damped single DOF system

3.2.2 Complex Number Representation of Vectors

It is convenient in many situations where harmonic excitation and response are in place to use the complex number representation of vectors. Figure 3.3 shows the one-to-one mapping that connects the classical representation of a vector and the one utilizing complex numbers. Considering the rod in Figure 3.3 rotates at constant angular velocity ω, the projections of point P on the Cartesian frame axes are:

$$\begin{cases} x = R\cos(\omega t) \\ y = R\sin(\omega t) \end{cases} \tag{3.12}$$

because the angle that positions the rotating vector is $\theta = \omega t$ (ω being the constant angular speed).

The velocity components are the time derivatives of x and y of Equation (3.12), and therefore the total velocity is:

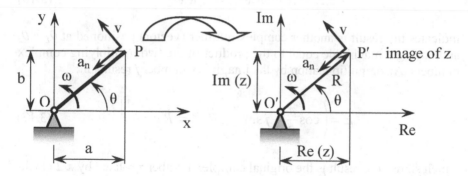

Figure 3.3 Classical planar representation of a vector versus complex-number representation of the same vector

$$v = \sqrt{v_x^2 + v_y^2} = \sqrt{\left(\frac{dx}{dt}\right)^2 + \left(\frac{dy}{dt}\right)^2} = \omega R \qquad (3.13)$$

Similarly, the normal acceleration is found by using the x- and y-components and its well-known value is:

$$a_n = \sqrt{a_x^2 + a_y^2} = \omega^2 R \qquad (3.14)$$

The one-to-one mapping of the rotating vector of Figure 3.3 into the complex number representation is warranted by the fact that a complex number is defined by a real component and an imaginary one. When the x- and y-projections of a vector are identical to the real and imaginary parts of a complex number, respectively, a vector in a plane is mapped into the image of a complex number in the complex plane. The complex number that is the map of the rotating vector in Figure 3.3 can be expressed in algebraic form, as well as in trigonometric and exponential forms (the latter due to Euler's formula), namely:

$$z = x + jy = R\left[\cos(\omega t) + j\sin(\omega t)\right] = Re^{j\omega t} \qquad (3.15)$$

The exponential form of a complex number is compact and is used in problems involving harmonic amounts. A few properties of exponential-form complex numbers are illustrated next. Multiplying two complex numbers, namely:

$$z_1 z_2 = R_1 e^{j\theta_1} R_2 e^{j\theta_2} = R_1 R_2 e^{j(\theta_1 + \theta_2)} \tag{3.16}$$

indicates the result is another complex number (vector) positioned at $\theta_1 + \theta_2$ and having a magnitude equal to the product of the two multiplying complex numbers. Also, multiplication by the imaginary number j results in:

$$jz = \left(\cos\frac{\pi}{2} + j\sin\frac{\pi}{2} \right) R e^{j\theta} = R e^{j\left(\theta + \frac{\pi}{2}\right)} \tag{3.17}$$

which shows the result is the original complex number z rotated by $\pi/2$ clockwise. Similarly, division by j rotates a complex number by $-\pi/2$ (or counterclockwise) because:

$$\frac{z}{j} = -jz = \left(\cos\frac{3\pi}{2} + j\sin\frac{3\pi}{2} \right) R e^{j\theta} = R e^{j\left(\theta + \frac{3\pi}{2}\right)} = R e^{j\left(\theta - \frac{\pi}{2}\right)} \tag{3.18}$$

The velocity and acceleration of point P′ in the complex plane are found by taking the first and second time derivative of z, namely:

$$\begin{cases} v = \dfrac{dz}{dt} = i\omega e^{j\omega t} = j\omega z \\[2mm] a_n = \dfrac{d^2 z}{dt^2} = \dfrac{dv}{dt} = -\omega^2 e^{j\omega t} = -\omega^2 z \end{cases} \tag{3.19}$$

The first Equation (3.19) indicates the velocity vector is rotated $\pi/2$ in a clockwise direction with respect to the position vector R (Figure 3.3), whereas the normal acceleration is parallel to the displacement vector but has an opposite direction—both situations being well known properties of the constant angular velocity rotation.

3.2.3 Q-Factor

The Q-factor is a figure of merit that takes into consideration the various energy losses in a vibrating system. For an oscillator, it is generally defined as:

$$Q = 2\pi \frac{U_s}{U_d} \tag{3.20}$$

where U_s is the energy stored (in the absence of losses) and U_d is the energy dissipated during one oscillatory cycle.

For a single DOF mechanical system, as the one shown in Figure 3.1, the energy stored in an oscillatory cycle (when damping the energy loss source is disregarded) is due to the elastic spring and is expressed as:

$$U_s = \frac{1}{2} k X^2 \tag{3.21}$$

Considering the work done by the viscous damping force is fully converted into energy lost during one oscillatory cycle, and considering a linear system, as the one shown in Figure 3.1, the damping energy is computed as:

$$U_d = \int F_d dx = c \int \dot{x} dx \tag{3.22}$$

The Q-factor is formulated by considering the interaction between the vibratory system and a harmonic (sine or cosine) excitation, and in such a characterization, the Q-factor is a forced-response one. The damping of a system can also be judged based on the free response, which would remove any dependency on excitation. Both ways are briefly discussed next.

3.2.3.1 Forced-Response Q-Factor

When a sinusoidal force acts on the mass-dashpot system pictured in Figure 3.1, the solution is obtained by carrying out the integration of Equation (3.22) for one period, and the energy lost through damping during one oscillation cycle is:

$$U_d = \pi c \omega X^2 \tag{3.23}$$

Consequently, the Q-factor defined in Equation (3.20) becomes:

$$Q = \frac{k}{c\omega} = \frac{1}{2\beta\zeta} \tag{3.24}$$

where $\beta = \omega/\omega_r$. At resonance ($\omega = \omega_r$ and $\beta = 1$), the Q-factor reduces to:

$$Q_r = \frac{k}{c\omega_r} = \frac{1}{2\zeta} \tag{3.25}$$

Example 3.1
   Analyze the Q-factor corresponding to the underdamped translational vibrations of a micromechanical system modeled as a single DOF system under sinusoidal excitation.

Solution:
     Equation (3.41) is used for the plot of Figure 3.4 (a), whereas the plot of Figure 3.4 (b) is drawn based on Equation (3.25). Increasing the actuation frequency (which amounts to increasing β) and the damping ratio ζ results in smaller Q-factors (Figure 3.4 (a)). Similarly, by increasing the damping ratio, the Q-factor diminishes (Figure 3.4 (b)).

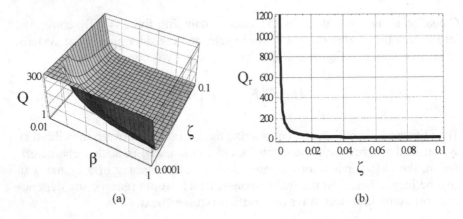

(a)   (b)

Figure 3.4 Quality factors: (a) regular quality factor; (b) resonant quality factor

Example 3.2
     Demonstrate that for a freely damped single DOF vibratory system the Q-factor can be defined as the number of oscillations required to reduce the system's energy to $1/e^{2\pi}$ (approximately $1/535$) of its original energy.

Solution:
     According to Equation (3.7), the maximum displacement is obtained as:

$$x_{max}(t) = Xe^{-\zeta\omega_r t} \tag{3.26}$$

If the energy of the system is the one stored in the spring, namely:

$$U = \frac{1}{2}kx_{max}^2 \tag{3.27}$$

and if n was the number of oscillations necessary to reduce the initial energy to the proportion mentioned in the problem, it means that:

$$\frac{U_0}{U} = \frac{1}{e^{-2\zeta\omega_r nT_r}} = e^{4\pi\zeta n} \tag{3.28}$$

where it has been considered that time t is expressed as:

$$t = nT_r$$ (3.29)

and the initial time is $t_0 = 0$. The relationship between the resonant period T_r and circular resonant frequency has also been considered:

$$T_r = \frac{2\pi}{\omega_r}$$ (3.30)

The problem's condition is:

$$\frac{U_0}{U} = e^{2\pi}$$ (3.31)

Comparing Equations (3.28) and (3.31) yields:

$$n = \frac{1}{2\varsigma} = Q_r$$ (3.32)

which demonstrates the problem assertion.

Free decaying underdamped vibrations can be evaluated by means of the *logarithmic decrement* δ, which is defined as the natural logarithm of the ratio of any two successive amplitudes, and, according to Equation (3.7), can be calculated as:

$$\delta = \ln \frac{x_n}{x_{n+1}} = \ln \frac{e^{-\varsigma\omega_r nT}}{e^{-\varsigma\omega_r (n+1)T}} = \varsigma\omega_r T$$ (3.33)

By taking into account Equation (3.6), which gives the relationship between the undamped and damped resonant frequencies, Equation (3.33) changes to:

$$\delta = \frac{2\pi\varsigma}{\sqrt{1-\varsigma^2}}$$ (3.34)

When using Equation (3.25), which expresses the resonant Q-factor in terms of the damping ratio, in conjunction with Equation (3.34), the resonant Q-factor results:

$$Q_r = \frac{1}{2}\sqrt{1 + 4\frac{\pi^2}{\delta^2}} \qquad (3.35)$$

Example 3.3

Determine the resonant Q-factor of a nano cantilever whose amplitude decays to $1/e^n$ after m free oscillations. Also determine the equivalent viscous damping.

Solution:

Considering that:

$$\frac{x_1}{x_m} = \frac{x_1}{x_2}\frac{x_2}{x_3}\cdots\frac{x_{m-2}}{x_{m-1}}\frac{x_{m-1}}{x_m} \qquad (3.36)$$

the natural logarithm of this relationship is applied, namely:

$$\ln\frac{x_1}{x_m} = \ln\frac{x_1}{x_2} + \ln\frac{x_2}{x_3} + \cdots + \ln\frac{x_{m-2}}{x_{m-1}} + \ln\frac{x_{m-1}}{x_m} = (m-1)\delta \qquad (3.37)$$

The problem statement is:

$$\frac{x_1}{x_m} = e^n \qquad (3.38)$$

or

$$\ln\frac{x_1}{x_m} = n \qquad (3.39)$$

Comparing Equations (3.37) and (3.39) results in:

$$\delta = \frac{n}{m-1} \qquad (3.40)$$

The resonant Q-factor becomes, by means of Equations (3.35) and (3.40):

$$Q_r = \frac{1}{2}\sqrt{1 + \frac{4\pi^2(m-1)^2}{n^2}} \qquad (3.41)$$

Figure 3.5 is the plot of Q_r as a function of m and n.

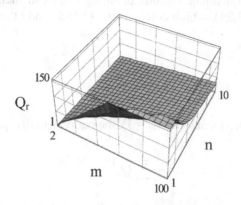

Figure 3.5 Resonant quality factor in terms of the number of cycles and amplitude ratio

Example 3.4

Consider the case in which several loss mechanisms act simultaneously on a MEMS, and that their individual Q-factors are known. Determine the total equivalent Q-factor, the corresponding equivalent damping ratio, as well as the resonant damping ratio, if the individual loss contribution superimpose linearly.

Solution:

The inverse of the Q-factor, as defined in Equation (3.20), is:

$$Q^{-1} = \frac{U_d}{2\pi U_s} \tag{3.42}$$

When several dissipation mechanisms are simultaneously present, the total loss energy can be expressed as:

$$U_d = \sum_i U_{d,i} \tag{3.43}$$

where $U_{d,i}$ are individual loss energy terms. Combination of Equations (3.42) and (3.43) results in:

$$Q_{eq}^{-1} = \frac{\sum_i U_{d,i}}{2\pi U_s} = \sum_i \left(\frac{U_{d,i}}{2\pi U_s} \right) = \sum_i Q_i^{-1} \tag{3.44}$$

In other words, the inverse of the total Q-factor is the sum of the individual Q-factor inverses. An equivalent viscous damping ratio can then be found, according to Equation (3.24), which connects the Q-factor and the equivalent damping ratio, namely:

$$\zeta_{eq} = \frac{1}{2\beta} \sum_i Q_i^{-1} \tag{3.45}$$

At resonance (when the frequency ratio $\beta = 1$), the equivalent damping ratio is:

$$\zeta_{eq,r} = \frac{1}{2} \sum_i Q_i^{-1} \tag{3.46}$$

3.2.3.2 Free-Response Q-Factor

A damped vibratory system can also be characterized in terms of energy efficiency by formulating a Q-factor corresponding to its free response. Unlike the customary approach to the Q-factor where a harmonic force is applied to the mechanical system, the free-response Q-factor is defined based on the initial conditions of free vibrations. In the case in which an initial velocity applies to a single DOF underdamped system (the initial displacement being assumed zero), the free response of the system, according to Equation (3.5), is:

$$x(t) = \frac{\dot{x}(0)}{\omega_d} e^{-\zeta \omega_r t} \sin(\omega_d t) \tag{3.47}$$

The damping energy lost during one oscillation cycle is of the form:

$$U_d = \int F_d dx = c \int \dot{x} dx \tag{3.48}$$

After taking the time derivative of $x(t)$ from Equation (3.47), by also considering the relationship between the damping coefficient c and the damping ratio ζ (Equation (3.3)), the damping energy of Equation (3.48) can be expressed as:

$$U_d = \frac{1}{2} \left(1 - e^{-\frac{4\pi\zeta}{\sqrt{1-\zeta^2}}} \right) m\dot{x}(0)^2 \tag{3.49}$$

As Equation (3.49) suggests, the damping energy is constant for specified system parameters and initial conditions, and is not cycle-dependent (as probably expected).

The elastic energy that is stored during one oscillation cycle (when considering there are no losses) is:

$$U_d = \frac{1}{2} m \dot{x}(0)^2 \qquad (3.50)$$

By using its definition of Equation (3.20), the free-response damping factor is:

$$Q = \frac{2\pi}{1 - e^{-\frac{4\pi\zeta}{\sqrt{1-\zeta^2}}}} \qquad (3.51)$$

Equation (3.51) gives the Q-factor of a freely vibrating system as a function of the damping ratio, and this relationship is plotted in Figure 3.6.

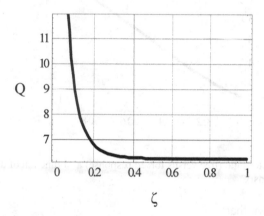

Figure 3.6 Quality factor as a function of damping ratio (underdamped case) in a free response

It can simply be shown that:

$$\begin{cases} \lim_{\zeta \to 0} Q = \infty \\ \lim_{\zeta \to 1} Q = 2\pi \end{cases} \qquad (3.52)$$

While the upper limit value is expected (the Q-factor would go to infinity when there are no energy losses, according to Equation (3.20)), the lower limit (2π) corresponding to the critically damped case shows that the entire original kinetic energy of the system is converted to damping energy (the two energies are equal, if the factor 2π is ignored).

Example 3.5

Compare the free-response Q-factor of Equation (3.51) to the resonant Q-factor corresponding to the forced response (Equation (3.25)).

Solution:

The two Q-factors can be compared by considering their ratio, which is:

$$r_Q = \frac{4\pi\zeta}{1 - e^{-\frac{4\pi\zeta}{\sqrt{1-\zeta^2}}}} \tag{3.53}$$

This ratio function is plotted in Figure 3.7.

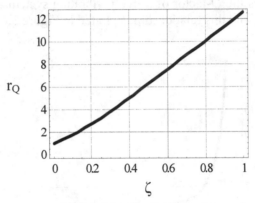

Figure 3.7 Quality factor ratio (free versus forced response) as a function of damping ratio for underdamped vibrations

It can also be shown that:

$$\begin{cases} \lim\limits_{\zeta \to 0} r_Q = 1 \\ \lim\limits_{\zeta \to 1} r_Q = 4\pi \end{cases} \tag{3.54}$$

Equation (3.54) indicates the two Q-factors are only equal in the absence of any viscous damping ($\zeta = 0$). In the opposite case (critically damped system, $\zeta = 1$), the Q-factor of the free response is 4π times larger than the classical, forced-response Q-factor.

As shown in subsequent sections of this chapter, an equivalent damping ratio can be formulated that incorporates all the damping sources, and therefore an equivalent (overall) Q-factor is obtained. Equation (3.51) can be used to express the damping ratio that corresponds to a given Q-factor, namely:

$$\zeta = \frac{\ln\left(1-\dfrac{2\pi}{Q}\right)}{\sqrt{16\pi^2 + \left[\ln\left(1-\dfrac{2\pi}{Q}\right)\right]^2}} \tag{3.55}$$

Example 3.6
    Express the Q-factor corresponding to the free response of a single DOF mass-dashpot system when the initial conditions consist of non-zero displacement $x(0)$. The initial velocity is assumed zero.

Solution:
    When an initial displacement is applied to a mass-dashpot system, the response to this initial condition is, according to the general solution of Equation (3.5):

$$x(t) = \left[\frac{\zeta}{1-\zeta^2}\sin(\omega_d t) + \cos(\omega_d t)\right]x(0)e^{-\zeta\omega_r t} \tag{3.56}$$

The velocity function can therefore be determined, together with the damping energy lost during one vibration cycle, as shown for the case with non-zero initial velocity. By taking into account that the elastically stored energy is:

$$U_s = \frac{1}{2}kx(0)^2 \tag{3.57}$$

the Q-factor can be expressed as:

$$Q = \frac{\pi\left(1-\zeta^2\right)\left(1+\coth\dfrac{2\pi\zeta}{\sqrt{1-\zeta^2}}\right)}{1+\left(2\sqrt{1-\zeta^2}-1\right)\zeta^2-\zeta^4} \tag{3.58}$$

Figure 3.8 plots the relative error between the Q-factor of Equation (3.58) and the one of Equation (3.51) for damping ratios not exceeding 0.1—which is a realistic upper limit for underdamped MEMS.

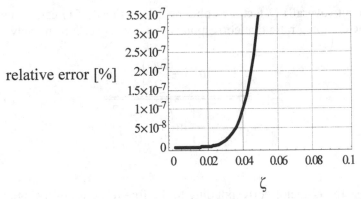

$$\text{relative error [\%]}$$

Figure 3.8 Relative errors between free-response quality factors: velocity versus displacement initial conditions

As the figure shows, the two models yield results that are in excellent agreement for the feasible domain of the damping ratio.

3.3 STRUCTURAL DAMPING

It has been shown that materials contribute to energy losses in driven systems, particularly in harmonically driven ones. In metallic materials, for instance, the energy dissipated per cycle is independent of frequency for a wide range (Thomson [1]), and is proportional to the square of the response amplitude, namely:

$$U_d = \alpha X^2 \tag{3.59}$$

Converting different forms of damping into viscous damping is advantageous from a computational standpoint because of the velocity dependency of the viscous damping. The energy loss through equivalent viscous damping during one oscillation cycle (period) is:

$$U_{d,eq} = \pi c_{eq} \omega X^2 \tag{3.60}$$

Equations (3.59) and (3.60) yield:

$$c_{eq} = \frac{\alpha}{\pi \omega} \tag{3.61}$$

The motion equation for a structurally damped system is therefore:

$$m\ddot{x} + \frac{\alpha}{\pi \omega}\dot{x} + kx = F\cos(\omega t) \tag{3.62}$$

If complex form is used again and when the following relationship is considered:

$$\dot{x} = j\omega x \qquad (3.63)$$

Equation (3.62) can be rewritten as:

$$m\ddot{x} + \left(k + j\frac{\alpha}{\pi} \right) x = Fe^{j\omega t} \qquad (3.64)$$

By using the notation (Thomson [1]):

$$\alpha = \pi k\gamma \qquad (3.65)$$

Equation (3.62) becomes:

$$m\ddot{x} + k\left(1 + j\gamma \right) x = Fe^{j\omega t} \qquad (3.66)$$

The quantity $k(1 + j\gamma)$ is called *complex stiffness* and the factor γ is the *structural damping factor*. When an exponential-form particular solution is sought for Equation (3.66), the real amplitude becomes:

$$X = \frac{F}{\sqrt{\left(k - m\omega^2 \right)^2 + \gamma^2 k^2}} \qquad (3.67)$$

At resonance, Equation (3.67) transforms into:

$$X_r = \frac{F}{\gamma k} \qquad (3.68)$$

For viscous damping, the resonant amplitude is:

$$X_r = \frac{F}{c\omega_r} = \frac{F}{2k\varsigma} \qquad (3.69)$$

Comparison of Equations (3.68) and (3.69) indicates that:

$$\gamma = 2\varsigma \qquad (3.70)$$

for equal resonant amplitudes. The output-input amplitude ratio (*transfer function*, as it will be shown with more detail in Chapter 4) is:

$$G(j\omega) = \frac{X}{F} = \frac{1}{k - m\omega^2 + j\gamma k} \tag{3.71}$$

whereas the *frequency response function* (also treated in Chapter 4) is:

$$H(j\omega) = \frac{kX}{F} = \frac{1}{1 - \beta^2 + j\gamma} = \frac{1 - \beta^2}{\left(1 + \beta^2\right)^2 + \gamma^2} - j\frac{\gamma}{\left(1 + \beta^2\right)^2 + \gamma^2} \tag{3.72}$$

Example 3.7
     The Q-factor is determined experimentally at resonance for a metal micro-cantilever whose length l, cross-sectional width w, and thickness t are known. Determine the structural loss coefficient α by assuming the experimental test is conducted in vacuum (such that friction losses can be neglected), at low temperature (to discard thermal damping effects), and when support losses are disregarded.

Solution:
     By taking into account Equations (3.25) and (3.70), the structural damping factor γ is expressed as:

$$\gamma = \frac{1}{Q_r} \tag{3.73}$$

which can be transformed by way of Equation (3.65) into:

$$\alpha = \frac{\pi k}{Q_r} \tag{3.74}$$

As known from elementary beam theory, the lumped-parameter stiffness of the cantilever at its free end is:

$$k = \frac{Ewt^3}{4l^3} \tag{3.75}$$

Consequently, the structural loss coefficient becomes:

$$\alpha = \frac{\pi Ewt^3}{4l^3 Q_r} \tag{3.76}$$

Equation (3.76) emphasizes that the loss coefficient is inversely proportional to the resonant Q-factor. It can also be seen that long (l large) and thin (w and t small) microcantilevers produce less structural damping.

3.4 SQUEEZE-FILM DAMPING

In squeeze-film damping, a MEMS plate-like member moves against a fixed surface and the gas in between generates a viscous damping resistance to motion, as sketched in Figure 3.9. Two particular gas regimes will be analyzed next together with their models, namely: the *continuum flow regime* and the *free molecular flow regime*.

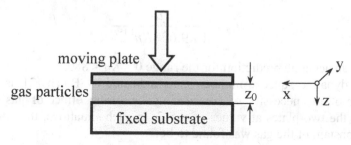

Figure 3.9 Squeeze-film damping

3.4.1 Continuum Flow Regime

In cases in which gas pressure is close to the normal (atmospheric) one and the gap is considerably larger than the free molecular path of gas molecules, the gas behaves as a continuum and approaches/results pertaining to continuum gas models are applicable.

Integration of the Poisson-type equation, which expresses the film pressure under isothermal conditions, was performed by Blech [2], for instance, who provided the following viscous damping coefficient for a rectangular plate of dimensions l and w ($l > w$):

$$c = \frac{64\sigma p l w}{\pi^6 z \omega} \sum_{m,odd} \sum_{n,odd} \frac{m^2 + r^2 n^2}{m^2 n^2 \left[\left(m^2 + r^2 n^2 \right)^2 + \frac{\sigma^2}{\pi^4} \right]} \quad (3.77)$$

where $r = w/l$ and:

$$\sigma = \frac{12\mu_{eff} w^2}{pz^2} \omega \quad (3.78)$$

is the *dynamic squeeze-number*. In the equations above ω is the frequency of the mobile plate, p is the atmospheric pressure, z is the channel gap and μ_{eff} is the effective dynamic viscosity. The last quantity is a corrected value of a regular dynamic viscosity number, when taking into account the relationship between the gas channel dimensions and the molecular mean free path, λ, (the distance between a molecule's two consecutive collisions). This is best expressed by the *Knudsen number*:

$$Kn = \frac{\lambda}{z} \tag{3.79}$$

Veijola et al. [3], for instance, suggest the effective dynamic viscosity:

$$\mu_{eff} = \frac{\mu}{1+9.638 Kn^{1.159}} \tag{3.80}$$

which is an accurate prediction for the range $0 < Kn < 880$.

The dynamic squeeze number σ, as indicated by Blech [2], is also an indicator of the necessity of considering the spring effect of gas trapped between the two plates at values of $\sigma > 3$. In such situations, the equivalent spring constant of the gas was found to be:

$$k = \frac{64\sigma^2 p l w}{\pi^8 z} \sum_{m,odd} \sum_{n,odd} \frac{m^2 + r^2 n^2}{m^2 n^2 \left[\left(m^2 + r^2 n^2\right)^2 + \frac{\sigma^2}{\pi^4} \right]} \tag{3.81}$$

Example 3.8

Compare the viscous damping coefficient of Equation (3.77) when the first four terms of the double series are retained with the viscous damping coefficient provided by Zhang et al. [4], namely:

$$c = \frac{96}{\pi^4} \frac{l^3 w^3}{l^2 + w^2} \frac{\mu}{z^3} \tag{3.82}$$

Consider that air pressure is $p = 100{,}000$ N/m^2, dynamic viscosity is $\mu = 1.85 \times 10^{-5}$ N-s/m^2, and use the first level of approximation.

Solution:

For $m = n = 3$, Equation (3.77) gives a damping coefficient denoted by c_{33}. When the plate's geometry is of interest, one can select an operating frequency and a gap distance, for instance, $f = 500{,}000$ Hz and $z = 10$ μm. By also using the other numerical values of this example, the plot of Figure 3.10 is obtained, which shows the relative errors between c_{33} and c, calculated as:

$$e = \frac{c - c_{33}}{c_{33}} \tag{3.83}$$

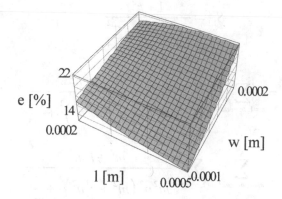

Figure 3.10 Relative errors in damping coefficients in terms of plate geometry in squeeze-film damping

For $l = 200$ μm and $w = 50$ μm, the plot of Figure 3.11 is obtained, which indicates the influence of vibration frequency and gap dimension on the same damping coefficient ratio.

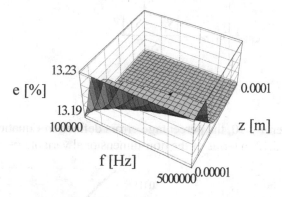

Figure 3.11 Relative error in damping coefficients in terms of vibration frequency and gap dimension in squeeze-film damping

Example 3.9

Analyze the precision of calculating the squeeze-film damping coefficient of Equation (3.77) as a function of the number of n and m terms in the corresponding infinite series.

Solution:

The squeeze number of Equation (3.77) is expressed as a series expansion in m and n. The double sum that defines it is expressed next by taking several levels of approximation, namely:

$$S_{1,1} = \frac{1+r^2}{\left(1+r^2\right)^2 + \dfrac{\sigma^2}{\pi^4}} \tag{3.84}$$

$$S_{3,3} = \sum_{m=1,\,m-odd}^{3} \sum_{n=1,\,n-odd}^{3} \frac{m^2 + r^2 n^2}{m^2 n^2 \left(m^2 + r^2 n^2\right)^2 + \dfrac{\sigma^2}{\pi^4}} \tag{3.85}$$

$$S_{5,5} = \sum_{m=1,\,m-odd}^{5} \sum_{n=1,\,n-odd}^{5} \frac{m^2 + r^2 n^2}{m^2 n^2 \left(m^2 + r^2 n^2\right)^2 + \dfrac{\sigma^2}{\pi^4}} \tag{3.86}$$

Equations (3.84), (3.85), and (3.86) are used to form the following relative error numbers:

$$\begin{cases} e_{11-33} = \dfrac{|S_{11} - S_{33}|}{S_{11}} \\[4mm] e_{11-55} = \dfrac{|S_{11} - S_{55}|}{S_{11}} \\[4mm] e_{33-55} = \dfrac{|S_{33} - S_{55}|}{S_{33}} \end{cases} \tag{3.87}$$

For a squeeze number $\sigma = 10$, the percentage errors defined in Equation (3.87) are plotted in Figure 3.12 in terms of the non-dimensional variable $r = w/l$.

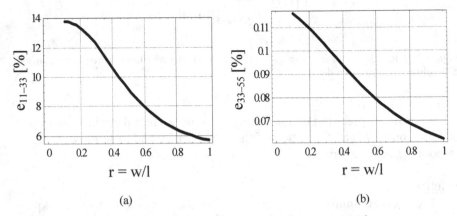

(a)   (b)

Figure 3.12 Relative errors in computing the squeeze number

As Figure 3.12 shows, the relative errors decrease sharply from the third level of approximation (when six terms are retained from the series; Equation (3.86)), which is compared to the second level of approximation (when four terms are retained from the series, as shown in Equation (3.85)). The relative error e_{11-55} of Equation (3.87) is very similar to the error e_{33-55} and was not plotted here. It is therefore safe, in terms of accuracy, to truncate the series expansion involved in calculating the viscous damping coefficient corresponding to the squeeze film phenomenon at $m = n = 3$.

Example 3.10

Analyze the squeeze number variation with respect to the width of a microplate that moves against a fixed plate and the spacing between the two plates in the case of air. Known are the following amounts: molecular free mean path $\lambda = 85$ μm, dynamic viscosity coefficient $\mu = 1.85 \times 10^{-5}$ N-s/m, pressure $p = 101{,}325$ N/m^2, frequency $f = 100$ MHz.

Solution:

With the numerical values given in this problem, and by taking into account that $\omega = 2\pi f$, the squeeze number becomes:

$$\sigma = \frac{1.375}{1 + 6.15e^{-8}z^{-1.159}}\left(\frac{w}{z}\right)^2 \tag{3.88}$$

Figure 3.13 shows the variation of the squeeze number in terms of the plate width and the spacing. Figure 3.14 gives the squeeze number dependency of the gap for a fixed width $w = 800$ μm.

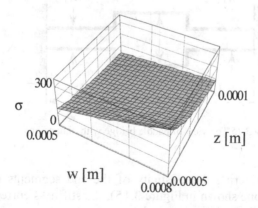

Figure 3.13 Squeeze number as a function of plate width w and gap z

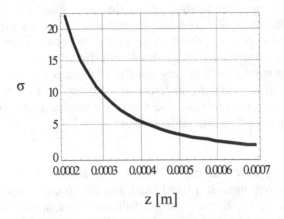

z [m]

Figure 3.14 Squeeze number as a function of gap *z*

Solving the equation $\sigma = 3$, where σ is given in Equation (3.78) results in $z = 541.5$ μm. Figures 3.13 and 3.14 indicate that σ decreases with the gap increasing and therefore, according to Blech's prescription, for gaps larger than 541.5 μm, air escapes the gap and the additional spring behavior is not manifested.

Example 3.11

Examine the influence of gas entrapping ($\sigma = 10$) in a paddle microbridge (Figure 3.15) on its bending-related resonant frequency. Assume the out-of-the-plane motion of the paddle segment is always parallel to the substrate.

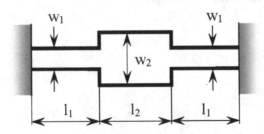

Figure 3.15 Top view of paddle bridge with dimensions

Solution:

When only considering the elasticity of the end segments of a paddle microbridge (as the one shown in Figure 3.15), the stiffness corresponding to out-of-the-plane bending (about the *y*-axis) is:

$$k_{b,y} = 2\frac{12EI_y}{l_1^3} = 2\frac{Ew_1t^3}{l_1^3} \qquad (3.89)$$

In the case air is entrapped between the out-of-the-plane vibrating middle plate and substrate, the corresponding spring effect is expressed by the stiffness given in Equation (3.81) and the total stiffness is the one of a spring parallel connection, namely:

$$k_{b,y}^{tot} = 2\frac{Ew_1t^3}{l_1^3} + \frac{64\sigma^2 pl_2w_2}{\pi^8 z} \sum_{m,odd}\sum_{n,odd} \frac{m^2 + r^2n^2}{m^2n^2\left[\left(m^2 + r^2n^2\right)^2 + \frac{\sigma^2}{\pi^4}\right]} \qquad (3.90)$$

The mass of the microbridge is considered to be provided by the middle segment, which, according to Figure 3.15, is:

$$m = \rho w_2 l_2 t \qquad (3.91)$$

When taking a one-term series approximation in Equation (3.90) ($m = n = 1$), the following resonant frequency percentage error can be formulated:

$$e_\omega = \frac{\omega_{b,y}^{total} - \omega_{b,y}}{\omega_{b,y}} = \frac{\sqrt{k_{b,y}^{total}} - \sqrt{k_{b,y}}}{\sqrt{k_{b,y}}} \qquad (3.92)$$

Equation (3.92) has been used to draw Figures 3.16 and 3.17. Figure 3.16 is plotted for $l_2 = 200$ μm and $w_2 = 100$ μm, whereas Figure 3.17 is plotted for $l_2 = 200$ μm and $w_1 = 50$ μm.

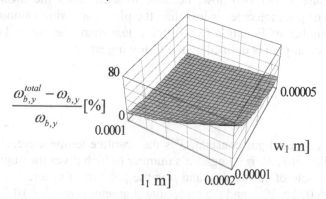

Figure 3.16 Relative errors in bending resonant frequency of a paddle microcantilever ($l_2 = 200$ μm, $w_2 = 100$ μm)

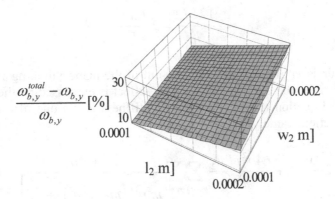

Figure 3.17 Relative errors in bending resonant frequency of a paddle microcantilever ($l_1 = 200$ μm, $w_1 = 50$ μm)

As Figure 3.16 indicates, the relative differences between the out-of-the-plane bending resonant frequency when air elasticity is taken into account and the one in which air elasticity is not considered, increase with the length of the mid-segment increasing and the width decreasing up to a maximum of 80% for the selected parameter ranges. The effects of length and width of the root segment (the flexible ones) on the same resonant frequency differences are shown in Figure 3.17, which indicates that differences do increase with both length and width, increasing up to a relative maximum of 25% for the parameters analyzed in the figure.

3.4.2 Free Molecular Flow Regime

The squeeze-film damping models presented thus far are accurate for situations in which the gas behaves as a continuum, and this condition is satisfied when the pressure is not very low, because in such cases the mean free molecular path of gas molecules is less than the plate gap (which amounts to the Knudsen number of Equation (3.69) being less than one, $Kn < 1$). The mean free molecular path is defined by the following equation:

$$\lambda = \frac{RT}{\sqrt{2}\pi d^2 N_A p}$$

(3.93)

where R is the universal gas constant, T is the absolute temperature, d is the gas molecule diameter, N_A is Avogadro's number (which gives the number of molecules in a mole of substance), and p is the pressure. For air, $R = 8.314$ J/mol-K, $N_A = 6.022 \times 10^{23}$, and the molecular diameter is $d = 3 \times 10^{-10}$ m. In case of a normal temperature of 300°K, the mean free molecular path of Equation (3.93) becomes:

$$\lambda = \frac{0.01}{p} \tag{3.94}$$

Under normal pressure conditions, such as $p = 100,000$ N/m^2, the mean free molecular path is $\lambda = 0.1$ μm, whereas for small pressures (almost vacuum), such as $p = 1$ N/m^2, the mean free molecular path is $\lambda = 1$ cm. Regular gaps in MEMS are of the order of micrometers, and therefore, for low (vacuum) pressures, the molecular mean free path is much smaller than the gap in a squeeze-film situation. Consequently, the continuum laws are no longer applicable and models pertaining to the free molecular domain are in place. By using notions of the momentum transfer, the Christian model [5], gives the following estimate of the Q-factor due to air damping at low pressures:

$$Q = \left(\frac{\pi}{2}\right)^{\frac{3}{2}} \frac{\rho t \omega}{p} \sqrt{\frac{RT}{M_m}} \tag{3.95}$$

where ρ is the gas mass density, t is the plate thickness, p is the pressure, and M_m is the molar weight of gas. This model assumed an infinitely large volume and considered the Maxwell-Boltzmann distribution of gas velocity. The results of Christian's model (which was derived for macroscale applications) indicated Q-factors larger than experimental measurements indicated. Kadar et al. [6] used a variant of the Maxwell-Boltzmann distribution, namely the Maxwell stream distribution, and proposed the following Q-factor:

$$Q = \frac{1}{\pi}\left(\frac{\pi}{2}\right)^{\frac{3}{2}} \frac{\rho t \omega}{p} \sqrt{\frac{RT}{M_m}} = \sqrt{\frac{\pi}{8}} \frac{\rho t \omega}{p} \sqrt{\frac{RT}{M_m}} \tag{3.96}$$

which is essentially π times smaller than Christian's model prediction.

Bao et al. [7] propose a similar Q-factor model that accounts for the plate dimensions and gap, namely:

$$Q = (2\pi)^{\frac{3}{2}} \frac{\rho t \omega}{p} \frac{z}{2(l+w)} \sqrt{\frac{RT}{M_m}} \tag{3.97}$$

where z is the gap and l and w are the dimensions of a rectangular plate. This model also considers the energy transfer from the oscillating plate to the gas and the reflection wall effects.

Example 3.12

A plate with $l = 180$ μm, $w = 40$ μm, and $t = 1$ μm is suspended by two single-wall carbon nanotubes (SWCNTs) 1.4 in diameter and oscillates out-of-the-plane against the substrate in air at normal temperature $T = 300°K$ and at a frequency of $f = 1$ MHz. Calculate the length of the SWCNT beams that will render the continuum-model and molecular-model Q-factors equal. Consider the following numerical properties: free gap $z = 10$ μm, pressure $p = 1000$ N/m².

Solution:

For air, the molecular weight is $M_m = 0.029$ kg/mol and the molecular diameter is $d = 3 \times 10^{-10}$ m; the gas constant is $R = 83145$ J/mol-K and the density at sea level and normal temperature is 1.2 kg/m³. With the numerical data of the example and with the aid of Equation (3.94), it is found that the molecular free path is $\lambda = 10$ μm, which is equal to the equilibrium gap, and therefore both the continuum and molecular models are likely to be valid. The continuum-model Q-factor is obtained from the corresponding damping coefficient as:

$$Q_c = \frac{k}{c\omega} \qquad (3.98)$$

It can be seen that this Q-factor depends on the spring stiffness, whereas the molecular-model Q-factor does not depend on any stiffness. Anyway, in order for the two models' Q-factors to be equal, the stiffness needs to have a specified value. By calculating the two models' Q-factors, an equation in k results, which gives $k = 17.27$ N/m. At the same time, it is known from mechanics of materials that the stiffness of a clamped-guided beam is:

$$k = \frac{3\pi d^4 E}{16 l_b^3} \qquad (3.99)$$

where E is the elastic modulus of the beam. An average value of $E = 10^{12}$ N/m² will be considered, and l_b is the unknown beam length. The total stiffness is twice the one given in Equation (3.99) because there are two beams supporting the plate, and therefore by equating that stiffness to the number found above, the resulting beam length is $l_b = 11.18$ nm.

Bao's model assumed constant gas particle velocity, and, in addition, the amplitude of oscillations was considered much smaller than the gap dimension. The time interval when a molecule is located between the resonator and the wall was assumed to be much smaller than the plate oscillation period. Hutcherson and Ye [8] proposed a Q-factor model that was two times smaller than the Q-factor, according to Bao's model prediction. This model allowed for variations in the gas particle velocity and was proved to be valid for situations in which the ratio of the gap to the plate length is around 1/200.

The Q-factor in squeeze-film damping is further affected by gas–surface interactions, of which out gassing from surfaces and gas molecule adsorption by the plates are the most important. A quantifier of these interactions is the *normal momentum accommodation coefficient* (NMAC), α_n, which ranges from 0 for no adsorption to 1 for full adsorption. Polikarpov et al. [9] proposed the following damping ratio, which took into account the gas–surface adsorption interaction:

$$\zeta = \frac{2-\alpha_n}{\rho t}\, p \sqrt{\frac{2m_m}{\pi k T}} \qquad (3.100)$$

where ρ is the plate's material mass density, t is the plate thickness (as previously mentioned), m_m is the molecular mass of the gas, k is Boltzmann's constant, and T is the absolute temperature. By taking Equation (3.24) into account, which defines the relationship between the damping ratio and the Q-factor, the latter can be expressed by also considering that:

$$\frac{m_m}{k} = \frac{M_m}{R} \qquad (3.101)$$

as:

$$Q = \left(\frac{\pi}{2}\right)^{\frac{3}{2}} \frac{2}{2-\alpha_n}\, \frac{\rho t f}{p} \sqrt{\frac{RT}{M_m}} \qquad (3.102)$$

Equation (3.102) also considered the relationship between the angular frequency ω and the normal one f, namely: $\omega = 2\pi f$.

3.4.3 Squeeze-Film Damping for Rotary-Motion Plates

Equivalent viscous damping coefficients expressed so far that involved squeeze-film damping referred to translation and therefore were related to linear-motion damping forces. In the case sketched in Figure 3.18, a plate rotates about a fixed pivot point and the gas is squeezed between the moving plate and the fixed substrate producing a damping torque, which opposes the motive angular velocity ω.

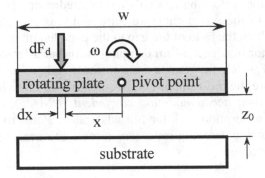

Figure 3.18 Squeeze-film damping and rotary plate

This topic has usually been approached by linearizing the Reynolds equation that expressed the variable pressure about the x-direction under the assumption that the angular motions of the plate are small compared to the static gap z_0. The linearized equation is subsequently solved, and its solution is used to determine the total resistive damping torque (Darling et al. [10], Dotzel et al. [11], Pan et al. [12] and Bao et al. [13], to cite just a few of the work dedicated to this topic). Veijola et al. [14] presented a simple model that yielded the damping coefficient pertaining to a plate rotating at an angular velocity ω, which is connected to the corresponding torque as:

$$M_d = c_r \omega \qquad (3.103)$$

The model starts by considering the linearized, temperature-independent Reynolds equation in one dimension (the x-direction):

$$\frac{z_0^2}{12\,\mu_{\mathit{eff}}} \times \frac{d^2 p(x)}{dx^2} = \frac{\omega x}{z_0} \qquad (3.104)$$

where the linear velocity of the plate at a distance x measured from the central pivot point is simply $\omega \times x$, and p_0 is the ambient (constant) pressure. The solution to Equation (3.104) is a third-order polynomial and its two integration constants are determined by applying the trivial boundary conditions:

$$p(w/2) = p(-w/2) = 0 \qquad (3.105)$$

The pressure is therefore:

$$p(x) = c_1\left(4\frac{x^3}{w^3} - \frac{x}{w}\right) \qquad (3.106)$$

where:

$$c_1 = \frac{\mu_{eff} \omega w^3}{2z_0^3} \tag{3.107}$$

The elementary damping force acting on an elementary surface area $dA = dxdy$ and opposing the plate rotation can be expressed as:

$$dF_d = p(x)dxdy \tag{3.108}$$

and this force produces an elementary damping torque:

$$dM_d = xdF_d = xp(x)dxdy \tag{3.109}$$

The total damping torque is found by integrating Equation (3.109) over the whole plate area, namely:

$$M_d = \int_A dM_d = \int_{-l/2}^{l/2} \left[\int_{-w/2}^{w/2} xp(x)dx \right] dy \tag{3.110}$$

After integration and consideration of Equation (3.103), which gives the relationship between damping torque and angular velocity, the torsional damping coefficient is expressed as:

$$c_r = \frac{\mu_{eff} w^5 l}{60 z_0^3} \tag{3.111}$$

Example 3.13
   A plate is suspended at its ends by two serpentine springs that are clamped at their opposite ends to the substrate, as sketched in Figure 3.19. Compare the squeeze-film damping that is generated when the plate moves out-of-the plane parallel to the substrate to the damping corresponding to the small-angle rotation of the plate about the x-axis.

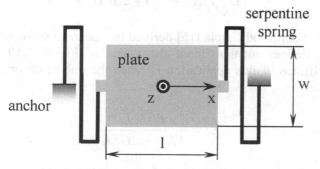

Figure 3.19 Plate with two end serpentine springs

Solution:

The damping comparison is carried out by means of the resonant Q-factors, which, according to the definition of Equation (3.25), depend on inertia, damping, and stiffness. For the translatory motion about the z-axis, the Q-factor is:

$$Q_z = \frac{\sqrt{MK_z}}{c_z} \tag{3.112}$$

where M is the mass of the plate, c_z (Equation (3.82)) is the damping coefficient of the z-axis translation, and K_z is the stiffness of the spiral springs corresponding to the same motion. Similarly, the Q-factor defining damping due to squeeze-film effects accompanying the resonant rotary vibrations of the plate can be expressed as:

$$Q_r = \frac{\sqrt{K_r J_x}}{c_r} \tag{3.113}$$

where J_x is the plate mechanical moment of inertia about the x-axis, c_r (Equation (3.111)) is the damping coefficient of this motion and K_r is the torsional stiffness of the two serpentine springs. The plate's moment of inertia depends on mass as:

$$J_x = \frac{M}{12}\left(w^2 + t_p^2\right) \tag{3.114}$$

where t_p is the plate thickness.

The stiffness of a serpentine spring expressing translatory motion about the z-axis in Figure 3.19 was given by Lobontiu and Garcia [15], and because there are two springs in parallel in this application, the corresponding stiffness is twice the one of an original spring, namely:

$$K_z = \frac{3EGI_y I_t}{3EI_y l_1 l_2(l_1 + 3l_2) + 2GI_t(l_1^3 + 2l_2^3)} \tag{3.115}$$

Similarly, Lobontiu and Garcia [15] derived the torsional stiffness of a spiral spring that defines rotation about the x-axis of Figure 3.19. The total torsional stiffness of this application is twice the one of an original spiral spring, namely:

$$K_r = \frac{EGI_y I_t}{EI_y l_2 + 2GI_t l_1} \tag{3.116}$$

In Equations (3.115) and (3.116), E and G are the longitudinal and shear modulii of the spring material, whereas l_1 is the length of half of a long leg defining the spiral spring and l_2 is the length of a spring short leg, as also shown in a previous example in Chapter 1. Figure 3.20 shows the cross-section of a spiral spring, which is assumed constant and identical for both the long and short legs.

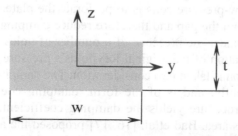

Figure 3.20 Cross-sectional dimensions of the spiral spring

Considering the cross-section is thin, the moments of inertia are related as:

$$I_t = 4I_y = \frac{wt^3}{3} \tag{3.117}$$

All the amounts that are necessary to compute the Q-factors of Equations (3.112) and (3.113) are now available. The non-dimensional variables c_1, c_2, c_w, and c_t are introduced and defined as follows: $l_1 = c_1 \times l$, $l_2 = c_2 \times l$, $w = c_w \times l$, $t_p = c_p \times l$. The plots of Figure 3.21 show the variation of the rotation-to-translation Q-factor ratio. In Figure 3.21 (a), $c_1 = 0.5$ and $c_2 = 0.1$, whereas in Figure 3.21 (b), $c_w = 0.2$ and $c_t = 0.01$.

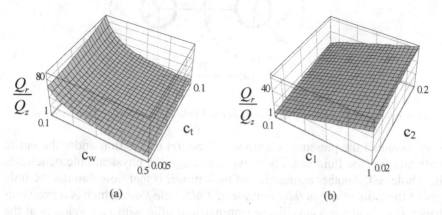

Figure 3.21 Numerical simulation for rotary-to-translatory quality factor ratio for squeeze-film damping: (a) $c_1 = 0.5$, $c_2 = 0.1$; (b) $c_w = 0.2$, $c_t = 0.01$

3.4.4 Squeeze-Film Damping: Translatory Perforated Plates

Proper operation of MEMS/NEMS under squeeze-film damping conditions is often times hampered by the relatively high damping coefficients (particularly for thick plates), which reduce the device Q-factor. An alternative to using slide-film damping instead of squeeze-film damping or to packaging the device in low-pressure cells is to perforate the plate, which allows air to flow through from the gap and therefore reduce damping. Finding the damping coefficient under the presence of a number of holes in the original plate implies modification of the original Reynolds's pressure equation so that the perforation region is taken into consideration. The damping force can be computed subsequently, and is of the form: damping coefficient times plate velocity; this procedure yields the damping coefficient. Of the many contributions to this area, Bao et al. [16, 17] proposed the approach of dividing the plate into cells with holes at centers, as shown, for instance, in Figure 3.22.

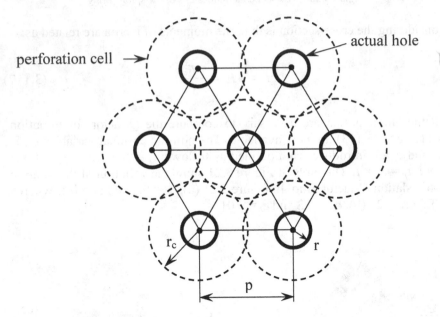

Figure 3.22 Portion of hole-plate showing actual hole and perforation cell arrangement

They assumed the pressure is a smooth function of position under the entire plate and that the fluid actually flowing through the physical hole penetrates the whole cell. Another assumption of their model is that flow through the hole (about the z-direction) is *fully developed Poiseuille flow*, which is a pressure-driven flow, and has a curvilinear symmetric profile with zero velocity at the edges and maximum velocity at the hole center. In doing so, the actual area of a rectangular plate, for instance, is transformed into an equivalent smaller area where pressure acts uniformly, as indicated in Figure 3.23.

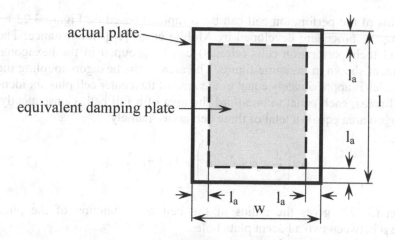

actual plate

equivalent damping plate

Figure 3.23 Top view of actual and equivalent damping plate

By applying this procedure, the following damping coefficient is obtained under the additional assumption that at least three holes exist across the plate in any direction:

$$c = \frac{12\mu}{z_0^3} l_a^2 \left(w - 2l_a\right)\left(l - 2l_a\right) \tag{3.118}$$

where w and l are the plate in-plane dimensions, z_0 is the original (static) gap, μ is the dynamic viscosity (which for slip and transition regimes can be substituted by μ_{eff} [Equation (3.80)], according to Veijola et al. [3]), and l_a is the *attenuation length* (Bao et al. [16, 17]), which is computed as:

$$l_a = \sqrt{\frac{2z_0^3 t_{eff}\eta(\beta)}{3\beta^2 r^2}} \tag{3.119}$$

In Equation (3.119), t_{eff} is the effective thickness of the plate, which takes into account additional flow resistance at the perforation ends (particularly when the hole radius r compares to the plate thickness t), and is calculated as:

$$t_{eff} = t + \frac{3\pi r}{8} \tag{3.120}$$

The coefficient β in Equation (3.119) is the ratio between the hole radius r and the perforation cell radius r_c, namely: $\beta = r/r_c$. In the same Equation (3.119):

$$\eta(\beta) = 1 + \frac{3r^4\left(4\beta^2 - \beta^4 - 4\ln\beta - 3\right)}{16tz_0^3} \tag{3.121}$$

The radius of the perforation cell can be computed based on Figure 3.22 by following the procedure developed by Mohite et al. [18], for instance. The holes and their perforation cells (circles) can be grouped in the hexagonal arrangement shown in the same figure. The area of the hexagon coupling the seven circles is approximately equal to the area of the center cell plus six identical cell areas, each equal to one-third the area of a full cell. Consequently, the hexagon area equals a total of three cell areas, namely:

$$6 \times \left(\frac{1}{2} \times p \times \frac{\sqrt{3}}{2} p \right) = 3 \times \left(\pi r_c^2 \right) \tag{3.122}$$

Equation (3.122) gives the radius of the cell as a function of the pitch distance p between two adjacent plate holes:

$$r_c = 0.525 p \tag{3.123}$$

Example 3.14
    Study the influence of the pitch dimension p on the attenuation length and of the squeeze-film damping of a rectangular plate with holes in it. The plate's dimensions are $l = 500$ μm, $w = 100$ μm, $t = 2$ μm. The initial gap is $z_0 = 10$ μm and the air's dynamic viscosity is $\mu = 1.73 \times 10^{-5}$ N-s/m^2.

Solution:
    By using the given numerical data, Equations (3.118) through (3.123) are used to express l_a and c in terms of only the pitch distance p and the hole radius r. Figure 3.24 shows the two functions plotted against p and r. Both the attenuation length and the damping coefficient increase with the pitch distance in a quasi-linear fashion, as shown in Figure 3.24, whereas the magnitude of the hole radius has a nonlinear influence, which is, however, smaller compared to the influence of the pitch distance.

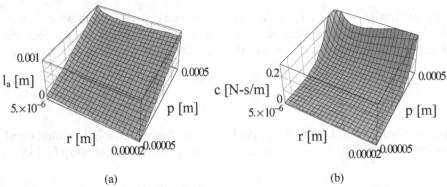

(a)   (b)

Figure 3.24 Damping characteristics for a plate with holes: (a) attenuation length; (b) damping coefficient

3.5 SLIDE-FILM DAMPING

When the motion of a plate takes place parallel to another fixed plate (such as the substrate in a MEMS device), shearing of the fluid underneath and above the moving plate will generate viscous damping resistance through the relative fluid-structure sliding. The slide-film damping is sketched in Figure 3.25.

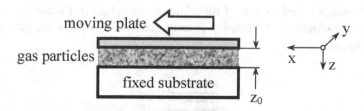

Figure 3.25 Slide-film damping

Depending on the Knudsen number, which compares the mean free molecular path to the fluid path, there are normally four different flow types, and the damping coefficients are determined by different models. For Knudsen numbers smaller than 0.001, which means the free molecular path is at least three orders of magnitudes smaller than the fluid gap, the viscous damping can be assessed by macro-scale, conventional methods pertaining to *continuum models*. Larger values of the *Kn* number indicate the mean free molecular path and gap dimension become comparable and micro/nano phenomena such as gas rarefaction and gas–surface interactions have to considered. As such, when $0.001 < Kn < 0.1$, the flow is known as *slip flow*, and slip velocity boundary conditions have to be accounted for. To determine the damping coefficients, Navier-Stokes equations are solved for both flow categories. For $0.1 < Kn < 10$, *transition flow* conditions are set up, while for $Kn > 10$ the flow is *free molecular*. These cases are discussed next.

3.5.1 Continuum Flow Regime

In the case of Knudsen numbers that are less than 0.001, the flow is governed by macro-scale laws and the viscous damping coefficient is determined as follows. The boundary conditions are considered fixed, namely:

$$\begin{cases} v_x(0) = 0 \\ v_x(z_0) = v \end{cases} \tag{3.124}$$

The shearing stresses present between two adjacent fluid layers in one-dimensional flow are expressed (e.g., see Landau and Lifschitz [19]) as:

$$\tau = \mu \frac{dv_x(z)}{dz} \tag{3.125}$$

One model that enables predicting the damping coefficient is the *Couette model*, according to which the velocity profile of the fluid between the two plates varies linearly from 0 (at the fixed plate) to the mobile plate velocity at the interface with it. The fluid above the moving plate is assumed to displace with a velocity equal to that of the plate, as sketched in Figure 3.26 below. As a direct consequence, damping is only generated by the fluid between the two plates.

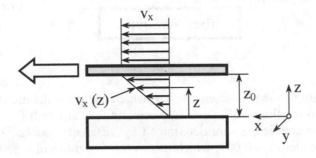

Figure 3.26 Velocities in Couette slide-film damping

For Couette-type flow, the linear velocity profile of Figure 3.26 is expressed as:

$$v_x(z) = \frac{z}{z_0} v_x \tag{3.126}$$

By combining Equations (3.125) and (3.126), the shearing stress becomes:

$$\tau = \mu \frac{v_x}{z_0} \tag{3.127}$$

which indicates the stress is constant over the two plates gap. The damping force produced at the plate–fluid interface can be calculated by multiplying the shear stress to the mobile plate area, namely:

$$F_d = \tau A = \frac{\mu A}{z_0} v_x \tag{3.128}$$

A linear damping force is the product of a damping coefficient to the velocity, and therefore Equation (3.128) yields the following viscous damping coefficient owing to Couette-type slide-film effects:

$$c_C = \frac{\mu A}{z_0} \tag{3.129}$$

The corresponding Q-factor of Equation (3.24) can be calculated by means of c_C from Equation (3.129) as:

$$Q_C = \frac{k z_0}{\mu A \omega} \tag{3.130}$$

where k is the spring stiffness associated with the mobile plate (which is elastically supported over the substrate) and ω is the frequency of the sinusoidal force that drives the mobile plate. At resonance, when the driving force frequency equals the plate-spring resonant frequency, the Couette-type Q-factor becomes:

$$Q_{C,r} = \frac{z_0}{\mu A} \sqrt{km} \tag{3.131}$$

where m is usually the mass of the moving plate. Equation (3.130) indicates that the Q-factor increases by reducing the dynamic viscosity of the gas, as well as the plate area, and by increasing the gap, together with the mass and stiffness of the plate-spring system.

Example 3.15
    A plate microresonator, as the one sketched in Figure 3.27, is driven at resonance by a comb-drive actuator. The plate is elastically supported by two spiral springs. Design a plate-spring system that will have a specified Q-factor at resonance for given air viscosity and plate gap. Assume also that the plate area is known and that the plate-to-spring thickness ratio and the spring's leg length ratio (the spring is shown in Figure 3.28) are specified as well.

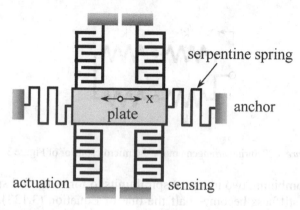

Figure 3.27 Electrostatically actuated and sensed microplate resonator with spiral springs (top view)

Solution:
    Equation (3.131) can be rewritten as:

$$Q_{C,r} = \frac{z_0}{\mu}\sqrt{\frac{\rho kt}{A}} \qquad (3.132)$$

where ρ is the plate's mass density and t is its thickness. The total stiffness of this microresonator is twice the stiffness of a single spring because there are two springs here acting in parallel. The stiffness of a spiral spring can be found as the inverse of the compliance given by Lobontiu and Garcia [15] as:

$$k = \frac{3EI_z}{2l_1^2\left(2l_1 + 3l_2\right)} \qquad (3.133)$$

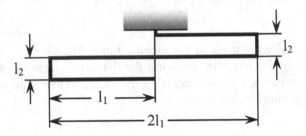

Figure 3.28 Geometry of spiral spring unit

The geometry of a spiral spring unit is shown in Figure 3.28, but each of the springs shown in Figure 2.27 is formed of two serially connected spiral spring units, and the total stiffness is therefore the equivalent stiffness of the spring arrangement shown in Figure 3.29.

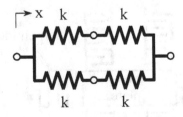

Figure 3.29 Spring arrangement for the microresonator of Figure 3.28

By serially combining two identical spiral units to form one full spiral spring, the resulting stiffness becomes half the one of Equation (3.133). By further combining the two resulting springs in parallel, the final equivalent stiffness is twice the one of a serial chain, and therefore equal to the one of the original spring unit of Equation (3.133). By considering the spring cross-section is square with a thickness t_s, the Q-factor of Equation (3.132) becomes:

$$Q_{C,r} = c \frac{t^{5/2}}{l_2^{3/2}}$$

(3.134)

where:

$$c = \frac{z_0 c_t^2}{2\mu c_l} \sqrt{\frac{\rho E}{2A(3 + 2c_l)}}$$

(3.135)

has the dimension of length to the $- (2/5)$ power and with: $l_1 = c_l l_2$, and $t_s = c_t t$. Equation (3.134) suggests that the Q-factor can be improved by increasing the plate and spring thickness and by shortening the legs of the spiral spring. By selecting one parameter, for instance l_2, the other unknown can be computed as:

$$t = \left(\frac{Q_{C,r}}{c} \right)^{2/5} l_2^{3/5}$$

(3.136)

3.5.2 Slip Flow Regime

Slip flow regimes are set up for Knudsen numbers $0.001 < Kn < 0.1$, when the mean free molecular path-to-gap ratio increases and when the gas velocity is not zero at the gas-fixed surface interface due to some gas molecular motion. The flow velocity is still linear (the Couette flow). For relatively slow motion of the plate (when gas inertia is not accounted for), as well as for fast vibrating plates (where inertia of gas is a factor), different damping coefficients can be obtained analytically by solving the Navier-Stokes equation in conjunction with using slip velocity boundary conditions, as shown in the following.

3.5.2.1 Frequency-Independent Damping

One modality of deriving the damping coefficient is expressing the maximum fluid shear stress, which occurs at the fluid–plate boundary:

$$\tau = \mu_{eff} \frac{v_x}{z_0}$$

(3.137)

where μ_{eff} is the effective dynamic viscosity and is determined according to various assumptions and contains corrections (mainly in terms of the Knudsen number) that will be discussed a bit later in this section. From Equation (3.137) one can calculate the damping force as:

$$F_d = \tau A = \frac{\mu_{eff} A}{z_0} v_x \qquad (3.138)$$

Because this force is equal to the damping coefficient multiplied by velocity, it follows that the damping coefficient is, according to Equation (3.138):

$$c = \frac{\mu_{eff} A}{z_0} \qquad (3.139)$$

3.5.2.2 Frequency-Dependent Damping: Stokes Model

For large Reynolds numbers, where inertia effects are larger than viscosity effects, damping is dependent on frequency because gas velocity distribution becomes dependent on time, and different models are in place (Veijola and Turowski [20]). The Navier-Stokes equation, which describes the diffusion problem with no pressure gradient in one dimension, is:

$$\frac{\partial v_x(z,t)}{\partial t} = \upsilon \frac{\partial^2 v_x(z,t)}{\partial z^2} \qquad (3.140)$$

where υ is the kinematic viscosity. The generic solution to this partial-derivative differential equation can be obtained in the frequency domain, when the velocity of the plate is:

$$v_x(z,t) = V_x(z)\sin(\omega t) \qquad (3.141)$$

It can be shown that the generic amplitude of Equation (3.141) is of the form:

$$V_x(z) = C_1 \sin h\left(\sqrt{\frac{j\omega}{\upsilon}} z\right) + C_2 \cos h\left(\sqrt{\frac{j\omega}{\upsilon}} z\right) \qquad (3.142)$$

For fixed boundary conditions, Equation (3.142) reduces to:

$$V_x(z) = V \frac{\sin h\left(\sqrt{\dfrac{j\omega}{\upsilon}}z\right)}{\sin h\left(\sqrt{\dfrac{j\omega}{\upsilon}}z_0\right)} \tag{3.143}$$

3.5.2.2.1 Above-the-plate model

The viscous damping coefficient of the motion between an oscillating plate and the unbounded fluid (gas) can be determined by integrating the same Navier-Stokes equation, Equation (3.140). The integration can be performed for fixed boundary conditions (according to the continuum model) or by considering various order slip at the boundaries.

Fixed boundary conditions (continuum model)
For fixed boundary conditions (no velocity slip at the plate interface; see Kundu [21]) the following equations apply:

$$\begin{cases} v_x(0,t) = V_x \cos(\omega t) \\ v_x(\infty,t) = bounded \end{cases} \tag{3.144}$$

Solutions of the type are sought:

$$v_x(z,t) = e^{j\omega t} V_x(z) \tag{3.145}$$

where all the involved functions are complex functions. (It should be mentioned that z is measured from the plate towards the fluid.) By substituting Equation (3.145) into Navier-Stokes Equation (3.140) and by using the two boundary conditions of Equation (3.144), the fluid velocity (which is a real quantity) can be expressed as:

$$v_x(z,t) = v_x e^{-\sqrt{\frac{\omega}{2\upsilon}}z} \, \mathrm{Re}\left(e^{j\omega t} e^{-(1+j)\sqrt{\frac{\omega}{2\upsilon}}z} \right) = v_x e^{-\sqrt{\frac{\omega}{2\upsilon}}z} \cos\left(\omega t - \sqrt{\frac{\omega}{2\upsilon}}z \right) \tag{3.146}$$

Equation (3.146) resembles the solution to a wave propagation problem, but in actuality there are no restoring forces participating in this motion, and therefore, it represents a diffusion problem (derived from solving the Navier-Stokes original diffusion equation). The amplitude of the fluid motion indicated in Equation (3.146) is:

$$V_x(z) = v_x e^{-\sqrt{\frac{\omega}{2\upsilon}}z} \tag{3.147}$$

and varies with z as shown in Figure 3.30. As z increases the influence of the boundary condition-generated vibration diminishes. For a value of:

$$\delta_d = 4\sqrt{\frac{\upsilon}{\omega}} \tag{3.148}$$

which is known as the *diffusion length* or parameter, the amplitude is:

$$V_x(\delta_d) = V_x(0)e^{-\frac{4}{\sqrt{2}}} = 0.05v_x \tag{3.149}$$

which is 5% of the wall velocity. A parameter similar to the diffusion length is the *penetration depth*, δ, which is the distance where the motion amplitude reduces by a factor of e, which means that:

$$v_x e^{-\sqrt{\frac{\omega}{2\upsilon}}\delta} = v_x e^{-1} \tag{3.150}$$

and therefore:

$$\delta = \sqrt{\frac{2\upsilon}{\omega}} \tag{3.151}$$

The diffusion length and penetration depth are related as:

$$\delta_d = 2\sqrt{2}\delta \tag{3.152}$$

The viscous damping coefficient at the moving plate–fluid interface is found by first determining the shear stress at that interface through application of Newton's law of viscosity (Equation (3.125)). The velocity amplitude derivative at the interface is:

$$\left.\frac{dV_x(z)}{dz}\right|_{z=0} = -v_x\sqrt{\frac{\omega}{2\upsilon}} \tag{3.153}$$

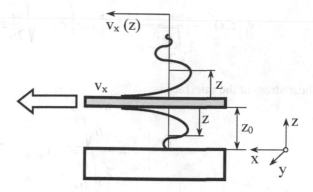

Figure 3.30 Velocities in Stokes slide-film damping

By ignoring the minus sign in Equation (3.153), the shear stress of Equation (3.125) can be expressed, and therefore the corresponding shear force corresponding to the plate surface's A is:

$$F_d = \mu \sqrt{\frac{\omega}{2\upsilon}} A v_x \qquad (3.154)$$

A typical viscous damping force is equal to damping coefficient times velocity and, consequently, the damping coefficient is:

$$c_d = \mu \sqrt{\frac{\omega}{2\upsilon}} A = \rho A \sqrt{\frac{\upsilon \omega}{2}} = A \sqrt{\frac{\rho \omega \mu}{2}} \qquad (3.155)$$

First-order slip boundary conditions

For rarefied gas, the continuum-model boundary condition at the moving plate needs to be amended, as the gas velocity will differ from the plate's velocity. First-order slip boundary condition (e.g., see Kundu [21]) are expressed as:

$$\begin{cases} v_x(0,t) = V_x - \lambda \dfrac{\partial v_x}{\partial z}\bigg|_{z=0} \\ v_x(\infty,t) = bounded \end{cases} \qquad (3.156)$$

where λ is the free molecular path. By carrying out the procedure that has been detailed for the continuum model, the following gas velocity is found at the moving plate boundary:

$$v_x(z,t) = \frac{v_x}{\lambda\sqrt{\dfrac{\omega}{2\upsilon}} - 1} e^{-\sqrt{\frac{\omega}{2\upsilon}}z} \cos\left(\omega t - \sqrt{\frac{\omega}{2\upsilon}}z\right) \qquad (3.157)$$

The shear stress at the interface is:

$$\tau = \mu\frac{\partial v_x}{\partial x}\bigg|_{z=0} = \frac{\mu\sqrt{\dfrac{\omega}{2\upsilon}}}{\lambda\sqrt{\dfrac{\omega}{2\upsilon}} - 1}V_x \qquad (3.158)$$

The total force at the gas–plate interface is determined by multiplying the stress of Equation (3.158) by the plate's area. Because the resulting force is the product of plate velocity amplitude V_x by a coefficient, this being the standard form of a damping force, the corresponding damping coefficient is:

$$c = \frac{\mu\sqrt{\dfrac{\omega}{2\upsilon}}A}{\lambda\sqrt{\dfrac{\omega}{2\upsilon}} - 1} \qquad (3.159)$$

By taking into account that the dynamic viscosity is mass density times kinematic viscosity, the damping coefficient of Equation (3.159) can also be written as:

$$c = \frac{\upsilon\rho A\left(\lambda\omega + \sqrt{2\upsilon\omega}\right)}{\lambda^2\omega - 2\upsilon} \qquad (3.160)$$

The penetration depth in this situation is found from its definition equation, namely:

$$\frac{v_x}{\lambda\sqrt{\dfrac{\omega}{2\upsilon}} - 1} e^{-\sqrt{\frac{\omega}{2\upsilon}}\delta} = v_x e^{-1} \qquad (3.161)$$

which results in:

$$\delta = \lambda - \sqrt{\frac{2\upsilon}{\omega}} \qquad (3.162)$$

3.5.2.2.2 Between-the-plates model

Again, fixed boundary conditions, as well as slip boundary ones will be considered here.

Fixed boundary conditions (continuum model)

To determine the damping coefficient at the moving plate–fluid interface for the space enclosed between the moving and the fixed plates, a solution to the Navier-Stokes equation is the one suggested by Landau and Lifshitz [19]:

$$v_x(z,t) = [A\sin(\beta z) + B\cos(\beta z)]e^{-j\omega t} \qquad (3.163)$$

where:

$$\beta = (1+j)\sqrt{\frac{\omega}{2\upsilon}} \qquad (3.164)$$

and z is the variable length parameter that ranges within the $[0, z_0]$ interval, and is measured from the mobile plate (when $z = 0$), as suggested in Figure 3.30. By using the continuum-model (no-slip) boundary conditions of this problem:

$$\begin{cases} v_x(0) = V_x e^{-j\omega t} \\ v_x(z_0) = 0 \end{cases} \qquad (3.165)$$

where the first boundary condition used the complex-number notation to denote the mobile plate velocity, the space-dependent portion of Equation (3.163) becomes:

$$v_x(z) = V_x \frac{\sin[\beta(z_0 - z)]}{\sin(\beta z)} \qquad (3.166)$$

The shear stress at the moving plate–fluid interface is found by using Newton's law as:

$$\tau = \mu \frac{dv_x(z)}{dz}\bigg|_{z=0} = \mu \beta V_x \cot(\beta z_0) \qquad (3.167)$$

Equation (3.167) gives the complex number form of the shear stress, but only its real part accounts for the actual shear stress, which is:

$$\tau = \mu \beta_1 V_x \frac{\sin h\left(2\beta_1 z_0\right) + \sin\left(2\beta_1 z_0\right)}{\cos h\left(2\beta_1 z_0\right) - \cos\left(2\beta_1 z_0\right)}$$

(3.168)

with:

$$\beta_1 = \sqrt{\frac{\omega}{2\upsilon}}$$

(3.169)

Because, again, the damping force is shear stress times area but is also damping coefficient times velocity, the damping coefficient can be computed from Equation (3.167) as:

$$c = \mu \beta A \frac{\sin h\left(2\beta_1 z_0\right) + \sin\left(2\beta_1 z_0\right)}{\cos h\left(2\beta_1 z_0\right) - \cos\left(2\beta_1 z_0\right)}$$

(3.170)

Example 3.16

Evaluate the total damping coefficient generated due to air friction above plate that undergoes vibrations parallel to the substrate, as well as between the plate and the substrate, by considering the continuum model does apply (with fixed boundary conditions).

Solution:

Equations (3.155) and (3.170) give the damping coefficients outside the plate and between the plate and the substrate, respectively. The total loss in case several independent effects superimpose can be evaluated by means of the inverse of an equivalent Q-factor, which is obtained by adding up the inverses of the individual Q-factors, as shown by Equation (3.44). At the same time, according to Equation (3.24), the Q-factor is proportional to the inverse of the damping coefficient. Consequently, the total damping coefficient is the sum of individual contributions from above-the-plate and between-the-plates damping coefficients, namely:

$$c = c_a + c_b$$

(3.171)

where c_a is given in Equation (3.155) and c_b in Equation (3.170). The plot of Figure 3.31 shows the variation of the c_a/c ratio.

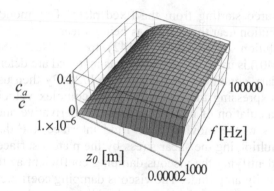

$\dfrac{c_a}{c}$ 0.4 100000

0 f [Hz]

$1. \times 10^{-6}$

z_0 [m]

0.00002 1000

Figure 3.31 Above-the-plate damping coefficient as a fraction of the total damping coefficient

For relatively small gaps and driving frequencies, the damping produced above the plate is less significant in the overall damping coefficient, as shown in the figure, cases in which the damping owing to fluid–structure interaction predominates. For higher gaps and frequencies, the fraction of damping produced above the plate increases and approaches the one generated between the plates.

First-order slip boundary conditions

The model described in the previous subsection together with its corresponding damping coefficient equation account for fixed boundary conditions, according to the continuum model. When slip boundary conditions are assumed (which account for flows in the slip and/or transition flow regimes), a different coefficient is produced, as shown next.

The *slip boundary conditions* account for gas rarefaction at low pressures and very small gaps that result in a gas velocity that is non-zero at the fixed wall (the slip velocity) and is different from the oscillating wall velocity at that interface. These changes in fluid velocity distribution alter the damping coefficient resulting from the rarefied fluid-moving plate interaction. The *first-order slip boundary conditions* (Burgdorfer [22]), also known as Maxwell's boundary conditions, are:

$$\begin{cases} v_x(0) = \lambda \dfrac{\partial v_x}{\partial z}\bigg|_{z=0} \\[2mm] v_x(z_0) = V_x - \lambda \dfrac{\partial v_x}{\partial z}\bigg|_{z=z_0} \end{cases} \tag{3.172}$$

where v_x is the gas slip velocity (at either the fixed plate or the oscillating one), V_x is the plate velocity amplitude, λ is the mean free molecular path, and z is measured starting from the fixed plate. This model considers the velocity distribution near the surface as being linear.

If the solution of Equation (3.142) to the Navier-Stokes equation (Equation (3.140)) is used, the two constants involved are determined by using the slip boundary conditions of Equation (3.172). By then using the normal procedure of expressing the real part of the complex velocity distribution, followed by calculation of its z-dependent first derivative and calculation of the shear stress at the moving plate–fluid interface, the damping force is obtained by multiplying the shear stress by the plate's surface area. This last step enables identifying the viscous damping coefficient as the multiplier of the plate's velocity amplitude. The viscous damping coefficient, as shown by Veijola and Turowski [20], for instance, is:

$$c = \mu \beta_1 A \frac{\sinh(2\beta_1 z_0) + \sin(2\beta_1 z_0) + c_1}{\cosh(2\beta_1 z_0) - \cos(2\beta_1 z_0) + c_2} \tag{3.173}$$

where the two terms, c_1 and c_2, are the additions to the damping coefficient corresponding to non-slip boundary conditions (Equation (3.170)) and are, according to Veijola and Turowski [20]:

$$\begin{cases} c_1 = 4\beta_1\lambda\left[\left(1+\beta_1^2\lambda^2\right)\cosh(2\beta_1 z_0)+\left(1-\beta_1^2\lambda^2\right)\cos(2\beta_1 z_0)\right] \\ \quad + 6\beta_1^2\lambda^2\left[\sinh(2\beta_1 z_0)-\sin(2\beta_1 z_0)\right] \\ c_2 = 4\beta_1\lambda\left[\left(1+2\beta_1^2\lambda^2\right)\sinh(2\beta_1 z_0)+\left(1-2\beta_1^2\lambda^2\right)\sin(2\beta_1 z_0)\right] \\ \quad + 4\beta_1^2\lambda^2\left[\left(2+\beta_1^2\lambda^2\right)\cosh(2\beta_1 z_0)-\left(2-\beta_1^2\lambda^2\right)\cos(2\beta_1 z_0)\right] \end{cases} \tag{3.174}$$

Higher-order Slip Boundary Conditions

The first-order boundary slip model captures the linear velocity distribution at boundaries for vibration frequencies that are not so high. In the case of higher frequencies, the linear character of velocity distribution near the oscillating plate needs to be replaced by a higher-order distribution. A second-order slip boundary condition, as shown by Beskok and Karniadakis [23], Beskok et al. [24], Bahukudumbi et al. [25], Park et al. [26] or Veijola and Turowski [20], requires:

$$\begin{cases} v_x(0) = \lambda \left.\frac{\partial v_x}{\partial z}\right|_{z=0} \\ v_x(z_0) = V_x - \frac{\lambda}{1-b\lambda}\left.\frac{\partial v_x}{\partial z}\right|_{z=z_0} \end{cases} \tag{3.175}$$

where the higher-order factor, according to Beskok and Karniadakis [23], is computed as:

$$b = \frac{\left.\dfrac{\partial^2 v_x}{\partial z^2}\right|_{z=z_0}}{2\left.\dfrac{\partial v_x}{\partial z}\right|_{z=z_0}}$$

(3.176)

where v_x was selected to be the velocity distribution according to the continuum model (Equation (3.146)). For $b = 0$, Equation (3.175) is identical to Equation (3.172), which describes the first-order slip model. By applying the same procedure indicated to the first-order slip model, the damping coefficient can be obtained. Veijola and Turowski [20] propose the following complex number form damping coefficient:

$$c = \frac{\mu A \beta_2 \left[2 + \beta_2 \lambda \tan h\left(\beta_2 z_0\right) - \beta_2^2 \lambda^2 \tan h^2\left(\beta_2 z_0\right)\right]}{4\beta_2 \lambda + \left(2 + \beta_2^2 \lambda^2\right)\tan h\left(\beta_2 z_0\right) - \beta_2 \lambda \tan h^2\left(\beta_2 z_0\right)}$$

(3.177)

where $\beta_2 = (2j\beta_1)^{1/2}$. A variant of modeling higher-order slip velocity boundary conditions at the moving wall is proposed by Bahukudumbi et al. [25], who propose the following boundary condition:

$$\begin{cases} v_x(0) = \lambda \left.\dfrac{\partial v_x}{\partial z}\right|_{z=0} \\[2mm] v_x(z_0) = V_x - \alpha\lambda \left.\dfrac{\partial v_x}{\partial z}\right|_{z=z_0} \end{cases}$$

(3.178)

where α is the slip coefficient, which is expressed, according to the same reference, as:

$$\alpha = 1.2977 + 0.71851\tan^{-1}\left[-1.17488\left(\frac{\lambda}{z_0}\right)^{0.58642}\right]$$

(3.179)

For a slip coefficient of $\alpha = 1$, the classical Maxwell, first-order slip velocity boundary condition is retrieved.

3.5.3 Free Molecular Flow Regime and Unifying Theories

For very large Knudsen numbers, Kn, the free molecular path exceeds by far the gap dimensions, and the flow is in the free regime. The shear stress on a plate that displaces at a velocity V_x can be expressed, according to Kogan [27], for instance, as:

$$\tau_\infty = \rho_0 V_x \sqrt{\frac{2RT}{\pi}} \tag{3.180}$$

where R is the universal gas constant and T is the temperature of the mobile plate. The damping coefficient corresponding to the free molecular flow regime is therefore:

$$c_\infty = \rho_0 A \sqrt{\frac{2RT}{\pi}} \tag{3.181}$$

Attempts have been made by researchers in this domain to generate closed-form analytical equations that would be applicable for the whole range of Knudsen numbers, covering the domain from continuum to free molecular. Corrections expressing slip-wall effects can be incorporated in effective values of either the shear stress or the dynamic viscosity, and therefore the simple linear Couette model can be used to generate damping coefficients by means of Equations (3.129), (3.130), and (3.131) at the beginning of this section treating the slide-film damping.

Veijola and Turowski [20], for instance, suggest using the following effective damping coefficient:

$$\mu_{eff} = \frac{\mu}{1 + 2\dfrac{\lambda}{z_0}} \tag{3.182}$$

where μ is the actual dynamic viscosity and λ is the free mean molecular path. The same reference proposes as an alternative a different effective dynamic viscosity, which is derived from normalizing the shear stress by the free molecular path as:

$$\mu_{eff} = \frac{\mu}{1 + 2K_n + 0.2K_n^{0.788} e^{-0.1K_n}} \tag{3.183}$$

which represents a curve fit obtained from numerical results of an integro-differential equation derived by Cercignani and Pagani [28] based on a linearized Boltzmann equation. First-order slip boundary conditions have been taken into account in these models. Having the effective dynamic viscosity enables calculation of the shear stress by means of Newton's law (Equation (3.125)). Similarly, Bahukudumbi et al. [25] propose the following shear stress ratio obtained from curve fitting of numerical results:

$$\frac{\tau}{\tau_\infty} = \frac{a(Kn)^2 + 2bKn}{a(Kn)^2 + cKn + b} \tag{3.184}$$

where $a = 0.530$, $b = 0.603$, $c = 1.628$ (values rounded to three decimal points). Equation (3.184) enables expressing the shear stress for any value of the Knudsen number, from 0 (continuum model) to infinity (free molecular regime). The damping force is obtained by multiplying the shear stress by the moving plate area, which enables determining the damping coefficient.

3.6 THERMOELASTIC DAMPING

The coupling between a strain/deformation field in an elastic body and the temperature field is best illustrated by the equation connecting the elongation of a bar, Δl, to the temperature variation, ΔT as:

$$\Delta l = \alpha l \Delta T \tag{3.185}$$

by means of the linear coefficient of thermal expansion α. An oscillating elastic body is out of equilibrium state, but local variations of its strain field interact with the thermal variations, according to Equation (3.185). This connection provides a mechanism for energy dissipation towards regain of equilibrium. This is actually a relaxation process consisting of an irreversible heat flow. For a vibrating beam, for instance, where the upper fibers contract and the lower fibers extend at a given moment in time, a heat flow loss occurs from the heated fibers (the extended ones) to the cooled fibers (the ones that contract). This energy loss phenomenon is known as TED.

Roszhart [29] derived a model for cantilever beams which predicts the following Q-factor:

$$Q = \frac{16\omega^2 t^4 c^2 \rho^2 + \kappa^2}{4\omega t^2 \alpha^2 \kappa E T} \tag{3.186}$$

where c is the specific heat, κ is the heat capacity, E is Young's modulus, ρ is the mass density, T is the temperature, t is the thickness, and ω is the vibration frequency.

In the 1930s, Zener [30] proposed an approximation to the Q-factor pertaining to thermoelastic losses, which is still considered operationally valid. Zener's model, also known as the *standard model of the anelastic solid*, considers that damping can be approximated by a relaxation process in which thermal diffusion occurs across the thickness of a vibrating beam. The standard model has at its core a generalization of Hooke's law, which is expressed as:

$$\sigma + \tau_\varepsilon \frac{d\sigma}{dt} = E_r \left(\varepsilon + \tau_\sigma \frac{d\varepsilon}{dt} \right) \tag{3.187}$$

where E_r is the relaxed modulus of elasticity, whereas τ_ε and τ_σ are the constant-strain relaxation time and constant-stress time, respectively. By assuming that the stress and strains vary harmonically, which in complex notation is:

$$\begin{cases} \sigma = \sigma_0 e^{j\omega t} \\ \varepsilon = \varepsilon_0 e^{j\omega t} \end{cases} \tag{3.188}$$

Equation (3.187) becomes:

$$\sigma_0 = E_{r,c} \varepsilon_0 \tag{3.189}$$

which is similar to the normal Hooke's law and where $E_{r,c}$ is a complex modulus, expressed as:

$$E_{r,c} = E_r \frac{1 + j\omega\tau_\sigma}{1 + j\omega\tau_\varepsilon} \tag{3.190}$$

The inverse of the Q-factor can be defined as the ratio of the imaginary part to the real part of the complex modulus, and after transforming the complex modulus of Equation (3.190) in its standard, complex-number form:

$$E_{r,c} = \frac{1 + \omega^2 \tau_\sigma \tau_\varepsilon}{1 + \omega^2 \tau_\varepsilon^2} + j \frac{\omega(\tau_\sigma - \tau_\varepsilon)}{1 + \omega^2 \tau_\varepsilon^2} \tag{3.191}$$

the inverse of the Q-factors can be expressed as:

$$Q^{-1} = \frac{E_u - E_r}{\sqrt{E_u E_r}} \times \frac{\omega_r \tau_{th}}{1 + (\omega_r \tau_{th})^2} \tag{3.192}$$

E_u is the unrelaxed modulus and is defined as:

$$E_u = \frac{\tau_\sigma}{\tau_\varepsilon} E_r$$ (3.193)

where the *thermal relaxation time* is:

$$\tau_{th} = \frac{t^2 \rho c}{\pi^2 k}$$ (3.194)

with c being the specific heat (at either constant pressure or constant volume [the differences are small between the two conditions]), ω_r being the cantilever resonant frequency, T being the temperature, t being the cantilever cross-sectional thickness, ρ being the mass density, and k being the thermal conductivity.

The Q-factor accounting for thermal damping losses, according to Zener's model and when considering the following approximation (Lifshitz and Roukes [31]):

$$\frac{E_u - E_r}{\sqrt{E_u E_r}} \approx \frac{E_{ad} - E}{\sqrt{E^2}} = \frac{E\alpha^2 T}{c}$$ (3.195)

is:

$$Q = \frac{c}{E\alpha^2 T} \times \frac{1 + (\omega_r \tau_{th})^2}{\omega_r \tau_{th}}$$ (3.196)

In Equation (3.195), E_{ad} is the adiabatic (unrelaxed) modulus of elasticity, whereas E is the regular, isothermal (relaxed) modulus of elasticity.

Example 3.17
Analyze the Q-factor due to thermal damping in terms of the relaxation constant $\omega_r \tau_{th}$.

Solution:
The inverse of the Q-factor, Q^{-1}, which is proportional to the losses incurred by a system, is plotted in a non-dimensional form, as shown in Figure 3.32, based on Equation (3.196). As Figure 3.32 shows, the non-dimensional damping has a maximum for $\omega_r \tau_{th} = 1$, which indicates that the resonant frequency and the relaxation rate (the inverse of the relaxation time) should be approximately equal for the maximum peak damping to occur.

When $\omega_r \gg 1/\tau_{th}$, the material does not have the necessary time to relax because the vibration is too fast. When, on the contrary, $\omega_r \ll 1/\tau_{th}$, the vibration is very slow and the system is in equilibrium technically, with little energy being lost.

Lifshitz and Roukes [31] derived the exact closed-form solution to the Q-factor corresponding to a thin cantilever by using the equations of linear thermoelasticity:

$$Q = \frac{c\zeta^2}{6E\alpha^2 T}\left[1 - \frac{\sinh\zeta + \sin\zeta}{\zeta(\cosh\zeta + \cos\zeta)}\right]^{-1} \tag{3.197}$$

with:

$$\xi = t\sqrt{\frac{\omega_r \rho c}{2k}} \tag{3.198}$$

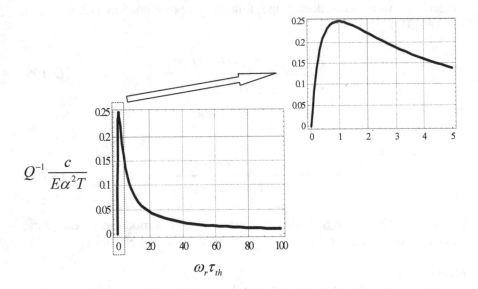

Figure 3.32 Non-dimensional thermal damping as a function of the relaxation parameter

3.7 OTHER INTRINSIC LOSSES

Other intrinsic losses are *phonon-mediated* and they include the ones produced by phonon–electron interactions as well as those generated by the interaction between phonons and the mechanical vibration of a microdevice. The first category of losses is a viscous drag exerted by the free electrons on oscillating ions and is mostly significant in metallic MEMS because it is based on material high electrical conductivity (Czaplewski et al. [32]).

The latter loss category is defined by the propagating mechanical vibration in a MEMS device, which causes the phonons to thermally readjust and reach a different equilibrium state, this state alteration being the channel for energy losses. The Q-factor in this case (Czaplewski et al. [32]) is of the Zener type, namely:

$$Q = \frac{\rho v_l^2}{CT\gamma^2} \times \frac{1 + \left(\omega_r \tau_{ph}\right)^2}{\omega_r \tau_{ph}} \tag{3.199}$$

where C is the heat capacity per unit volume, v_l is the longitudinal wave (sound) velocity and the phonon relaxation time is:

$$\tau_{ph} = \frac{3\kappa}{\rho C_p v_D^2} \tag{3.200}$$

κ being the material thermal conductivity, C_p being the constant-pressure heat capacity, and v_D being the Debye sound velocity defined as:

$$v_D = v_l v_t \sqrt[3]{\frac{3}{2v_l^3 + v_t^3}} \tag{3.201}$$

It is known that the longitudinal velocity is:

$$v_l = \sqrt{\frac{E}{\rho}} \tag{3.202}$$

whereas the transverse wave velocity (or the group velocity) is:

$$v_t = \frac{\lambda}{T} = \frac{\lambda\omega}{2\pi} \tag{3.203}$$

The constant γ in Equation (3.199) is the *Gruneisen's constant* (e.g., see Braginski et al. [33] and Burakowsky and Preston [34]), whose definition is:

$$\gamma = \frac{\alpha E}{\rho C_v} \tag{3.204}$$

Burakowsky and Preston [34] propose the following equation to determining γ, which depends only on material density:

$$\gamma = \frac{1}{2}c_1\rho^{1/3} + c_2\rho^{c_3} \tag{3.205}$$

where c_1, c_2, and c_3 are constants that can be evaluated individually for various materials, and the same reference gives those values for 20 metallic materials.

Another internal dissipation mechanism is due to *defects* or *disorder* in a material, and relaxation (damping) is provided by reconfiguration/reordering between equilibrium states that usually occurs through motion by atoms, vacancies, impurities, or dislocations. The Q-factor of a MEMS oscillator subjected to this particular type of loss is of the generic Zener-type, namely:

$$Q = c\frac{1 + (\omega_r\tau)^2}{\omega_r\tau} \tag{3.206}$$

where the constant c depends on the type of defect and its intensity. The relaxation time (Czaplewski et al. [32]) can be expressed for these processes as an Arrhenius-type equation:

$$\frac{1}{\tau} = \frac{1}{\tau_0}e^{-\frac{E_a}{RT}} \tag{3.207}$$

where R is the universal gas constant ($R = 8.31$ J/(mol K)), τ_0 is the characteristic atomic vibration period and is of the order of 10^{-13} s, and E_a is the activation energy of the relaxation process. Usually this energy is equal to the self-diffusion energy and is of the order of 1–2 eV per mol.

Example 3.18

Evaluate the relaxation time involved in the losses due to defect motion if the characteristic atomic vibration period is $\tau_0 = 2.5 \times 10^{-13}$ s and the activation energy is $E_a = 1.5$ eV. Considering a microresonator whose resonant frequency is 10,000 Hz is defined by a Q-factor $Q = 1000$, which is solely due to defect motion losses, determine the constant c of Equation (3.206).

Solution:

By taking into account that 1 eV $= 1.6 \times 10^{-9}$ J and by considering the numerical data of this problem, the relaxation time can be expressed as a function of temperature, and Figure 3.33 plots this relationship. As Figure 3.33 indicates, the relaxation time decreases when the temperature increases, as expected, but is largely determined by the value of the characteristic atomic vibration period τ_0. The constant c is simply found from Equation (3.206) as:

$$c = \frac{2\pi f_r \tau}{1 + \left(2\pi f_r \tau\right)^2} Q \qquad (3.208)$$

and its numerical value is $c = 6.28 \times 10^{-6}$.

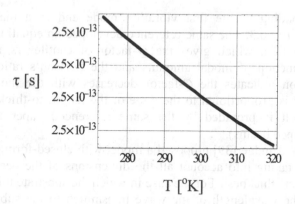

Figure 3.33 Relaxation time as a function of temperature

3.8 SUBSTRATE (ANCHOR) LOSSES

Vibration of micro/nano mechanical devices, particularly the resonant ones, is transmitted to the substrate and dissipates into it; the corresponding losses are known as support (or substrate, or anchor) losses. Reduction of these losses, as indicated by Mihailovich and MacDonald [35], for instance, can be achieved by either utilizing symmetry in designing resonant structures such that zero (desirably) forces/moments are transmitted to the support or by interposing a mass between the oscillating structure and the substrate such that the energy of the mass is small compared to the energy of the original structure.

Modeling and quantifying losses to the substrate by a micro/nano oscillator is generally performed by assessing the vibration energy, which is transmitted through the anchor regions by the shearing forces and bending or torsional moments generated locally by the vibrating structures. Park and Park [36] developed a methodology enabling evaluation of the anchor losses by means of a modified Fourier semi-analytic technique involving numeric solutions. Osaka et al. [37] considered cantilever beams of infinite width that are attached to a semi-infinite substrate and gave the following Q-factor corresponding to anchor losses:

$$Q \approx 2.17 \frac{l^3}{t^3} \qquad (3.209)$$

where l is the beam length and t is its thickness. Hao et al. [38] derived a very similar Q-factor, namely:

$$Q \approx 2.09 \frac{l^3}{t^3} \tag{3.210}$$

value corresponding to the first vibration mode and to a material with Poisson's ratio $\mu = 0.3$. The same reference developed an equation requiring numeric integration, which gives the Q-factor of cantilevers for various modes as a function of modal amounts and the Poisson's ratio. The respective equation indicates the Q-factor decreases with the mode number increasing and is proportional to the cube of the length-to-thickness ratio. A similar result is provided by the same referenced paper for bridges (clamped-clamped beams).

Photiadis and Judge [39] proposed a model with closed-form Q-factor of cantilevers by taking into account all the dimensions of the beam together with the substrate thickness. For the case in which the substrate thickness t_s is smaller than the wavelength of the wave transmitted to the substrate, they derived the following Q-factor equation:

$$Q = 1.05 \frac{l}{w} \times \frac{t_s^2}{t^2} \tag{3.211}$$

where, in addition to the parameters already introduced here, w is the beam cross-sectional width. Equation (3.211) is a simplification of a more generic equation derived in the same reference—the equation mentioned here corresponds to cantilever designs with thicknesses far smaller than the substrate thickness and was shown valid for cases when $\lambda_s/3 < t_s < \lambda_s$. Figure 3.34 shows the dependency of the Q-factor on the l/w and t_s/t ratios.

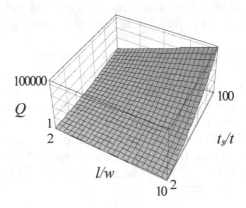

Figure 3.34 Quality factor due to anchor losses in terms of geometry

Example 3.19

Design a cantilever that would be able to produce an anchor loss-related Q-factor of $Q = 10,000$ when the substrate has a thickness $t_s = 500$ μm.

Solution:

Equation (3.211) enables expression of the cantilever thickness as:

$$t = 1.025\sqrt{\frac{1}{Q} \times \frac{l}{w}} \, t_s \qquad (3.212)$$

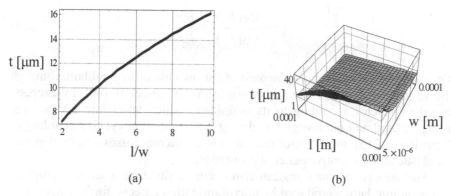

(a)   (b)

Figure 3.35 Cantilever thickness in terms of length and width

For the numerical data of the problem, the solution is: $t = 5.125 \, (l/w)^{1/2}$. The thickness is plotted in terms of the cantilever length and cross-sectional width in Figure 3.35. As Figure 3.35 shows it, the thickness increases with the length-to-width ratio for the specified values of the Q-factor and substrate thickness. For $l/w = 5$, the cantilever thickness becomes: $t = 11.6$ μm.

Photiadis and Judge [39] also analyzed designs for which the substrate thickness is larger compared to the wavelength of the transmitted vibration, and proposed the following Q-factor:

$$Q = 3.226\frac{l}{w} \times \frac{l^4}{t^4} \qquad (3.213)$$

Both Equations (3.211) and (3.213) were derived for a material with $\mu = 3$ (steel-type). It can be seen from Equation (3.213) that the substrate thickness does not affect the Q-factor for relatively thick substrates. There are also marked differences between the predictions of Equation (3.209) by Osaka et al. [37], Equation (3.210) by Hao et al. [38], on one side, and Equation (3.213) by Photiadis and Judge [39].

3.9 SURFACE LOSSES

The generic Zener model can also stand for dissipative processes different from the TED since by taking into account a complex modulus of the type:

$$E_c = E_1 + jE_2 \qquad (3.214)$$

where E_1 can be the real, conventional elastic modulus of a specific material, and E_2 stands for the dissipative part of that material (generated by lattice defects motion, for instance), the Q-factor is simply:

$$Q = \frac{\mathrm{Re}(E_c)}{\mathrm{Im}(E_c)} = \frac{E_1}{E_2} \qquad (3.215)$$

Equation (3.215) can actually be derived for the case of internal bulk (through the volume) dissipation (e.g., as shown by Yasumura et al. [40]) by expressing the Q-factor according to its definition, which takes into account the energy stored and the energy lost during one oscillation cycle. According to Equation (3.215), the Q-factor related to bulk internal losses strictly depends on elastic and dissipative material properties.

Surface loss mechanisms can also occur in situations such as disruption of the atomic lattice produced by microfabrication defects, for instance, or in cases of surface contamination (such as adsorbates on the surface). Yasumura et al. [40] derived the following Q-factor owing to surface losses for a cantilever having the cross-sectional width w and thickness t:

$$Q = \frac{wt}{2\delta(3w+t)} \times \frac{E_1}{E_{s,1}} Q_s \qquad (3.216)$$

where δ is the thickness of a thin dissipative layer for which the following complex modulus is used:

$$E_{s,c} = E_{s1} + jE_{s2} \qquad (3.217)$$

defined by a Q-factor:

$$Q_s = \frac{E_{s1}}{E_{s2}} \qquad (3.218)$$

For cantilevers with large widths compared to their thickness, $w \gg t$, Equation (3.216) simplifies to:

$$Q = \frac{t}{6\delta} \times \frac{E_1}{E_{s,1}} Q_s \qquad (3.219)$$

As indicated by Equations (3.216) and (3.219), the surface losses are affected by the cross-sectional dimensions, in addition to elastic and dissipative material properties, and are not influenced by the cantilever length, as one would expect by taking into account that length defines the longitudinal area, $w \times l$.

Problems

Problem 3.1
    The differential equation expressing the free damped vibrations of a single DOF microresonator is: $\ddot{x} + 20\dot{x} + 1,000,000x = 0$. Determine the mass m, damping coefficient c, and stiffness k, as well as the damped resonant frequency ω_d of this lumped-parameter system.

Problem 3.2
    Repeat Problem 3.2 in the case of the following differential equation: $\ddot{x} + 500\dot{x} + 6,250,000x = 0$.

Problem 3.3
    The free damped vibrations of a two DOF mechanical microsystem are characterized by the following mass, damping and stiffness matrices:
$[M] = \begin{bmatrix} a & 0 \\ 0 & a \end{bmatrix}; [C] = \begin{bmatrix} b & 0 \\ 0 & b \end{bmatrix}; [K] = \begin{bmatrix} d_1 & -d_2 \\ -d_2 & d_1 \end{bmatrix}$, where the real coefficients a, b, d_1, and d_2 are specified. Identify a mechanical microsystem that possesses these properties by drawing a schematic of the microsystem and by also calculating its individual physical parameters.

Problem 3.4
    Repeat Problem 3.3 in the case the microsystem's matrices are:
$[M] = \begin{bmatrix} a & 0 \\ 0 & a \end{bmatrix}; [C] = \begin{bmatrix} b & 0 \\ 0 & b \end{bmatrix}; [K] = \begin{bmatrix} d_1 & -d_2 \\ -d_2 & d_1 \end{bmatrix}$.

Problem 3.5
    The free damped response of the mechanical microfilter of Figure 3.36 is determined experimentally, consisting of the logarithmic decrements δ_1 and δ_2, and the two damped resonant frequencies ω_{d1} and ω_{d2}. By also knowing the mass m of the two rigid oscillators, evaluate the stiffnesses k_1 and k_2.

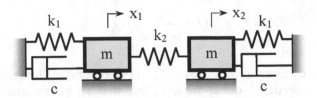

Figure 3.36 Two DOF microfilter with damping

Problem 3.6

Calculate the bending resonant frequency ω_r of a resonator for which the logarithmic decrement δ and damped frequency ω_d are known.

Problem 3.7

A paddle microcantilever is tested at resonance by using out-of-the-plane bending and torsion. It is determined the torsion-to-bending damped resonant frequency ratio is r. By using a lumped-parameter model, evaluate the overall losses corresponding to these motions. By using a lumped-parameter model, evaluate the loss corresponding to torsion as a function of the loss produced through bending.

Problem 3.8

Solve Problem 3.7 by considering a paddle microbridge instead of a paddle microcantilever.

Problem 3.9

A paddle microcantilever is tested at resonance by monitoring its damped, out-of-the-plane response. The resonant Q-factor Q_r and the damped resonant frequency ω_d are determined experimentally. By knowing all geometric and inertia parameters of the microbridge, use a lumped-parameter model (with the paddle rigid and the root massless and compliant) to evaluate the elasticity (Young's) modulus E of the microbridge material.

Problem 3.10

Solve Problem 3.9 by considering a paddle microbridge instead of a paddle microcantilever.

Problem 3.11

Establish a relationship between the resonant (forced) Q-factor and the free-response Q-factor (with non-zero initial velocity) for a single DOF damped microresonator when the logarithmic decrement δ is known.

Problem 3.12

A constant rectangular cross-section microbridge with $E = 155$ GPa, $\rho = 2300$ kg/m^3 and length $l = 100$ μm is displaced by 5 μm at its midpoint and then let to freely vibrate. After $t = 50$ s, its midpoint vibration amplitude is 30 nm. Evaluate the Q-factor corresponding to the overall losses.

Problem 3.13

A paddle microcantilever with both segments contributing to compliance and inertia is vibrated in out-of-the-plane bending in a vacuum environment to evaluate structural losses, and the corresponding Q-factor is determined experimentally. By ignoring other losses of this microsystem, and by considering all geometric and material parameters of the microbridge are known, calculate the structural loss coefficient α.

Problem 3.14

A Q-factor of 7,800 is experimentally determined for a trapezoid cantilever whose minimum width, maximum width, thickness, and length are 20 μm, 80 μm, 1 μm, and 300 μm, respectively. The structural damping roughly represents 80% of the total losses. Determine the elastic modulus of this cantilever's material. It is also known that $\alpha = 0.0005$.

Problem 3.15

A microbridge formed of a central plate of 90 μm length, 10 μm width, and 1 μm thickness, and two side CNTs, each 20 μm long and 50 nm in diameter, vibrates normally to the substrate. Knowing the bridge–substrate initial gap is 3 μm and that the Q-factor owing to squeeze-film damping is 9,300, calculate the dynamic viscosity. Known are also $\omega = 6000$ rad/s and $E = 50$ GPa. Use Zhang's continuum-gas model.

Problem 3.16

A 300 μm × 50 μm × 2 μm plate vibrates normal to the substrate at 10 kHz. The plate is supported by two end rectangular cross-section beams, each 50 μm long, 10 μm wide, and 1 μm thick. The initial gap is 12 μm, the gas pressure is 0.001 atm, and the dynamic viscosity is 1.8×10^{-5} N-s/m^2. Calculate the squeeze-film damping coefficient.

Problem 3.17

A plate 200 μm long and 40 μm wide is used to determine the nature of an unknown gas by monitoring the plate's vibratory response against a substrate (the initial gap is 8 μm). The plate vibrates at 8,000 Hz at normal temperature and pressure. Considering the Q-factor corresponding to squeeze-film damping is 7,600, calculate the molecular mass of the gas.

Problem 3.18

The plate shown in Figure 3.37 is supported by two CNT beams and can vibrate in out-of-the-plane translation and rotary motion. Knowing that $l = 400$ μm, $w = 70$ μm, $l_b = 300$ μm, $d = 60$ nm (d is the CNT diameter), and also that the Q-factors owing to squeeze-film damping are $Q_t = 6,500$ for translation and $Q_r = 6,800$ for rotation, find the plate-substrate gap z_0 and the dynamic viscosity μ. Known is also that $f_t = f_r = 100$ Hz.

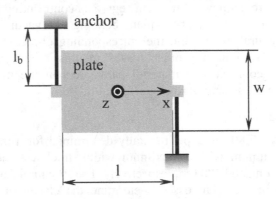

Figure 3.37 Bridge suspended on two carbon nanotube beams

Problem 3.19

A microresonator consists of a central square plate (of known mass m and dimensions $L \times L$), which is supported symmetrically on the midpoints of its sides by four identical CNT beams of known length l, diameter d, and modulus of elasticity E. At a very small pressure p, this microsystem can be used as a thermal sensor. The out-of-plane vibrations of the plate are monitored experimentally and the Q-factor Q is determined at $\omega - \omega_r$. By using Bao's molecular-flow model, and by also knowing the gas molecular mass M_m and the gas constant R, determine the gas temperature T.

Problem 3.20

A torsional micromirror is formed of a rectangular plate and two end beams. The inertia and geometric properties of the plate and beams are known. Compute the equivalent viscous damping ratio when the Q-factor corresponding to torsional vibrations of the micromirror is known. Consider only the losses produced through squeeze-film damping.

Problem 3.21

A paddle microbridge with the paddle rigid and the root segments massless and compliant (of constant rectangular cross-section) is used as a torsional oscillator to assess the dynamic viscosity coefficient μ of an unknown gas. All design and material properties being known, as well as the damped resonant frequency ω_d and resonant quality factor Q_r, devise an algorithm to determine μ. Known is also that $f_t = f_r = 100$ Hz.

Problem 3.22

Find the damped resonant frequency of the plate with 10 holes, as shown in Figure 3.38, by considering the squeezed-film stiffness. The plate is suspended by two end springs, each of 2 N/m stiffness. Known are the hole radius $r = 3$ μm, $l = 80$ μm, $w = 20$ μm, $p = 15$ μm, and $\mu = 1.7 \times 10^{-5}$ N-s/m^2, $z_0 = 8$ μm, $t = 2$ μm and $\rho = 2300$ kg/m^3.

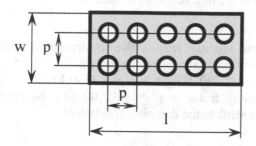

Figure 3.38 Plate with holes

Problem 3.23

Study the variation of c in Problem 3.22 as a function of the number of holes n.

Problem 3.24

A plate vibrates parallel to the substrate by maintaining a constant gap of 10 μm. The air density is $\rho = 1.1$ kg/m^3 and the dynamic viscosity is $\mu = 1.7 \times 10^{-5}$ N-s/m^2. By using the continuum model, determine the frequency at which the damping coefficient corresponding to the above-the-plate fluid–structure interaction is equal to the one between the moving plate and the substrate. Known are also $l = 200$ μm and $w = 100$ μm.

Problem 3.25

A plate is supported by two identical end springs and vibrates at 50,000 Hz parallel to the substrate. The area of the plate is 40,000 μm^2. Knowing the penetration depth is 80 μm, the Q-factor owing to above-the-plate fluid–structure interaction is 5,000, find the stiffness of the spring.

Problem 3.26

The losses due to friction with the fluid above the plate are 0.8 of the losses generated by air friction between the plate and the substrate. Considering the first-order slip boundary conditions, find the dynamic viscosity knowing the constant gap $z_0 = 15$ μm, vibration frequency $f = 65,000$ Hz, free molecular path $\lambda = 20$ μm, and plate area $A = 20,000$ μm^2.

Problem 3.27

A constant rectangular cross-section microcantilever of given length l, cross-sectional dimensions w and t, mass density ρ, specific heat c, coefficient of thermal expansion α, period T, frequency ω, and elastic modulus of elasticity E vibrates in vacuum. The thermal losses are monitored by means of the Q-factor Q, which is determined experimentally. Calculate the thermal conductivity k of the microcantilever material.

Problem 3.28

A 5 μm thick plate vibrates against the substrate at 100,000 Hz and $T = 300$ K. The properties of the plate material are: $\alpha = 3.5 \times 10^{-6}$ K^{-1}, $\kappa = 1,500$ W/(m × K), $c = 700$ J/(kg × K), $\rho = 2300$ kg/m^3, $E = 1.65$ GPa. Knowing the total quality factor is $Q = 130,000$ and the elastic support stiffness is $k = 4$ N/m, calculate the damping coefficient.

Problem 3.29

The Q-factor of a silicon trapezoid microresonator corresponding to defect motion is 8,000. Knowing the relaxation period is 2×10^{-13} s and the activation energy is 2 eV, as well as the microresonator's dimensions (length 200 μm, thickness 2 μm, base width 50 μm, and tip width 20 μm), determine the equivalent viscous damping ratio.

Problem 3.30

A rectangular microcantilever 300 μm long, 50 μm wide, and 2 μm thick, vibrates in air against the substrate at 16,000 Hz with an initial gap of 12 μm. The total losses due to squeeze-film damping and anchor losses are expressed by a Q-factor of 20,000. Evaluate the substrate thickness. The dynamic viscosity is $\mu = 1.8 \times 10^{-5}$ N-s/m^2.

References

1. W.T. Thomson, *Theory of Vibrations with Applications*, Third Edition, Prentice Hall, Englewood Cliffs, 1988.
2. J.J. Blech, On isothermal squeeze films, Journal of Lubrication Technology, 105, 1983, pp. 615–620.
3. T. Veijola, T. Tinttunen, H. Nieminen, V. Ermolov, T. Ryhanen, Gas damping model for a RF MEM switch and its dynamic characteristics, IEEE MTT-S International Microwave Symposium Digest, 2, 2002, pp. 1213–1216.
4. L. Zhang, D. Cho, H. Shiraishi, W. Trimmer, Squeeze film damping in microelectro-mechanical systems, ASME Micromechanical Systems, Dynamic Systems Measurements and Control, 40, 1992, pp. 149–160.
5. R.G. Christian, The theory of oscillating-vane vacuum gauges, Vacuum, 16, 1966, pp. 149–160.
6. Zs. Kadar, W. Kindt, A. Bossche, J. Mollinger, Quality factor of torsional resonators in the low-pressure region, Sensors and Actuators A, 53, 1996, pp. 299–303.

7. M. Bao, H. Yang, H. Yin, Y. Sun, Energy transfer model for squeeze-film damping in low vacuum, Journal of Micromechanics and Microengineering, 12, 2002, pp. 341–346.
8. S. Hutcherson, W. Ye, On the squeeze-film damping of microresonators in the free-molecular regime, Journal of Micromechanics and Microengineering, 14, 2004, pp. 1726–1733.
9. P.J. Polikarpov, S.F. Borisov, A. Kleyn, J.-P. Taran, Normal momentum transfer study by a dynamic technique, Journal of Applied Mechanics and Technical Physics, 44, 2003, pp. 298–303.
10. R. B. Darling, C. Hivick, J. Xu, Compact analytical models for squeeze film damping with arbitrary venting conditions, Transducers '97 International Conference on Solid State Sensors and Actuators, 2, 1997, pp. 1113–1116.
11. W. Dotzel, T. Gessner, R. Hahn, C. Kaufmann, K. Kehr, S. Kurth, J. Mehner, Silicon mirrors and micromirror arrays for spatial laser beam modulation, Transducers '97 International Conference on Solid State Sensors and Actuators, 1, 1997, pp. 81–84.
12 F. Pan, J. Kubby, E. Peeters, A.T. Tran, Squeeze film damping effect on the dynamic response of a MEMS torsion mirror, Journal of Micromechanics and Microengineering, 8, 1998, pp. 200–208.
13. M. Bao, Y. Sun, Y. Huang, Squeeze-film air damping of a torsion mirror at a finite tilting angle, Journal of Micromechanics and Microengineering, 16 (11), 2006, pp. 2330–2335.
14. T. Veijola, A. Pursula, P. Raback, Extending the valability of squeezed-film damper models with elongation of surface dimensions, Journal of Micromechanics and Microengineering, 15 (9), 2005, pp. 1624–1636.
15. N. Lobontiu, E. Garcia, Mechanics of Microelectromechanical Systems, Kluwer Academic Press, New York, 2004.
16. M. Bao, H. Yang, Y. Sun, Y. Wang, Squeeze-film air damping of thick hole plate, Sensors and Actuators A, 108, 2003, pp. 212–217.
17. M. Bao, H. Yang, Y. Sun, P.J. French, Modified Reynolds' equation and analytical analysis of squeeze-film air damping of perforated structures, Journal of Micromechanics and Microengineering, 13, 2003, pp. 795–800.
18. S.S. Mohite, H. Kesari, V.R. Sonti, R. Pratap, Analytical solutions for the stiffness and damping coefficients of squeeze films in MEMS devices with perforated back plates, Journal of Micromechanics and Microengineering, 15, 2005, pp. 2083–2092.
19. L.D. Landau, E.M. Lifshitz, Fluid Mechanics, Pergamon, London, 1959.
20. T. Veijola, M. Turowski, Compact damping for laterally moving microstructures with gas rarefaction effects, Journal of Microelectromechanical Systems, 10 (2), 2001, pp. 263–273.
21. P.K. Kundu, Fluid Mechanics, Academic Press, San Diego, 1990.
22. A. Burgdorfer, The influence of the mean free path on the performance of hydrodynamic gas lubricated bearings, Journal of Basic Engineering, 81, 1959, pp. 94–99.
23. A. Beskok, G.E. Karniadakis, Simulation of heat and momentum transfer in complex microgeometries, Journal of Thermophysics and Heat Transfer, 8 (4), 1994, pp. 647–655.
24. A. Beskok, G.E. Karniadakis, W. Trimmer, Rarefaction and compressibility effects in gas microflows, Journal of Fluids Engineering, 118, 1996, pp. 448–456.
25. P. Bahukudumbi, J.H. Park, A. Beskok, A unified engineering model for steady and quasi-steady shear-driven gas microflows, Microscale Thermophysical Engineering, 7, 2003, pp. 291–315.
26. J.H. Park, P. Bahukudumbi, A. Beskok, Rarefaction effects on shear driven oscillatory gas flows: a direct simulation Monte Carlo study in the entire Knudsen regime, Physics of Fluids, 16 (2), 2004, pp. 317–330.
27. M.N. Kogan, Rarefied Gas Dynamics, Plenum, New York, 1969.
28. C. Cercignani, C.D. Pagani, Variational approach to boundary-value problems in kinetic theory, The Physics of Fluids, 9, 1966, pp. 1167–1173.

29. T.W. Roszhart, The effect of thermoelastic internal friction on the Q of micromachined silicon resonators, Technical Digest on Solid-State Sensor and Actuator Workshop, 1990, pp. 13–16.
30. C. Zener, Elasticity and Anelasticity of Metals, University of Chicago Press, Chicago, 1948.
31. R. Lifshitz, M.L. Roukes, Thermoelastic damping in micro and nanomechanical systems, Physical Review B, 61 (8), 2000, pp. 5600–5609.
32. D.A. Czaplewski, J.P. Sullivan, T.A. Friedmann, D.W. Carr, B.E. Keeler, J.R. Wendt, Mechanical dissipation in tetrahedral amorphous carbon, Journal of Applied Physics, 97, 2005, pp. 023517, 1–023517, 10.
33. V. B. Braginski, V.P. Mitrofanov, V.I. Panov, Systems with Small Dissipation, University of Chicago Press, Chicago, 1985.
34. L. Burakowsky, D.L. Preston, An analytical model of the Gruneisen parameter at all densities, Journal of Physical Chemistry and Solids, 65, 2004, pp. 1581–1595.
35. R.E. Mihailovich, N.C. MacDonald, Dissipation measurements of vacuum operated single-crystal silicon microresonators, Sensors and Actuators A, 50, 1995, pp. 199–207.
36. Y.-H Park, K.C. Park, High-fidelity modeling of MEMS resonators – Part I: Anchor loss mechanisms through substrate, Journal of Microelectromechanical Systems, 13 (2), 2004, pp. 238–247.
37. H. Osaka, K. Itao, S. Kuroda, Damping characteristics of beam-shaped micro-oscillators, Sensors and Actuators A, 49, 1995, pp. 87–95.
38. Z. Hao, A. Erbil, F. Ayazi, An analytical model for support loss in micromachined beam resonators with in-plane flexural vibrations, Sensors and Actuators A, 109, 2003, pp. 156–164.
39. D. Photiadis, J.A. Judge, Attachment losses of high Q oscillators, Applied Physics Letters, 85 (3), 2004, pp. 482–484.
40. K.Y. Yasumura, T.D. Stowe, E.M. Chow. T. Pfafman, T.W. Kenny, B.C. Stipe, D. Rugar, Quality factors in micron- and submicron-thick cantilevers, Journal of Microelectromechanical Systems, 9 (1), 2000, pp. 117–125.

Chapter 4

FREQUENCY AND TIME RESPONSE OF MEMS

4.1 INTRODUCTION

Microelectromechanical systems are discussed in this chapter by taking into account the forcing factor and therefore the forced response. When actuation is produced by harmonic (sinusoidal, cosinusoidal) factors, the frequency response needs to be analyzed, which, essentially, consists of characterizing the response amplitude and phase shift over the excitation frequency range. The Laplace transform and the transfer function approach are used to study topics such as transmissibility, coupling, mechanical-electrical analogies, as well as applications such as microgyroscopes and tuning forks. When excitation is not harmonical, the time response of MEMS has to be addressed. The Laplace transform method, the state-space approach, and time-stepping schemes are discussed in connection with the time response of MEMS. Non-linear problems, such as those generated by large deformations, and dedicated modeling/solution methods, such as time-stepping schemes or the approximate iteration method are presented, all in the context of MEMS applications.

4.2 FREQUENCY RESPONSE OF MEMS

Harmonic excitation of microelectromechanical systems is often used for actuating/sensing purposes. When there is no damping in the system, the response is a vibration having the same frequency as the excitation force/moment and in phase with it. When damping is present (particularly of a viscous nature), the system response is a vibration whose amplitude is proportional to the excitation amplitude, having the same oscillation frequency and out-of-phase. Analyzing the frequency response of a harmonically driven or sensed system implies analyzing the response amplitude and phase angle (only for damped vibrations) over the whole frequency range. This section analyzes the frequency response of single and multiple DOF microelectromechanical systems that are modeled through the lumped-parameter technique using the transfer function approach based on the Laplace transform method.

4.2.1 Frequency Response

The response characteristics of a mechanical system under harmonic (sinusoidal/cosinusoidal) excitation are introduced by using a single DOF system that is studied in the absence and then in the presence of viscous damping.

4.2.1.1 Undamped Frequency Response

An undamped, single DOF mechanical system under harmonic excitation is sketched in Figure 4.1.

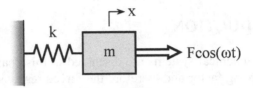

Figure 4.1 Single DOF mass-spring system under harmonic excitation

The equation of motion, which governs the vibrations of the system, shown in Figure 4.1 is:

$$m\ddot{x} + kx = F\cos(\omega t) \tag{4.1}$$

The complex number notation that has been introduced in Chapter 3 can be used in connection to Equation (4.1) considering that both the excitation function and the system response are complex numbers, with the understanding that, however, both the excitation and the response are the real parts of their corresponding complex number representations. This means Equation (4.1) can be expressed as:

$$m\ddot{x} + kx = Fe^{j\omega t} \tag{4.2}$$

and therefore its solution is of the form:

$$x = Xe^{j\omega t} \tag{4.3}$$

which, substituted in Equation (4.2), results in:

$$X = \frac{F}{-m\omega^2 + k} \tag{4.4}$$

Consequently, the complex form solution of Equation (4.2) is:

$$x = \frac{F}{-m\omega^2 + k} e^{j\omega t} \tag{4.5}$$

Because only the real part should be retained from the complex form, the solution of Equation (4.1) is:

$$x = \frac{F}{-m\omega^2 + k} \cos(\omega t) \tag{4.6}$$

Equation (4.4) allows formulating the following ratio:

$$G = \frac{X}{F} = \frac{1}{k - m\omega^2} \tag{4.7}$$

which is known as *transfer function* and is extremely useful in describing the *steady-state response*, and will be used in this section. Another parameter, the *frequency response function*, is defined as:

$$H = kG = \frac{kX}{F} = \frac{1}{1 - \beta^2} \tag{4.8}$$

and it actually represents the ratio of the amplitudes of the elastic (spring) force and the excitation force. The variable β in Equation (4.8) is the ratio of the excitation frequency to the resonant frequency, namely:

$$\beta = \frac{\omega}{\omega_r} \tag{4.9}$$

where:

$$\omega_r = \sqrt{\frac{k}{m}} \tag{4.10}$$

is the resonant (natural) frequency of the system. When the excitation frequency $\omega = \omega_r$, the transfer function, the frequency response function and the response amplitude all go to infinity, as can be checked from Equations (4.4), (4.7), and (4.8). A system with zero losses is an idealized situation, and definitely irreparable damage will incur to such a system under resonant conditions. Real systems, however, are characterized by energy losses (which, in general, and as shown in the Chapter 3, can be expressed in the form of equivalent viscous damping), and therefore the resonant response amplitude has a finite value.

4.2.1.2 Damped Frequency Response

A damped, single DOF mechanical system under harmonic excitation is sketched in Figure 4.2.

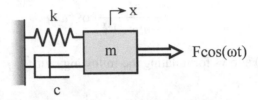

Figure 4.2 Single DOF mass-dashpot system under harmonic excitation

A mass-dashpot system's response to harmonic excitation, in the case where viscous damping is present, is made up of two components: one represents the system's reaction at its own resonant frequency and the other is the system's response at the excitation and its frequency input. The former component dies out in time eventually due to damping and is important in characterizing the *transient response* (the first phase in time), whereas the latter defines the *steady-state response*. The system sketched in Figure 4.2 is characterized by the dynamic equation:

$$m\ddot{x} + c\dot{x} + kx = Fe^{j\omega t} \tag{4.11}$$

where, again, the complex representation of the excitation has been used. The solution to Equation (4.11) is the sum of two terms: one, the *complementary solution*, is the solution to the homogeneous equation (where there is no excitation) and it characterizes the *transitory response*; the second solution term is the *particular solution*, which defines the steady-state response when time goes to infinity and after the transient effects have generally vanished altogether. The frequency response characterization of a system regards its steady-state response and the related particular solution. As such, an exponential solution as the one of Equation (4.3) is sought here, too. After taking the two time derivatives and substituting them into Equation (4.11), the particular solution's amplitude is:

$$X = \frac{F}{-m\omega^2 + k + jc\omega} = \frac{F(k - m\omega^2)}{(k - m\omega^2)^2 + c^2\omega^2} - j\frac{Fc\omega}{(k - m\omega^2)^2 + c^2\omega^2} \tag{4.12}$$

By taking into account the following relationship:

$$x - jy = \sqrt{x^2 + y^2}\, e^{-j\tan^{-1}(y/x)} \tag{4.13}$$

Equation (4.12) can be rewritten as:

$$X = \frac{F}{\sqrt{(k - m\omega^2)^2 + c^2\omega^2}}\, e^{-j\tan^{-1}(c\omega/(k - m\omega^2))} \tag{4.14}$$

which means the particular solution of Equation (4.11) is:

$$x_p = \frac{F}{\sqrt{(k - m\omega^2)^2 + c^2\omega^2}}\, e^{j\left[\omega t - \tan^{-1}(c\omega/(k - m\omega^2))\right]} \tag{4.15}$$

The transfer function is determined by means of Equation (4.7) as:

$$G(j\omega) = \frac{X}{F} = \frac{1}{k - m\omega^2 + jc\omega} \tag{4.16}$$

The modulus (magnitude) of this transfer function is:

$$|G(j\omega)| = \frac{1}{\sqrt{(k - m\omega^2)^2 + c^2\omega^2}} \tag{4.17}$$

By taking into account only the real parts of the excitation and response signals, it follows that:

$$x_p = X_p \cos(\omega t + \varphi) \tag{4.18}$$

with:

$$\begin{cases} X_p = |G(j\omega)|\, F \\ \varphi = -\tan^{-1}\left(\dfrac{c\omega}{k - m\omega^2}\right) \end{cases} \tag{4.19}$$

Equations (4.18) and (4.19) indicate that the response amplitude is proportional to the excitation amplitude F (the proportionality factor is the modulus of the transfer function) and that the response is a harmonic one having the same vibration frequency with the excitation. In other words, defining the frequency response of a single DOF viscously damped system is equivalent to defining the modulus of the transfer function and the phase angle between input (excitation) and output (response).

Example 4.1
A paddle microbridge, as the one of Figure 2.5, is driven electrostatically in out-of-the-plane vibration by a cosinusoidal force at $\omega = 0.8\ \omega_r$ (the resonant frequency). By using a lumped-parameter model with the paddle being rigid and the root segments being flexible and massless, determine the thickness and length of the root segments such that an amplitude of 1 μm is achieved when a maximum force of 100 μN is applied. The microbridge material has a modulus of elasticity of 160 GPa and the width of its roots is equal to 50 μm. Consider the viscous damping coefficient is $\zeta = 0.2$.

Solution:
By using the definitions of the damping ratio (Equation (3.3)) and frequency ratio (Equation (3.28)), the modulus of the transfer function can be expressed as:

$$|G(j\omega)| = \frac{1}{k\sqrt{1+2\left(2\varsigma^2-1\right)\beta^2+\beta^4}} \tag{4.20}$$

The stiffness k of the paddle microbridge, according to the lumped-parameter that uses the inertia of the paddle only and the compliance of the two root segments, is similar to the one of Equation (2.9), namely:

$$k = 2Ew_1\left(\frac{t}{l_1}\right)^3 \tag{4.21}$$

By taking into account Equation (4.16), as well as Equations (4.19) and (4.20), the stiffness can be expressed as:

$$k = \frac{F}{U_z\sqrt{1+2\left(2\varsigma^2-1\right)\beta^2+\beta^4}} \tag{4.22}$$

where F is the amplitude of the cosinusoidal force and U_z is the amplitude of the microbridge deflection at its midpoint. By combining Equations (4.21) and (4.22), the following relationship is produced:

$$\frac{t}{l_1} = \sqrt[3]{\frac{F}{2Ew_1U_z\sqrt{1+2(2\varsigma^2-1)\beta^2+\beta^4}}} \tag{4.23}$$

The numerical data of this example result in $t/l = 0.018$. For a thickness of $t = 1$ μm, for instance, the root length is approximately $l_1 = 56$ μm.

4.2.2 Transfer Function Approach to Frequency Response

As shown in the previous subsection, the steady-state response of a damped system under harmonic excitation has an amplitude proportional to the excitation amplitude, the proportionality factor being the modulus of the transfer function. The concept of *transfer function*, as well as the corresponding modeling approach, will be introduced in this section by means of the *Laplace transform method*.

4.2.2.1 Laplace Transform and the Transfer Function

The Laplace transform maps a real function depending on a real variable into a complex function depending on a complex variable. In dynamics problems, the real variable is time, and the definition of the Laplace transform is:

$$\mathcal{L}[u(t)] = \int_0^\infty u(t)e^{-st}dt = U(s) \tag{4.24}$$

where $u(t)$ is the original function and $U(s)$ is the Laplace-transformed function. The Laplace transform of a function exists if the function is *continuous* (or *piecewise continuous* for finite intervals) for the zero-to-infinity time interval. The other condition for a function to be Laplace transformable is that the function is of exponential order (see specialized texts, such as those of Ogata [1] or Nise [2], for more details) requiring:

$$\lim_{t\to\infty}\left(|u(t)|e^{-\sigma t}\right) = \begin{cases} 0, \sigma \geq \sigma_c \\ \infty, \sigma < \sigma_c \end{cases} \tag{4.25}$$

where σ is a real and positive number and σ_c is the *abscissa of convergence*. Equation (4.25) basically requires that the time-dependent function grows slower than an exponential function $e^{\sigma t}$. It can be shown that polynomial or harmonic functions, for instance, are Laplace-transformable. Knowledge of the Laplace transform $U(s)$ enables finding the original, time-dependent function by means of the *inverse Laplace transform* as:

$$u(t) = \mathcal{L}^{-1}[U(s)] \tag{4.26}$$

Tables are usually provided that give the Laplace transforms of elementary functions, such that by knowing a transformed function, its time-dependent counterpart can easily be found by simple tabular inspection. Table 4.1 shows a few Laplace-transformed function pairs necessary here (more pairs can be found in Ogata [1]).

Table 4.1 Elementary functions paired through Laplace transformation

$u(t)$	$U(s)$
t^n	$\dfrac{n!}{s^{n+1}}$
$\sin(\omega t)$	$\dfrac{\omega}{s^2 + \omega^2}$
$\cos(\omega t)$	$\dfrac{s}{s^2 + \omega^2}$
e^{at}	$\dfrac{1}{s-a}$

The Laplace transform is a linear operator, which means that given the set of functions $u_i(t)$ and constants c_i, the following relationship holds true:

$$\mathcal{L}\left[\sum_{i=1}^{n}\left(c_i u_i(t)\right)\right] = \sum_{i=1}^{n}\left(c_i U_i(s)\right) \tag{4.27}$$

with $U_i(s)$ being the Laplace transform of $u_i(t)$.

The most powerful feature, probably, of the Laplace transform approach is the one that enables transforming ordinary linear differential equations with time-unvarying coefficients into algebraic equations by implicitly incorporating the initial conditions into the transformation. The resulting algebraic equation can be solved for $U(s)$, and the unknown original function $u(t)$ can ultimately be determined from $U(s)$ by using the inverse Laplace transform. The following equation provides the Laplace transform of the n-th order differential of a function $u(t)$:

$$\mathcal{L}\left[\frac{d^n u(t)}{dt^n}\right] = s^n U(s) - \sum_{i=1}^{n}\left(s^{n-i}\left.\frac{d^{i-1}u(t)}{dt^{i-1}}\right|_{t=0}\right) \tag{4.28}$$

where the sum in the right-hand side collects the initial conditions of the time-dependent dynamic model. The MEMS dynamic models usually extend no higher than the second-order differential, such that of interest here are the following two transformation rules, which result from the generic Equation (4.28):

$$\begin{cases} \mathcal{L}[\dfrac{d^2u(t)}{dt^2}] = \mathcal{L}[\ddot{u}(t)] = s^2U(s) - su(0) - \dot{u}(0) \\[4mm] \mathcal{L}[\dfrac{du(t)}{dt}] = \mathcal{L}[\dot{u}(t)] = sU(s) - u(0) \end{cases} \qquad (4.29)$$

Example 4.2

Find the solution to the following second-order differential equation: $\ddot{u} + 4a\dot{u} + 3a^2u = b\cos(\omega t)$ for a and b being positive real constants and with zero initial conditions, namely: $du/dt\,(0) = 0$ and $u(0) = 0$.

Solution:

By combining the linearity feature of the Laplace transform operator (Equations (4.27) and (4.29) and Table 4.1), the Laplace transform is applied to this example's differential equations, which results in:

$$U(s) = \frac{bs}{(s^2 + 4as + 3a^2)(s^2 + \omega^2)} = \frac{A_1}{s + a} + \frac{A_2}{s + 3a} + \frac{A_3}{s - j\omega} + \frac{A_4}{s + j\omega} \qquad (4.30)$$

The constants A_1 to A_4, which correspond to simple-fraction expansion, can be found by elementary algebra. A_1, for instance, is found as:

$$A_1 = \frac{bs}{(s + 3a)(s^2 + \omega^2)}\bigg|_{s=-a} = -\frac{b}{2(\omega^2 + a^2)} \qquad (4.31)$$

The other constants are similarly found as:

$$A_2 = \frac{3b}{2(\omega^2 + 9a^2)}; A_3 = \frac{b}{2(3a + j\omega)}; A_4 = \frac{b}{2(3a - j\omega)} \qquad (4.32)$$

The original function $u(t)$ can thus be obtained as:

$$u(t) = A_1e^{-at} + A_2e^{-3at} + A_3e^{j\omega} + A_4e^{-j\omega} \qquad (4.33)$$

By substituting the constants A_1 to A_4 from Equations (4.31) and (4.32) into Equation (4.33) and after some more algebraic manipulation, the original time-dependent function becomes:

$$u(t) = \frac{b}{2}\left(\frac{e^{-3at}}{\omega^2 + 9a^2} - \frac{e^{-at}}{\omega^2 + a^2}\right) + \frac{3ab}{\omega^2 + 9a^2}\cos\left(\omega t - \tan^{-1}\frac{\omega}{3a}\right) \quad (4.34)$$

When time goes to infinity, Equation (4.34) yields the particular solution, which describes the steady-state behavior of the single DOF system, and which can be expressed with an equation similar to Equation (4.18) as:

$$u_p(t) = U_p(\omega)\cos\left[\omega t + \varphi(\omega)\right] \quad (4.35)$$

where:

$$\begin{cases} U_p(\omega) = \dfrac{3ab}{\omega^2 + 9a^2} \\[2ex] \varphi(\omega) = -\tan^{-1}\dfrac{\omega}{3a} \end{cases} \quad (4.36)$$

Example 4.3
By using the Laplace direct and inverse transforms, derive the time response of the single DOF mechanical microsystem shown in Figure 4.3, which is actuated electrostatically by a comb driver being operated by a cosinusoidal voltage. Consider the initial displacement and initial velocity have zero values.

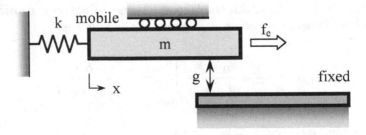

Figure 4.3 Mass-spring system and electrostatic longitudinal (comb-drive) actuation

Solution:
In longitudinal (or comb-drive) actuation, an electrostatic force is generated between the mobile plate of mass m and the fixed substrate, which is expressed (e.g., see Lobontiu and Garcia [3]) as:

$$f_e = \frac{\varepsilon l}{2g}v^2 \quad (4.37)$$

where ε is the medium (usually air) electric permittivity, l is the two plates common dimension perpendicular to the drawing plane, g is the gap that is maintained constant between the two plates, and v is the voltage, which here is assumed to vary harmonically as:

$$v = V \cos(\omega t) \qquad (4.38)$$

The equation of motion for the plate of Figure 4.3 is:

$$m\ddot{x} + kx = f_e \qquad (4.39)$$

By applying the direct Laplace transform to Equation (4.39), the counterpart $x(t)$, the function $X(s)$, is obtained as:

$$X(s) = \frac{\varepsilon l}{4g} V^2 \left[\frac{1}{s(ms^2 + k)} - \frac{s}{(s^2 + 4\omega^2)(ms^2 + k)} \right] \qquad (4.40)$$

Equation (4.40) can be expanded in simple fractions as:

$$X(s) = \frac{A_1}{s} + \frac{A_2 s + A_3}{ms^2 + k} + \frac{A_4 s + A_5}{s^2 + 4\omega^2} \qquad (4.41)$$

After determining the coefficients A_1 to A_5 by identifying the corresponding terms of Equations (4.40) and (4.41), the inverse Laplace transform is applied to Equation (4.41), and the time-domain position of the plate is obtained as:

$$x(t) = \frac{F}{k} \left[1 - \frac{2(1 - 2\beta^2)\cos(\omega_r t) - \cos(2\omega t)}{1 - 4\beta^2} \right] \qquad (4.42)$$

with:

$$F = \frac{\varepsilon l}{4g} V^2 \qquad (4.43)$$

Equation (4.42) shows that the solution is the superposition of the static spring force and two other forces: one that combines the natural response with the excitation frequency and the other, which is the result of excitation.

Example 4.4

Determine the voltage amplitude that needs to be applied at resonance to the microelectromechanical system of Example 4.3 so that the mobile plate travels a maximum distance $x_{max} = 15$ µm. Known are the following amounts: $\omega_r = 10,000$ rad/s, $k = 0.001$ N/m, $g = 1$ µm, $l = 800$ µm, and the air permittivity $\varepsilon = 8.8 \times 10^{-12}$ F/m.

Solution:
    At resonance, the excitation frequency ω is equal to the resonant frequency ω_r, and Equation (4.42) simplifies to:

$$x_r(t) = \frac{F}{k_r} \tag{4.44}$$

with F given in Equation (4.43) and k_r calculated as:

$$k_r = \frac{3k}{4\left[2 + \cos(\omega_r t)\right]\sin^2\left(\dfrac{\omega_r t}{2}\right)} \tag{4.45}$$

Maximizing x_r of Equation (4.44) is equivalent to minimizing k_r and by combining Equations (4.44) and (4.45), the voltage amplitude is calculated as:

$$V = 2\sqrt{\frac{gx_{r,max}k_{r,min}}{\varepsilon l}} \tag{4.46}$$

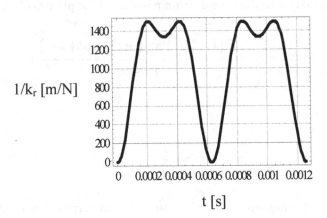

$1/k_r$ [m/N]

t [s]

Figure 4.4 Equivalent resonant compliance in terms of time

Figure 4.4 plots the inverse of k_r from Equation (4.45), which is actually an equivalent compliance, and it can be seen that the maximum value of $1/k_r$ is 1,500, which means $k_{r,min} = 1/1,500$. By using this value together with the other numerical data, it is found that $V = 75$ Volts.

The Laplace transform can also be used to solve systems of linear differential equations with time-unvarying coefficients. Consider a system of n second-order differential equations, whose matrix form is:

$$[M]\{\ddot{u}(t)\}+[C]\{\dot{u}(t)\}+[K]\{u(t)\}=\{f(t)\} \qquad (4.47)$$

where $[M]$, $[C]$, and $[K]$ are the inertia, damping and stiffness matrices; $\{u(t)\}$ and $\{f(t)\}$ are the displacement vector and forcing vectors:

$$\begin{cases} \{u(t)\}=\{u_1(t) \quad u_2(t) \quad \dots \quad u_n(t)\}^t \\ \{f(t)\}=\{F\}\cos(\omega t)=\{F_1 \quad F_2 \quad \dots \quad F_n\}^t \cos(\omega t) \end{cases} \qquad (4.48)$$

It has been assumed here that the excitation vector is of harmonic form, and that all excitation components differ only by their amplitudes.

By taking the Laplace transform to Equation (4.47), the Laplace-domain unknown vector $\{U(s)\}$ is obtained as:

$$\{U(s)\}=[A(s)]^{-1}\{B(s)\} \qquad (4.49)$$

where:

$$\begin{cases} [A(s)]=s^2[M]+s[C]+[K] \\ \{B(s)\}=\{F\}\dfrac{s}{\omega^2+s^2}+s[M]\{u(0)\}+[M]\{\dot{u}(0)\}+[C]\{u(0)\} \end{cases} \qquad (4.50)$$

with $\{u(0)\}$ being the initial displacement vector and the similar dotted vector being the initial velocity vector. The original vector $\{u(t)\}$ can now be found by taking the inverse Laplace transform to Equation (4.49). More details and utilization of the Laplace approach to multi DOF systems with generic excitation will be given in the section dedicated to the time response.

4.2.2.2 Transfer Function Approach

A system, be it mechanical, electrical, electromechanical, or of a different nature (but whose behavior can be modeled by means of differential equations) is often times considered, in the control domain especially, of the simple form indicated in Figure 4.5.

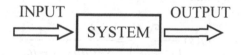

Figure 4.5 System with input and output signals

In a single DOF mass-dashpot mechanical system, which is subjected to a time-dependent force and which is described by the differential equation:

$$m\ddot{x} + c\dot{x} + kx = f(t) \tag{4.51}$$

the input is the forcing term $f(t)$ and the output is the displacement $x(t)$.

The concept of transfer function has been introduced in a Chapter 3 in an intuitive manner as the ratio of the displacement amplitude to the ratio of the forcing amplitude. Formally, the transfer function is defined as the ratio of the Laplace transform of the output to the Laplace transform of the input with zero initial conditions; the function is denoted by $G(s)$, and can be expressed as:

$$G(s) = \frac{X(s)}{F(s)}\bigg|_{\substack{x(0)=0 \\ \dot{x}(0)=0}} = \frac{1}{ms^2 + cs + k} \tag{4.52}$$

The generic concept of transfer function will be used more in subsequent sections of this chapter treating modeling in the time domain; for frequency response systems, a specialized transfer function operates, as shown next.

When cosinusoidal excitation is considered of the single DOF mechanical system shown in Figure 4.2, the corresponding differential equation is:

$$m\ddot{x} + c\dot{x} + kx = F\cos(\omega t) \tag{4.53}$$

By applying the Laplace transform to Equation (4.53), the transformed $X(s)$ becomes:

$$X(s) = \frac{Fs}{\left(ms^2 + cs + k\right)\left(s^2 + \omega^2\right)} = \frac{A_1}{s - s_1} + \frac{A_2}{s - s_2} + \frac{A_3}{s - j\omega} + \frac{A_4}{s + j\omega} \tag{4.54}$$

where s_1 and s_2 are the roots of the polynomial $ms^2 + cs + k$. The *theory of stability* (e.g., see Ogata [1]), shows that the real parts of s_1 and s_2 should be negative in order for the system's response to be stable. Assuming this condition is satisfied, as also shown in the previous example, it follows that the first two fractions of Equation (4.54) go to zero when time goes to infinity, which means that:

$$G(s)\frac{Fs}{\left(s^2 + \omega^2\right)} = \frac{A_3}{s - j\omega} + \frac{A_4}{s + j\omega} \tag{4.55}$$

The two constants of Equation (4.55) are:

$$\begin{cases} A_3 = G(s)\dfrac{Fs}{s+j\omega}\bigg|_{s=j\omega} = \dfrac{F}{2}G(j\omega) \\[4mm] A_4 = G(s)\dfrac{Fs}{s-j\omega}\bigg|_{s=-j\omega} = \dfrac{F}{2}G(-j\omega) \end{cases} \quad (4.56)$$

It is also known from complex number analysis that:

$$\begin{cases} G(j\omega) = |G(j\omega)|e^{j\varphi} \\[2mm] G(-j\omega) = |G(j\omega)|e^{-j\varphi} \end{cases} \quad (4.57)$$

where φ is the angle made by the position line of $G(j\omega)$ with the real (x) axis, as shown in Figure 4.6.

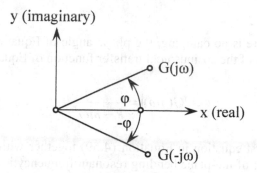

Figure 4.6 Transfer function as a complex number

The original function's particular solution is determined by combining Equations (4.54), (4.56), and (4.57) as:

$$x_p(t) = F|G(j\omega)|\frac{e^{j(\omega t+\varphi)} + e^{-j(\omega t+\varphi)}}{2} = F|G(j\omega)|\cos(\omega t + \varphi) \quad (4.58)$$

Equation (4.58) indicates that the response of a single DOF mechanical system to harmonic (cosinusoidal) excitation consists of a harmonic function whose amplitude is equal to the product of the excitation signal amplitude to the modulus of the *cosinusoidal transfer function* $G(j\omega)$. The frequency of the response function is separated from the one of the input signal by a *phase angle* φ. Consequently, the frequency response parameters are defined by means of the cosinusoidal transfer function as:

$$
\begin{cases}
\left| G(j\omega) \right| = \sqrt{\operatorname{Re}^2\left(G(j\omega)\right) + \operatorname{Im}^2\left(G(j\omega)\right)} \\[2mm]
\varphi = -\tan^{-1} \dfrac{\operatorname{Im}\left(G(j\omega)\right)}{\operatorname{Re}\left(G(j\omega)\right)}
\end{cases}
\tag{4.59}
$$

The minus sign of the phase angle indicates that the response signal is behind the excitation one. For a single DOF damped mechanical system, the actual expressions for the modulus and phase angle of the cosinusoidal transfer function are given in Equations (4.17) and (4.19).

Example 4.5
Analyze the out-of-the-plane bending frequency response of a paddle microcantilever vibrating in vacuum and fabricated of polysilicon with $E = 165$ GPa and $\rho = 2300$ kg/m^3. The microcantilever is defined by the following geometric parameters: $l_1 = l_2 = 200$ μm, $w_1 = 100$ μm, $w_2 = 20$ μm, and $t = 1$ μm. Use the lumped-parameter model by considering the paddle is rigid and provides the equivalent mass fully.

Solution:
Because there is no damping, the phase angle of Equation (4.59) is zero and the modulus of the cosinusoidal transfer function of Equation (4.57) is:

$$
\left| G(j\omega) \right| = \frac{1}{k - m\omega^2}
\tag{4.60}
$$

By using the first equation in Equation (4.59) together with the data of this example, the out-of-the-plane bending resonant frequency is found to be $\omega_r = 47{,}348$ rad/s. Figure 4.7 plots the frequency response function H of Equation (4.8) as a function of the frequency ratio β as defined in Equation (4.9). It can be seen that at resonance the frequency response function (and therefore the cosinusoidal transfer function, which is related to the frequency response function according to Equation (4.8)) goes to infinity.

Example 4.6
A paddle microbridge is used as a torsional resonator under harmonic excitation and with monitoring of the frequency response. By using the lumped-parameter model, evaluate the elastic properties of the microbridge material, as well as the overall losses (which are considered to be produced by viscous damping) for $l_1 = l_2 = l = 250$ μm, $w_1 = 30$ μm, $w_2 = 150$ μm, $t = 0.8$ μm, and $\rho = 2400$ kg/m^3. Known is also the resonant frequency $f_r = 22{,}000$ Hz and the phase angle $\varphi = 16.88°$ at $\omega = 100{,}000$ rad/s.

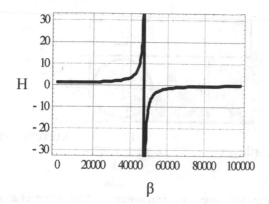

Figure 4.7 Frequency response function in terms of frequency ratio

Solution:

Equation (2.13), which gives the torsion resonant frequency of a microbridge, simplifies for this particular example to:

$$\omega_t^* = 2.82 \frac{t}{l} \sqrt{\frac{Gw_1}{\rho w_2 \left(t^2 + w_2^2\right)}}$$ (4.61)

As discussed in Chapter 3, the damped resonant frequency can be found from the undamped one by means of the equation:

$$\omega_t = \sqrt{1 - \zeta^2} \, \omega_t^*$$ (4.62)

The absolute value of the phase angle is:

$$\varphi = \tan^{-1} \frac{2\zeta\beta}{1 - \beta^2}$$ (4.63)

where β is the frequency ratio of Equation (4.9). Equations (4.61), (4.62), and (4.63) contain the unknowns G and ζ, which are found as: $G = 64$ GPa and $\zeta = 0.1$.

Example 4.7

The constant cross-section microcantilever of Figure 4.8 is used to sense an external magnetic field B by means of a metallic circular loop embedded in it that is ran by a current $i = I\cos(\omega t)$. Determine the magnetic field by using a lumped-parameter model. Known are the length of the cantilever l, its cross-section dimensions w and t, the current amplitude I, the actuation frequency ω, and the maximum tip deflection U_z, which is measured experimentally.

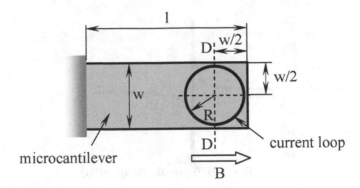

Figure 4.8 Microcantilever with loop carrying a current for harmonic excitation detection of external magnetic field

Solution:

The interaction between an electric current and a magnetic field generates a Lorentz-type force. As shown by Lobontiu and Garcia [3], who analyzed a similar example in the static domain, the result of the interaction between the current passing through the loop and the external magnetic field that is parallel to the microcantilever's length is a bending moment about the D-D axis (as shown in Figure 4.8), which is expressed as:

$$m_b = ABi = \pi R^2 BI \cos(\omega t) = M \cos(\omega t) \qquad (4.64)$$

As suggested by Equation (4.64), the bending moment changes its direction because of the current's changing sign, and the result is an alternating bending of the microcantilever. When damping is ignored, the lumped-parameter model of the vibrating cantilever is the one sketched in Figure 4.9.

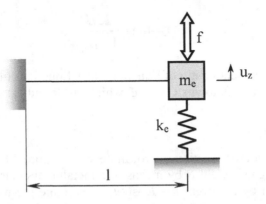

Figure 4.9 Equivalent lumped-parameter model of a microcantilever

To obtain the alternating tip force necessary in the equivalent translation model of Figure 4.9, it is assumed that the force f would produce the same tip deflection as the actual bending moment m_b. The equivalent force can simply be evaluated by using the beam theory as:

$$f = \frac{3m_b}{2l} = \frac{3M}{2l} \cos(\omega t) = F \cos(\omega t) \qquad (4.65)$$

The equivalent, lumped-parameter mass and stiffness (e.g., see Lobontiu [4]) are:

$$\begin{cases} k_e = \dfrac{3EI_y}{l^3} \\ m_e = \dfrac{33}{140} m \end{cases} \qquad (4.66)$$

where m is the microcantilever mass. By using the tip force of Equation (4.65) as well as the stiffness and mass fractions of Equation (4.66), the following differential equation describing the motion of the lumped mass m_e is obtained:

$$m_e \ddot{u}_z + k_e u_z = f \qquad (4.67)$$

When considering the input is f and the output is u_z, the transfer function corresponding to Equation (4.67) is:

$$G(s) = \frac{U_z(s)}{F(s)} = \frac{1}{m_e s^2 + k_e} \qquad (4.68)$$

The modulus of the cosinusoidal transfer function is actually the ratio of the real-valued amplitudes of the output and input, namely:

$$|G(j\omega)| = \frac{|U_z(j\omega)|}{|F(j\omega)|} \qquad (4.69)$$

The amplitude of the output is U_z, whereas the amplitude of the input is determined from Equations (4.64) and (4.65) as:

$$|F(j\omega)| = \frac{3\pi R^2 BI}{2l} \qquad (4.70)$$

By combining Equation (4.68) (with $j\omega$ instead of s), (4.69), and (4.70), the unknown value of B is found to be:

$$B = 0.012 \frac{\rho E t^4 w^2}{l R^2 I} U_z \omega^2 \qquad (4.71)$$

Example 4.8

The torsion micromirror of Figure 4.10 is set into motion by the interaction between the external magnetic field B and the alternating current $i = I \cos(\omega t)$, which passes through the circular metallic loop embedded in the central plate. Determine the maximum current amplitude such that the supporting circular cross-section beams do not yield.

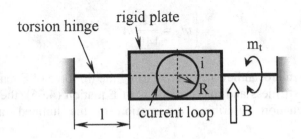

Figure 4.10 Electromagnetically actuated torsion micromirror with torsion hinges

Solution:

The result of the interaction between the magnetic field and the loop current is an alternating moment directed along the hinges axis. The moment is:

$$m_t = ABi = \pi R^2 BI \cos(\omega t) = M_t \cos(\omega t) \qquad (4.72)$$

where M_t is the torsion moment amplitude.

The torsional vibration lumped-parameter model is described by the equation:

$$J\ddot{\theta} + k\theta = m_t \qquad (4.73)$$

where J is the moment of inertia of the rigid plate about the hinges axis, and k is the torsion stiffness of the two hinges, which is equal to:

$$k = 2\frac{GI_p}{l} = \frac{\pi d^4 G}{16l} \qquad (4.74)$$

where d is the hinge diameter. By applying the Laplace transform to Equation (4.73), the following transfer function is obtained:

$$G(s) = \frac{1}{Js^2 + k} \tag{4.75}$$

when considering the torsion moment is the input and the rotation angle is the output. The amplitude of the output Θ is equal to the modulus of the cosinusoidal transfer function multiplied by the input amplitude M_t, namely:

$$\theta_{max} = \Theta = |G(j\omega)| M_t = \frac{\pi R^2 BI}{k - J\omega^2} \tag{4.76}$$

Equations (4.72) and (4.75)—with $j\omega$ instead of s—have been used to derive Equation (4.76).

It is known from mechanics of materials that the torsion angle and shear stress are expressed as:

$$\begin{cases} \theta = \dfrac{M_t l}{GI_p} \\[3mm] \tau = \dfrac{M_t d}{2I_p} \end{cases} \tag{4.77}$$

By combining the two equations in Equation (4.77), the maximum rotation angle can be obtained as a function of the maximum (admissible) shear stress as:

$$\theta_{max} = \frac{2l}{Gd} \tau_{max} \tag{4.78}$$

Equations (4.76) and (4.78) represent the same amount, and by equalizing the right-hand sides of these equations, and by also taking account the stiffness Equation (4.74), one can express the maximum current as:

$$I_{max} = 0.04 \frac{\pi Gd^4 - 16 J l \omega^2}{GdR^2 B} \tau_{max} \tag{4.79}$$

Figure 4.11 plots the maximum current I_{max} as a function of the hinge diameter d for the following numerical parameters: $G = 66$ GPa, $J = 3.2 \times 10^{-19}$ kg-m^2, $l = 100$ µm, $\tau_{max} = 10^8$ N/m^2, $B = 5$T, $\omega = 20,000$ rad/s. For large hinge

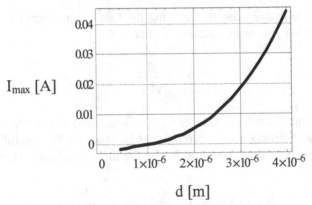

Figure 4.11 Maximum current versus hinge diameter

diameters, such as $d = 4$ μm, relatively large currents, of the order of 40 mA are needed. For smaller diameters, such as $d = 1$ μm, the maximum current is only 8.5 μA.

4.2.2.3 Transmissibility

Forces/moments or displacements in MEMS can be transmitted from a source (input) to an actuator/sensor (output), and if the input signal is harmonic, the concept of cosinusoidal transfer function can be used to relate the amplitudes of the input and output signals. Figure 4.12 is the schematic representation of a mechanical microsystem, which receives a harmonic input displacement y to be transmitted as an output displacement x through a mass-dashpot.

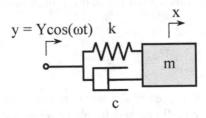

Figure 4.12 Displacement transmissibility principle

The equation of motion for the body can be formulated by Newton's law as:

$$m\ddot{x} + c(\dot{x} - \dot{y}) + k(x - y) = 0 \qquad (4.80)$$

which is rearranged as:

$$m\ddot{x} + c\dot{x} + kx = c\dot{y} + ky \qquad (4.81)$$

By applying the Laplace transform to Equation (4.81) where, again, the input is y and the output is x, the transfer function is:

$$G(s) = \frac{X(s)}{Y(s)} = \frac{cs+k}{ms^2 + cs + k} \tag{4.82}$$

The *transmissibility* is defined as the ratio of the modulus of the complex-form output to the modulus of the complex-form input, which is actually the ratio of the two signals amplitudes. For the mechanical system of Figure 4.12, the transmissibility is:

$$T = |G(j\omega)| = \frac{|X(j\omega)|}{|Y(j\omega)|} = \frac{|c\omega j + k|}{|-m\omega^2 + c\omega j + k|} = \frac{\sqrt{k^2 + c^2\omega^2}}{\sqrt{(k - m\omega^2)^2 + c^2\omega^2}} \tag{4.83}$$

Equation (4.83) can be formulated in terms of the damping ratio ζ and frequency ratio β as:

$$T = \sqrt{\frac{1 + 4\zeta^2\beta^2}{(1 - \beta^2)^2 + 4\zeta^2\beta^2}} \tag{4.84}$$

Figure 4.13 is the plot of T as a function of β for three different values of the damping ratio.

Simple calculations, as also shown in the plot of Figure 4.13, indicate that the transmissibility is equal to 1 for $\beta = 0$ and $\beta = \sqrt{2}$, which means that the

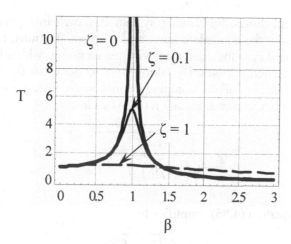

Figure 4.13 Transmissibility as a function of the frequency ratio

input signal is fully transmitted for these two values of the frequency ratio. For very small damping, the transmitted amplitude is large at resonance (when $\beta = 1$), as also suggested in Figure 4.13. For very large values of the excitation frequency (when $\beta \gg 1$), the transmitted amplitude is smaller than the input amplitude, and goes to zero as β tends to infinity.

Example 4.9
Out-of-the-plane bending of a microcantilever whose support is harmonically driven by a known signal y (as sketched in Figure 4.14 (a)) is used to evaluate the overall losses of the microsystem by monitoring the deflections of the microcantilever's free tip. Determine the quality factor (Q-factor) in terms of the system's parameters, excitation frequency, and transmissibility.

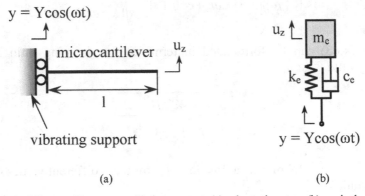

(a)   (b)

Figure 4.14 Microcantilever on oscillating support: (a) schematic setup; (b) equivalent lumped-parameter model

Solution:
The equivalent, lumped-parameter system is shown in Figure 4.14 (b), where m_e and k_e are the equivalent mass and stiffness fractions, provided in Equation (4.66) and c_e is the equivalent viscous damping, which is assumed to express all the energy losses. By taking into account that the Q-factor is expressed as a function of the damping ratio and frequency ratio as $Q = 1/(2\beta\zeta)$, as shown in Chapter 3, Equation (4.84) yields:

$$Q = \sqrt{\frac{1-T^2}{\left(1-\beta^2\right)^2 T^2 - 1}}$$

(4.85)

At resonance, Equation (4.85) simplifies to:

$$Q_r = \sqrt{T_r^2 - 1}$$

(4.86)

where T_r is the transmissibility at resonance and is expressed as:

$$T_r = \frac{\sqrt{1+4\zeta^2}}{2\zeta} \qquad (4.87)$$

Equations (4.86) and (4.87) give the quantitative relationship between the Q-factor, transmissibility, the microcantilever's parameters and the excitation frequency—the last two factors being incorporated into the frequency ratio β.

4.2.2.4 Coupled Systems

In transduction microsystems, which combine an actuation subsystem with a detection/sensing one, the two systems interact and are coupled. This interaction can be studied by means of the individual transfer functions and their combination. Assume that n subsystems, each defined by its own transfer function, interact in the form indicated in Figure 4.15.

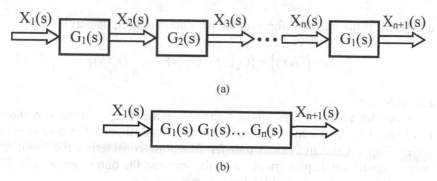

(a)

(b)

Figure 4.15 Coupled system: (a) serial interaction of n subsystems; (b) equivalent system

when the output signal from one subsystem is the input signal to the adjacent system, and if this coupling connection applied to all subsystems, then the following is true:

$$\frac{X_{n+1}(s)}{X_1(s)} = \frac{X_{n+1}(s)}{X_n(s)} \times \frac{X_n(s)}{X_{n-1}(s)} \times ... \times \frac{X_3(s)}{X_2(s)} \times \frac{X_2(s)}{X_1(s)} \qquad (4.88)$$
$$= G_n(s)G_{n-1}(s)...G_2(s)G_1(s)$$

By considering the ratio $X_{n+1}(s)/X_1(s)$ is an equivalent transfer function $G(s)$, it follows from Equation (4.88) that:

$$G(s) = G_1(s)G_2(s)...G_n(s) \qquad (4.89)$$

which shows that the overall (equivalent) transfer function of a coupled system formed of n subsystems is the product of subsystem transfer functions, which is indicated in Figure 4.15 (b). It should be mentioned that the approach discussed only applies when the subsystems do not load each other.

Equations (4.88) and (4.89) can also be expressed in matrix form as:

$$\{O(s)\} = [G(s)]\{I(s)\} \tag{4.90}$$

where $\{O(s)\}$ is the output vector, defined as:

$$\{O(s)\} = \{X_2 \quad X_3 \quad ... \quad X_{n+1}\}^t \tag{4.91}$$

$\{I(s)\}$ is the input vector, defined as:

$$\{I(s)\} = \{X_1 \quad X_2 \quad ... \quad X_n\}^t \tag{4.92}$$

and $[G(s)]$ is the diagonal-form transfer-function matrix, namely:

$$diag[G(s)] = \{G_1(s) \quad G_2(s) \quad ... \quad G_n(s)\} \tag{4.93}$$

Example 4.10

Consider a MEMS filter whose lumped-parameter representation is shown in Figure 4.16, and which is formed on n mass-dashpot units connected in series. Express the individual transfer functions considering the input and output signals are displacements, and also express the output amplitude. The system's input is a cosinusoidal displacement x_1.

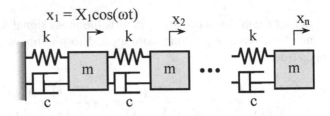

Figure 4.16 Serial n DOF mechanical microoscillator with harmonic input

Solution:

Actual systems do couple, however, in a slightly different manner from the one just presented. Consideration of the n DOF mechanical microsystem (which can be used to filter an input signal) of Figure 4.16, indicates the last mass is only connected at one end. For this serial system the output displacement from one subsystem is the input displacement to the subsequent one, and therefore the rule just presented of the overall transfer function being the product of individual transfer functions should apply. However, by writing the dynamic equations for the n masses, it is clear that three signals in the Laplace domain are related to one mass (DOF), while for transfer function formulation only two signals would be needed.

Consider the mass i in the sequence, which is shown in Figure 4.17 and whose equation of motion is:

$$m\ddot{x}_i + c\left(\dot{x}_i - \dot{x}_{i-1}\right) + c\left(\dot{x}_i - \dot{x}_{i+1}\right) + k\left(x_i - x_{i-1}\right) + k\left(x_i - x_{i+1}\right) = 0 \quad (4.94)$$

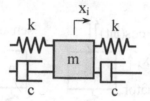

Figure 4.17 Generic mass with dashpots in a serial mechanical microsystem

The Laplace transform of Equation (4.94) with zero initial conditions results in:

$$-\left(cs + k\right)X_{i-1}(s) + \left(ms^2 + 2cs + 2k\right)X_i - \left(cs + k\right)X_{i+1}(s) = 0 \quad (4.95)$$

which shows three signals are involved with the generic mass subsystem. The last mass in the sequence, however, only relates two signals: the input one, $X_{n-1}(s)$ and the output one, $X_n(s)$. Its transfer function is therefore:

$$G_n(s) = \frac{X_n(s)}{X_{n-1}(s)} = \frac{cs + k}{ms^2 + cs + k} \quad (4.96)$$

Equation (4.96) enables expressing $X_n(s)$ as a function of $X_{n-1}(s)$, and generally it can be stated that:

$$G_{i+1}(s) = \frac{X_{i+1}(s)}{X_i(s)} \quad (4.97)$$

and

$$G_i(s) = \frac{X_i(s)}{X_{i-1}(s)} = \frac{cs + k}{ms^2 + 2cs + 2k - \left(cs + k\right)G_{i+1}(s)} \quad (4.98)$$

Equation (4.98) is a recurrence relationship that enables back calculation of the individual transfer functions for the serial mechanical system of Figure 4.16. In the end, the output amplitude is calculated as:

$$|X_n(j\omega)| = X_1 |G_1(j\omega)G_2(j\omega)...G_i(j\omega)...G_n(j\omega)| \qquad (4.99)$$

Example 4.11
Figure 4.18 is the schematic representation of an experiment involving harmonic excitation of a mass-spring system (m_a, k_a), which is to be per-formed on an optical table defined by m_t, c, and k_t. A sensing microdevice, defined by m_s, k_s is placed on the table. Knowing the minimum signal amplitude the sensor can detect is X_{min}, find the necessary amplitude of the actuator.

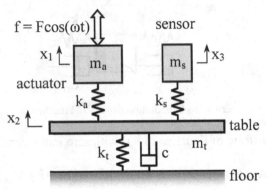

Figure 4.18 Optical table with actuator and sensing microdevices

Solution:
The equations of motion for the actuator, table, and sensor are:

$$\begin{cases} m_a\ddot{x}_1 + k_a(x_1 - x_2) = f \\ m_t\ddot{x}_2 + c\dot{x}_2 + k_t x_2 + k_a(x_2 - x_1) + k_s(x_2 - x_3) = 0 \qquad (4.100) \\ m_s\ddot{x}_3 + k_s(x_3 - x_2) = 0 \end{cases}$$

By Laplace-transforming Equation (4.100) with zero initial conditions yields:

$$\begin{cases} (m_a s^2 + k_a)X_1(s) - k_a X_2(s) = F(s) \\ -k_a X_1(s) + (m_t s^2 + cs + k_a + k_t + k_s)X_2(s) - k_s X_3(s) = 0 \qquad (4.101) \\ -k_s X_2(s) + (m_s s^2 + k_s)X_3(s) = 0 \end{cases}$$

Equations (4.101) can be expressed in matrix form as:

$$[A(s)]\{X(s)\} = \{F_1(s)\} \tag{4.102}$$

with:

$$\begin{cases} \{X(s)\} = \{X_1(s) \quad X_2(s) \quad X_3(s)\}' \\ \{F_1(s)\} = \{F(s) \quad 0 \quad 0\}' \end{cases} \tag{4.103}$$

and:

$$[A(s)] = \begin{bmatrix} m_a s^2 + k_a & -k_a & 0 \\ -k_a & m_t s^2 + cs + k_a + k_t + k_s & -k_s \\ 0 & -k_s & m_s s^2 + k_s \end{bmatrix} \tag{4.104}$$

Equation (4.102) allows solving for $\{X(s)\}$ as:

$$\{X(s)\} = [A(s)]^{-1}\{F_1(s)\} \tag{4.105}$$

The three variables $X_1(s)$, $X_2(s)$, and $X_3(s)$ can be expressed in terms of $F(s)$ by Equation (4.105). By only combining $X_1(s)$ and $X_3(s)$, and considering the harmonic nature of the excitation and the cosinusoidal transfer function approach, the minimum amplitude of the actuators is:

$$X_1 = |X_1(j\omega)| = \frac{X_{min}}{|G_{13}(j\omega)|} \tag{4.106}$$

where G_{13} is the transfer function between the actuator and sensor, and its modulus is:

$$|G_{13}(j\omega)| = \frac{k_a k_s}{\sqrt{\omega^2 c^2 (k_s - m_s \omega^2)^2 + \left[-k_s^2 + (k_s - m_s \omega^2)(k_a + k_t + k_s - m_t \omega^2)\right]^2}} \tag{4.107}$$

4.2.3 Electrical Microsystems

Microtransduction, which comprises actuation and sensing, is largely achieved in MEMS/NEMS by subsystems that involve (mostly) electrostatic means. Moreover, as shown in a subsequent section, mechanical components have electrical analogous counterparts, and therefore electromechanical micro-systems can be formulated unitarily in either the mechanical domain or the electrical one. A brief review of the main notions of electrical circuits that are needed in the modeling and analysis of electromechanical micro/nano systems will be given here.

As known from basic electrical circuit theory, the principal elements are the *resistor*, the *capacitor*, and the *inductor*. The *electrical impedance* is a central concept, which is defined as the ratio of the *voltage phasor* across an element to the *current phasor* through that element, namely:

$$Z = \frac{\tilde{E}}{\tilde{I}} \tag{4.108}$$

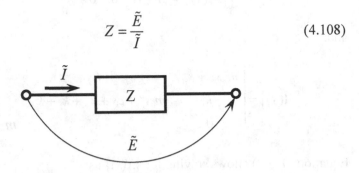

Figure 4.19   Schematic for electrical impedance definition

and its definition is symbolized as in Figure 4.19.

The phasor form of a harmonic signal is introduced to explain the electrical impedance definition mentioned above. Consider a harmonic signal:

$$x = X \cos(\omega t + \varphi) \tag{4.109}$$

which can also be expressed as the real part of the complex number:

$$x = \mathrm{Re}\left[X\left(\cos(\omega t + \varphi) + j\sin(\omega t + \varphi)\right)\right] = \mathrm{Re}\left[Xe^{j\varphi}e^{j\omega t}\right] \tag{4.110}$$

The following quantity:

$$\tilde{X} = Xe^{j\varphi} \tag{4.111}$$

is the phasor-form of the complex number. It can be seen that the phasor comprises full information (in exponential form) about a complex number which is defined by a modulus X and a phase angle φ. Equation (4.110) can be written now by means of Equation (4.111) as:

$$x = \text{Re}\left[\tilde{X}e^{j\omega t}\right] \tag{4.112}$$

Figure 4.20 shows the main electrical elements.

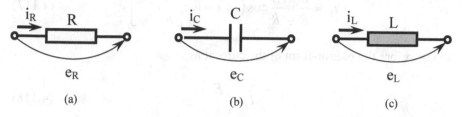

i_R R i_C C i_L L

e_R e_C e_L

(a)   (b)   (c)

Figure 4.20 Electrical elements with voltages: (a) resistor; (b) capacitor; (c) inductor

The voltages on the three components are expressed as:

$$\begin{cases} e_R = Ri_R \\ e_C = \dfrac{1}{C}\displaystyle\int i_C dt \\ e_L = L\dfrac{di_L}{dt} \end{cases} \tag{4.113}$$

whereas, conversely, the currents are expressed as:

$$\begin{cases} i_R = \dfrac{e_R}{R} \\ i_C = C\dfrac{de_C}{dt} \\ i_L = \dfrac{1}{L}\displaystyle\int e_L dt \end{cases} \tag{4.114}$$

Let us now formulate the impedances corresponding to the three electrical components by following the definition and when considering the following harmonic voltage is applied across each component:

$$e = E\cos(\omega t) = \text{Re}\left[Ee^{j\omega t}\right] \tag{4.115}$$

By comparing Equation (4.115) to the definition of a phasor (Equations (4.110) and (4.111)), it follows that the phasor form of the voltage is:

$$\tilde{E} = E \qquad (4.116)$$

Equation (4.114) allows expressing the current through the resistor as:

$$i_R = \frac{e}{R} = \frac{E}{R}\cos(\omega t) = \mathrm{Re}\left[\frac{E}{R}e^{j\omega t}\right] \qquad (4.117)$$

and therefore the phasor-form of the current as:

$$\tilde{I}_R = \frac{E}{R} \qquad (4.118)$$

By combining Equations (4.116) and (4.118), the electrical impedance corresponding to the resistor is:

$$Z_R = \frac{\tilde{E}_R}{\tilde{I}_R} = \frac{\tilde{E}}{\tilde{I}_R} = R \qquad (4.119)$$

Example 4.12
Derive the electrical impedances corresponding to a capacitor and an inductor when a cosinusoidal voltage of the type defined in Equation (4.115) is applied to each component.

Solution:
By dropping the voltage subscript, the current corresponding to the capacitor, as shown in the second Equation (4.114), is expressed as:

$$i_C = C\frac{de}{dt} = -\omega CE\sin(\omega t) = -\omega CE\cos\left(\omega t - \frac{\pi}{2}\right)$$
$$= \mathrm{Re}\left[-\omega CEe^{-j\frac{\pi}{2}}e^{j\omega t}\right] = \mathrm{Re}\left[j\omega CEe^{j\omega t}\right] \qquad (4.120)$$

Equation (4.120) shows that the capacity intensity phasor is:

$$\tilde{I}_C = j\omega CE \qquad (4.121)$$

By combining now Equations (4.116) and (4.121), the electrical impedance of a capacitor becomes:

$$Z_C = \frac{1}{j\omega C} = -\frac{j}{\omega C} \qquad (4.122)$$

To determine the impedance of the inductor, the current of Equation (4.114) is rewritten based on the cosinusoidal voltage (after dropping the subscript to the voltage) as:

$$i_L = \frac{1}{L}\int e\,dt = \frac{E}{\omega L}\sin(\omega t) = \frac{E}{\omega L}\cos\left(\omega t - \frac{\pi}{2}\right)$$

$$= \mathrm{Re}\left[\frac{E}{\omega L}e^{-j\frac{\pi}{2}}e^{j\omega t}\right] = \mathrm{Re}\left[-j\frac{E}{\omega L}e^{j\omega t}\right] \qquad (4.123)$$

which shows that:

$$\tilde{I}_L = -j\frac{E}{\omega L} \qquad (4.124)$$

Consequently, the impedance of an inductor is:

$$Z_L = \frac{\tilde{E}}{\tilde{I}_L} = -\frac{\omega L}{j} = j\omega L \qquad (4.125)$$

The three impedances just derived can very simply be determined by using the Laplace transform as shown next, and by introducing the notion of *Laplace-domain electrical impedance* as:

$$Z(s) = \frac{E(s)}{I(s)} \qquad (4.126)$$

where $E(s)$ is the Laplace transform of the voltage across an electrical component and $I(s)$ is the current transform through that component, both transforms being taken with zero initial conditions.

By applying the Laplace transform to the first Equation (4.114), it can be shown that:

$$Z_R(s) = \frac{E_R(s)}{I_R(s)} = \frac{E(s)}{I_R(s)} = R \qquad (4.127)$$

Similarly, the Laplace transform applied to the second Equation (4.114) results in:

$$Z_C(s) = \frac{E_C(s)}{I_C(s)} = \frac{E(s)}{I_C(s)} = \frac{1}{Cs} \qquad (4.128)$$

The inductor-related impedance is found from the third Equation (4.114) and its Laplace transform is:

$$Z_L(s) = \frac{E_L(s)}{I_L(s)} = \frac{E(s)}{I_L(s)} = Ls \qquad (4.129)$$

When substituting $j\omega$ for s in Equations (4.127), (4.128), and (4.129), the originally derived impedances of Equations (4.119), (4.122), and (4.125) are retrieved.

The Laplace-domain electrical impedance notion allows eliminating the step of passing from the time domain to the Laplace domain through the dedicated transformation, and enables working directly with algebraic amounts. Immediate advantages resulting from this shortcut regard the possibility of treating impedances, voltages, and currents in the Laplace domain as the ones related to resistors into the time domain. A series connection of n impedances for instance results in the following equivalent impedance:

$$Z_s(s) = \sum_{i=1}^{n} Z_i(s) \qquad (4.130)$$

whereas a parallel connection yields:

$$\frac{1}{Z_p(s)} = \sum_{i=1}^{n} \frac{1}{Z_i(s)} \qquad (4.131)$$

Example 4.13

Find the transfer function between the input voltage $e_i(t)$ and the output voltage $e_o(t)$ for the electrical circuit of Figure 4.21 by using the complex impedance approach. Determine the characteristics of the output voltage when time goes to infinity by considering the input is a cosinusoidal function of time.

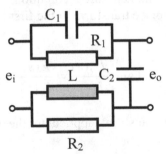

Figure 4.21 Electrical circuit with input and output voltage signals

Solution:

The actual electrical system of Figure 4.21 has a corresponding circuit that uses the impedances shown in Figure 4.22.

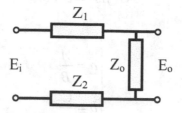

Figure 4.22 Equivalent electrical circuit with impedances

By considering the current is $I(s)$, the input and output Laplace-domain voltages can be expressed as:

$$\begin{cases} E_i(s) = \left[Z_1(s) + Z_2(s) + Z_o(s)\right] I(s) \\ E_o(s) = Z_o(s) I(s) \end{cases} \qquad (4.132)$$

Consequently, the transfer function is:

$$G(s) = \frac{E_o(s)}{E_i(s)} = \frac{Z_o(s)}{Z_1(s) + Z_2(s) + Z_o(s)} \qquad (4.133)$$

where:

$$\begin{cases} Z_1(s) = \dfrac{R_1}{1 + R_1 C_1 s} \\[2mm] Z_2(s) = \dfrac{L R_2 s}{R_2 + L s} \\[2mm] Z_o(s) = \dfrac{1}{C_2 s} \end{cases} \qquad (4.134)$$

By substituting Equation (4.134) into Equation (4.133), the transfer function becomes:

$$G(s) = \frac{L R_1 C_1 s^2 + \left(L + R_1 R_2 C_1\right) s + R_2}{L R_1 R_2 C_1 C_2 s^3 + L\left[R_1 C_1 + \left(R_1 + R_2\right) C_2\right] s^2 + \left[L + R_1 R_2\left(C_1 + C_2\right)\right] s + R_2} \qquad (4.135)$$

When $j\omega$ is used instead of s in Equation (4.135), the following complex number results:

$$G(j\omega) = a + bj \tag{4.136}$$

with:

$$\begin{cases} a = \dfrac{A}{B} \\[2mm] b = \dfrac{C}{B} \end{cases} \tag{4.137}$$

where:

$$\begin{cases} A = \omega^2 \left\{ L^2 \left[\omega^2 R_1^2 C_1 \left(C_1 + C_2 \right) + 1 \right] - L R_2^2 C_2 \left(\omega^2 R_1^2 C_1^2 + 1 \right) \right. \\[2mm] \qquad \left. + R_1^2 R_2^2 C_1 \left(C_1 + C_2 \right) \right\} + R_2^2 \\[2mm] B = 2\omega^4 L^2 R_1 R_2 C_2^2 + \omega^2 L^2 \left[\omega^2 R_1^2 \left(C_1 + C_2 \right)^2 + 1 \right] \\[2mm] \qquad + R_2^2 \left[\left(L C_2 \omega^2 - 1 \right)^2 + \omega^2 R_1^2 \left(C_1 + C_2 - L C_1 C_2 \omega^2 \right)^2 \right] \\[2mm] C = -\omega C_2 \left[\omega^2 L^2 \left(\omega^2 R_1^2 R_2 C_1^2 + R_1 + R_2 \right) + R_1 R_2^2 \right] \end{cases} \tag{4.138}$$

For a cosinusoidal input of amplitude E_i, the amplitude of the output is simply:

$$E_o = E_i \left| G(j\omega) \right| \tag{4.139}$$

with the modulus of the transfer function determined from Equations (4.137) and (4.138). Knowing the real and imaginary parts of $G(j\omega)$ also allows calculating the phase φ between the input and output voltages.

4.2.4 Mechanical-Electrical Analogy

Equations similar to the dynamic equation describing the motion of a single DOF mechanical system can be written for elementary electrical circuits. There are always two electrical systems that are analogous to a given mechanical system, as indicated in Figure 4.23.

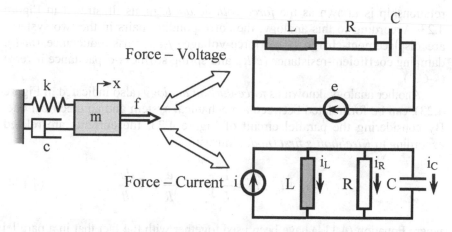

Figure 4.23 Force-voltage and force-current mechanical-electrical analogies

The equation of motion for the mass of Figure 4.23 is:

$$m\ddot{x} + c\dot{x} + kx = f \qquad (4.140)$$

For the series electrical circuit of the same figure, according to *Kirchhoff's second* (*loop* or *mesh*) *law*, the source voltage is the sum of voltages on the three elements, namely:

$$e = e_L + e_R + e_C = L\frac{di}{dt} + Ri + \frac{1}{C}\int i\,dt \qquad (4.141)$$

where Equation (4.113) has been applied for element voltages by also recognizing the current i is the same across all electrical components in a series circuit. It is also known that the current is expressed in terms of charge as:

$$i = \frac{dq}{dt} \qquad (4.142)$$

and therefore Equation (4.141) becomes:

$$L\ddot{q} + R\dot{q} + \frac{1}{C}q = e \qquad (4.143)$$

which is a second-order differential equation, similar to Equation (4.140), which defined the dynamics of the mechanical system of Figure 4.23. The source of motion is the force for the mechanical system, whereas for the electrical one, the source is voltage, and therefore the corresponding mechanical-electrical relationship is known as the *force-voltage analogy*, as illustrated in Figure 4.23. According to this analogy, the corresponding pairs in the two systems are displacement-charge $(x\text{-}q)$, force-voltage $(f\text{-}e)$, mass-inductance $(m\text{-}L)$, damping coefficient-resistance $(c\text{-}R)$, and spring stiffness-capacitance inverse $(k\text{-}C^{-1})$.

Another analogy, known as *force-current analogy* (also indicated in Figure 4.23), can be formulated between a mechanical system and an electrical one. By considering the parallel circuit of Figure 4.23, the currents are related according to *Kirchhoff's first (node) law* as:

$$i = i_L + i_R + i_C = \frac{1}{L}\int e\,dt + \frac{e}{R} + C\frac{de}{dt} \qquad (4.144)$$

where Equation (4.114) have been used together with the fact that in a parallel connection the voltage e is the same on each electrical component branch. It is known that the voltage and the magnetic flux ψ are related as:

$$e = \frac{d\psi}{dt} \qquad (4.145)$$

and therefore Equation (4.144) changes to:

$$C\ddot{\psi} + \frac{1}{R}\dot{\psi} + \frac{1}{L}\psi = i \qquad (4.146)$$

which is a second-order differential equation similar to Equation (4.140), which described the dynamics of a mechanical system; consequently, the mechanical system and the parallel electrical system are analogous. The following pairs describe the connection: displacement-magnetic flux $(x\text{-}\psi)$, force-current $(f\text{-}i)$, mass-capacitance $(m\text{-}C)$, damping coefficient-resistance inverse $(c\text{-}R^{-1})$, and stiffness-inductance inverse $(k\text{-}L^{-1})$.

Mechanical impedances can be formulated similarly to the electrical ones. The mechanical impedance of a mechanical element can be defined as the ratio of the force phasor to the resulting velocity phasor, namely:

$$Z = \frac{\tilde{F}}{\tilde{V}} \qquad (4.147)$$

By considering the following harmonic force:

$$f = F\cos(\omega t) = \mathrm{Re}\left[Fe^{j\omega t}\right] \qquad (4.148)$$

its related phasor, as discussed previously in this section, is:

$$\tilde{F} = F \tag{4.149}$$

If a mass element is acted upon by the force of Equation (4.149), the corresponding motion equation is defined by way of Newton's law as:

$$m\frac{dv_m}{dt} = f \tag{4.150}$$

where v_m is the velocity, which is found by integrating Equation (4.150), namely:

$$v_m = \frac{F}{m\omega}\sin(\omega t) = \frac{F}{m\omega}\cos\left(\omega t - \frac{\pi}{2}\right) = \mathrm{Re}\left[\frac{F}{m\omega}e^{-j\frac{\pi}{2}}e^{j\omega t}\right]$$

$$= \mathrm{Re}\left[-j\frac{F}{m\omega}e^{j\omega t}\right] \tag{4.151}$$

Equation (4.151) indicates that the velocity phasor is:

$$\tilde{V}_m = -j\frac{F}{m\omega} \tag{4.152}$$

By combining Equations (4.149) and (4.152), the mass-related mechanical impedance is:

$$Z_m = -\frac{\omega m}{j} = j\omega m \tag{4.153}$$

Example 4.14
    By using the phasor definition, derive the mechanical impedances for a spring of stiffness k and a damper of damping coefficient c. Consider that the harmonic force defined in Equation (4.148) is acting on each mechanical element.

Solution:
    In the case of a damper, the acting force and the opposing damping one balance out, and therefore:

$$cv_c = f \tag{4.154}$$

The velocity v_c can be expressed by using Equations (4.148) and (4.154) as:

$$v_c = \frac{F}{c}\cos(\omega t) = \text{Re}\left[\frac{F}{c}e^{j\omega t}\right] \qquad (4.155)$$

Equation (4.155) shows that the damping-related velocity phasor is:

$$\tilde{V}_c = \frac{F}{c} \qquad (4.156)$$

and, consequently, the corresponding mechanical impedance is:

$$Z_c = \frac{\tilde{F}}{\tilde{V}_c} = c \qquad (4.157)$$

For a spring element that is acted upon by the harmonic force of Equation (4.148) the force balance equation is:

$$kx = f \qquad (4.158)$$

By taking the time derivative of Equation (4.158) and by also considering Equation (4.148), the stiffness-related velocity is obtained as:

$$v_k = \frac{1}{k}\frac{df}{dt} = -\frac{F\omega}{k}\sin(\omega t) = -\frac{F\omega}{k}\cos\left(\omega t - \frac{\pi}{2}\right)$$

$$= \text{Re}\left[-\frac{F\omega}{k}e^{-j\frac{\pi}{2}}e^{j\omega t}\right] = \text{Re}\left[j\frac{F\omega}{k}e^{j\omega t}\right] \qquad (4.159)$$

Equation (4.159) indicates that the velocity phasor is:

$$\tilde{V}_k = j\frac{F\omega}{k} \qquad (4.160)$$

which means the stiffness-related mechanical impedance is:

$$Z_k = \frac{\tilde{F}}{\tilde{V}_k} = \frac{k}{j\omega} = -j\frac{k}{\omega} \qquad (4.161)$$

Example 4.15
    Derive the Laplace-domain mechanical impedances of a mass, damping, and spring elements. The Laplace-domain mechanical impedance is defined as the ratio of the Laplace transform of force to the Laplace transform of velocity.

Solution:

By applying the Laplace transform to Equation (4.150) and by using zero initial conditions one obtains:

$$ms V_m(s) = F(s) \tag{4.162}$$

and therefore the mass-related mechanical impedance is:

$$Z_m(s) = \frac{F(s)}{V_m(s)} = ms \tag{4.163}$$

The Laplace transform applied to Equation (4.154) results in:

$$c V_c(s) = F(s) \tag{4.164}$$

which yields:

$$Z_c(s) = \frac{F(s)}{V_c(s)} = c \tag{4.165}$$

By taking the time derivative to Equation (4.158) and then by applying the Laplace transform with zero initial conditions, the Laplace-domain, stiffness related mechanical impedance results, namely:

$$Z_k(s) = \frac{F(s)}{V_k(s)} = \frac{k}{s} \tag{4.166}$$

Equations (4.163), (4.165), and (4.166) are identical to Equations (4.153), (4.157), and (4.161), respectively, when using $j\omega$ instead of s.

Example 4.16

Determine an electrical system analogous to the mechanical microsystem, which is schematically shown by means of a lumped parameter model in Figure 4.24.

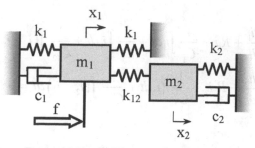

Figure 4.24 wo DOF mechanical system

Solution:
    The dynamic equations of the two DOF system are, according to Newton's law:

$$\begin{cases} m_1\ddot{x}_1 + c_1\dot{x}_1 + k_1x_1 + k_1x_1 + k_{12}(x_1 - x_2) = f \\ m_2\ddot{x}_2 + c_2\dot{x}_2 + k_2x_2 + k_{12}(x_2 - x_1) = 0 \end{cases} \qquad (4.167)$$

By using the force-voltage analogy, the equations corresponding to the electrical system are:

$$\begin{cases} L_1\ddot{q}_1 + R_1\dot{q}_1 + \dfrac{1}{C_1}q_1 + \dfrac{1}{C_1}q_1 + \dfrac{1}{C_{12}}(q_1 - q_2) = e \\ L_2\ddot{q}_2 + R_2\dot{q}_2 + \dfrac{1}{C_2}q_2 + \dfrac{1}{C_{12}}(q_2 - q_1) = 0 \end{cases} \qquad (4.168)$$

Equations (4.168) are actually Kirchhoff's second law, and the fact that there are two equations indicates the electrical system should have two loops. By inspection of Equation (4.168), the candidate electrical system, which is analogous to the mechanical one of Figure 4.24, is sketched in Figure 4.25.

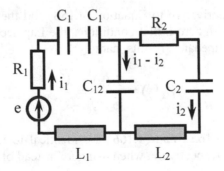

Figure 4.25 Two-loop electrical system, which is the analogous of the mechanical system of Figure 4.24

Kirchhoff's equations for the two loops of Figure 4.25 are, in their standard (current) form:

$$\begin{cases} L_1\dfrac{di_1}{dt} + R_1i_1 + \dfrac{1}{C_1}\int i_1dt + \dfrac{1}{C_1}\int i_1dt + \dfrac{1}{C_{12}}\int (i_1 - i_2)dt = e \\ L_2\dfrac{di_2}{dt} + R_2i_2 + \dfrac{1}{C_2}\int i_2dt + \dfrac{1}{C_{12}}\int (i_2 - i_1)dt = 0 \end{cases} \qquad (4.169)$$

By combining the known relationship between charge and current [Equation (4.142)] and Equation (4.169) yields Equation (4.168), which proves that, indeed, the electrical system proposed in Figure 4.25 is the electrical analogous of the mechanical system of Figure 4.24.

4.2.5 Electromechanical Microsystems with Variable Capacity

Transduction in MEMS/NEMS often times requires interaction between a mechanical subsystem and an electrical subsystem, such as a capacitive one. Comb-type driving or sensing is one modality of transduction that uses the variation of capacitance principle, as already seen in this chapter. The analogy between a mechanical system and an electrical one is very useful, as the mechanical parameters can be transformed/scaled into their electrical counterparts and therefore subsequent modeling of the MEMS in the electrical domain only becomes possible. Consider the MEMS of Figure 4.26, which is composed of a comb drive, a proof mass, and two beams that act as springs.

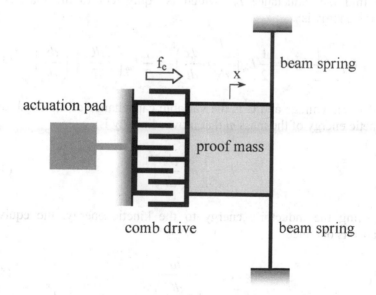

Figure 4.26 Comb-driven microelectromechanical system

Figure 4.27 (a) shows the electrical and mechanical models of the MEMS of Figure 4.26. Figure 4.27 (b) gives the full electrical schematic by using equivalent quantities L_m, R_m, and C_m that correspond to the actual mechanical quantities m (mass of the proof mass), c (damping coefficient), and k (beam spring stiffness). The problem of determining the electrical equivalent counterparts of the mechanical ones, which was studied by Lin et al. [5] or Bannon et al. [6], amongst others, can be approached considering the energy equivalence between mechanical and electrical elements with small/elementary variations of physical amounts.

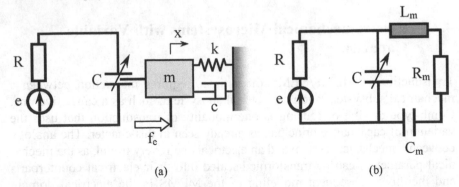

Figure 4.27 Schematics of the comb-driven MEMS: (a) electromechanical; (b) equivalent electrical

To find the inductance L_m, which is equivalent to the mass m, the inductor's energy is:

$$U_L = \frac{1}{2} L_m i^2 = \frac{1}{2} L_m \left(\frac{dq}{dx} \times \frac{dx}{dt} \right)^2 = \frac{1}{2} L_m \left(e \frac{dC}{dx} \right)^2 \left(\frac{dx}{dt} \right)^2 \quad (4.170)$$

where e is the voltage and C is the variable capacitance of the comb driver. The kinetic energy of the mass m that moves linearly is:

$$T = \frac{1}{2} m \left(\frac{dx}{dt} \right)^2 \quad (4.171)$$

By equating the inductor's energy to the kinetic energy, the equivalent inductance is obtained:

$$L_m = \frac{m}{\left(e \dfrac{dC}{dx} \right)^2} \quad (4.172)$$

The equivalent resistance R_m is determined by considering the electrical energy dissipated on the resistor and the mechanical energy dissipated through damping. The Joule-type electrical energy over a short time dt is:

$$U_R = R_m i^2 dt = R_m \left(\frac{dq}{dx} \times \frac{dx}{dt} \right)^2 dt = R_m \left(e \frac{dC}{dx} \right)^2 \left(\frac{dx}{dt} \right)^2 dt \quad (4.173)$$

whereas the energy dissipated through damping is equal to the work done by the damping force and can be expressed as:

$$U_d = c \frac{dx}{dt} dx \qquad (4.174)$$

By equating now the two energies of Equations (4.173) and (4.174), the equivalent resistance is obtained as:

$$R_m = \frac{c}{\left(e \frac{dC}{dx} \right)^2} \qquad (4.175)$$

The electrostatic energy that is stored on a capacitor can be expressed as:

$$U_C = \frac{dq^2}{2C_m} \qquad (4.176)$$

while the elastic potential energy stored in the actual spring is:

$$U = \frac{1}{2} k dx^2 \qquad (4.177)$$

By equating the energies expressed in Equations (4.176) and (4.177), and by also taking into account that:

$$\frac{dq}{dx} = e \frac{dC}{dx} \qquad (4.178)$$

the equivalent capacitance becomes:

$$C_m = \frac{\left(e \frac{dC}{dx} \right)^2}{k} \qquad (4.179)$$

It should be mentioned that in actual MEMS comb-type actuation the voltage e is the sum of a dc (bias) term and a harmonic component.

Example 4.17
Consider a purely cosinusoidal voltage is applied to actuate a MEMS as the one shown in Figure 4.26. Use the electrical equivalent system and the

complex impedance approach to determine the current through the mechanical loop of the equivalent electrical circuit.

Solution:
   By considering the current through the input (left) loop of Figure 4.27 is $I_i(s)$ and the one through the output (right) loop of the same figure is $I_o(s)$, the Kirchhoff's loop laws yield (by means of the complex impedance approach) the following equations:

$$\begin{cases} E_i(s) = I_i(s)R + \left[I_i(s) - I_o(s)\right]\dfrac{1}{Cs} \\[4mm] I_o(s)\left(L_m s + R_m + \dfrac{1}{C_m s}\right) = \left[I_i(s) - I_o(s)\right]\dfrac{1}{Cs} \end{cases} \tag{4.180}$$

Substitution of $I_i(s)$ between the two equations in Equation (4.180) produces the transfer function connecting the input voltage $E_i(s)$ and output current $I_o(s)$:

$$G(s) = \frac{I_o(s)}{E_i(s)} = \frac{C_m s}{L_m RCC_m s^3 + C_m\left(L_m + CRR_m\right)s^2 + \left(RC + RC_m + R_m C_m\right)s + 1} \tag{4.181}$$

By utilizing $j\omega$ for s in Equation (4.181), $G(j\omega)$ is obtained in the standard algebraic form of a complex number with the real and imaginary parts expressed as:

$$\begin{cases} \text{Re}\left[G(j\omega)\right] = \dfrac{A}{B} \\[4mm] \text{Im}\left[G(j\omega)\right] = \dfrac{C}{B} \end{cases} \tag{4.182}$$

with:

$$\begin{cases} A = \omega^2 C_m\left[C_m\left(R + R_m\right) - CR\left(\omega^2 L_m C_m - 1\right)\right] \\[2mm] B = \omega^2\{C_m[\omega^2 L_m^2 C_m\left(1 + \omega^2 R^2 C^2\right) + \omega^2 R^2 R_m^2 C^2 C_m \\[2mm] \quad -2\omega^2 L_m R^2 C\left(C + C_m\right) - 2L_m + 2R^2 C + \left(R + R_m\right)^2 C_m] + R^2 C^2\} + 1 \\[2mm] C = \omega C_m\left[1 - \omega^2 C_m\left(L_m + CRR_m\right)\right] \end{cases} \tag{4.183}$$

Again, knowledge of the real and imaginary parts of the cosinusoidal transfer function yields the modulus and the phase angle, which fully define the output current.

4.2.6 Microgyroscopes

Microgyroscopes are used as sensors in a large variety of applications, the best known being the ones in the automobile industry. They are based on the forces that act on a body moving with respect to a non-inertial reference frame.

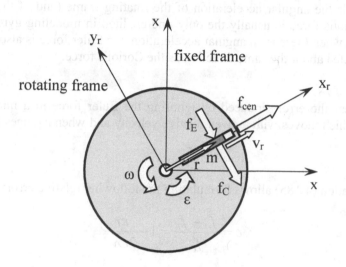

Figure 4.28 Forces acting on the point of mass m in a rotating reference frame

Consider, for instance, the reference frame x_rOy_r, which rotates (and is therefore a non-inertial reference frame) about the fixed (inertial) reference frame xOy with an angular velocity ω. A body of mass m moves with a velocity v_r measured in the non-inertial frame. It is known from mechanics that three forces need to be accounted for when referencing the motion in the rotating frame, namely: the centrifugal force f_{cen}, the Coriolis force f_C, and the Euler force f_E, which are shown in Figure 4.28 and are defined as:

$$\begin{cases} \overline{f}_{cen} = -m\overline{\omega} \times (\overline{\omega} \times \overline{r}) \\ \overline{f}_C = -2m\overline{\omega} \times \overline{v}_r \\ \overline{f}_E = -m\dfrac{d\overline{\omega}}{dt} \times \overline{r} \end{cases} \qquad (4.184)$$

These forces have the following magnitudes:

$$\begin{cases} f_{cen} = m\omega^2 r \\ f_C = 2m\omega v_r \\ f_E = m\dfrac{d\omega}{dt} r = m\varepsilon r \end{cases} \tag{4.185}$$

where ε is the angular acceleration of the rotating frame (and of the body). The Coriolis force is usually the only one credited in modeling gyroscopes, although when there is an angular acceleration, the Euler force is also present and directed about the same direction as the Coriolis force.

Example 4.18
Assess the errors induced by ignoring the Euler force in a microgyroscope, which moves with a given relative velocity and when r ranges between r_{min} and r_{max}.

Solution:
Equation (4.185) allows formulating the following relative error:

$$e = \frac{f_C - f_E}{f_C} = 1 - \frac{\varepsilon r}{2\omega v_r} \tag{4.186}$$

The errors can be reduced for small angular velocities and small relative velocities, as Equation (4.186) indicates, which also shows that large angular accelerations and large distances of the point mass from the center of rotation also reduce the errors. In other words, the error is minimum at r_{max} and maximum at r_{min}.

Consider now the microgyroscope's model of Figure 4.29. The body of mass m is constrained to move unidirectionally about the x-axis where harmonic driving is applied by a force f inside a massless carrier. At its turn, the carrier is constrained to move about a direction, which, initially, is perpendicular to the drive direction. The carrier is subjected to an angular velocity ω and an angular acceleration ε about an axis perpendicular to the drawing plane of Figure 4.29. As a result, a Coriolis force and an Euler force will act on the body about the sense direction. The equations of motion of the inside body about the drive and sense directions are:

$$\begin{cases} m\ddot{x} + c_d \dot{x} + k_d x = f \\ m\ddot{y} + c_s \dot{y} + k_s y = -m(2\omega\dot{x} + \varepsilon x) \end{cases} \tag{4.187}$$

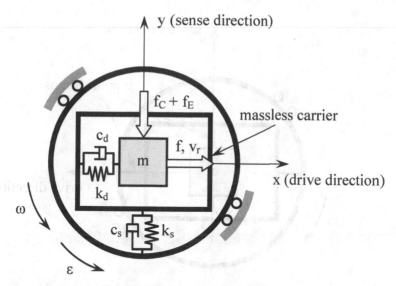

Figure 4.29 Microgyroscope model

where the Coriolis and Euler forces defined in Equation (4.185) have been used in the second Equation (4.187).

By applying the Laplace transforms with zero initial conditions to Equation (4.187), the following algebraic equations result:

$$\begin{cases} \left(ms^2 + c_d s + k_d\right) X(s) = F(s) \\ \left(ms^2 + c_s s + k_s\right) Y(s) = -m\left(2\omega s + \varepsilon\right) X(s) \end{cases} \quad (4.188)$$

Elimination of $X(s)$ between the two equations in Equation (4.188) produces the following transfer function connecting $F(s)$—the input—and $Y(s)$—the output:

$$G(s) = \frac{Y(s)}{F(s)} = -\frac{\left(2\omega s + \varepsilon\right) m}{\left(ms^2 + c_d s + k_d\right)\left(ms^2 + c_s s + k_s\right)} \quad (4.189)$$

Example 4.19

The microgyroscope of Figure 4.30 is encapsulated in a vacuum-like environment and is supported by two identical beams inside a carrier whose mass is neglected. The carrier is also suspended inside its enclosure by two beams that are identical to the ones suspending the central body (the four beams are of length *l*). Analyze the amplitude of the driving force as a

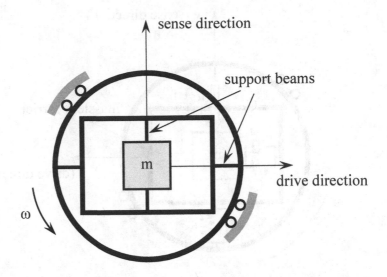

<figure>*Figure 4.30* Beam-suspended microgyroscope</figure>

function of the external angular velocity considering that $\varepsilon = 0$, $m = 2$ grams, $l = 100$ µm, $t = 5$ µm, $w = 1$ µm, $E = 160$ GPa, and the sensed amplitude is $Y = 1$ µm. Ignore Euler effects.

<u>Solution:</u>
    The force amplitude can be expressed as:

$$F = \frac{\left(k - m\omega^2\right)^2}{2m\omega^2} Y \tag{4.190}$$

where Y is the amplitude of the sensed signal and k is the stiffness of one pair of beams, which is equal to:

$$k = 2\frac{12EI_z}{l^3} = \frac{2Ew^3 t}{l^3} \tag{4.191}$$

By using the numerical data of this problem, the plot of Figure 4.31 is drawn, which shows the amplitude of the driving force as a function of the external angular velocity. It can be seen that the force amplitude varies nonlinearly with the circular frequency.

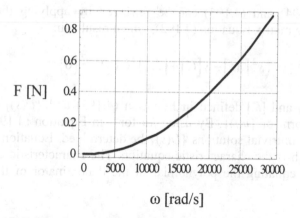

Figure 4.31 Driving force amplitude as a function of the external angular frequency

4.2.7 Resonant Frequencies and the Laplace Method

The characteristic equation of a multi DOF system can be determined in the Laplace domain. Consider, for instance, the microgyroscope of Figure 4.29. Its free-response dynamic equations can be set in matrix form as:

$$[M]\{\ddot{u}\}+[C]\{\dot{u}\}+[K]\{u\} = \{0\} \tag{4.192}$$

with:

$$\begin{cases} [M]=\begin{bmatrix} m & 0 \\ 0 & m \end{bmatrix} \\ [C]=\begin{bmatrix} c_d & 0 \\ 0 & c_s \end{bmatrix} \\ [K]=\begin{bmatrix} k_d & 0 \\ 0 & k_s \end{bmatrix} \end{cases} \tag{4.193}$$

being the mass, damping, and stiffness matrices and $\{u\} = \{x \quad y\}^t$ being the displacement vector. When solutions of the exponential type, $\{u\} = \{U\}$ $\sin(\omega t)$ are sought, the condition for nontrivial solutions to be found is:

$$\det\left(-\omega^2 [M]+\omega[C]+[K]\right)=0 \tag{4.194}$$

which is the characteristic equation, whose algebraic form is:

$$\left(-m\omega^2 +c_d\omega+k_d\right)\left(-m\omega^2 +c_s\omega+k_s\right)=0 \tag{4.195}$$

Equations (4.194) and (4.195) can be retrieved by applying the Laplace transform to the matrix Equation (4.192), which results in:

$$\left(s^2[M]+s[C]+[K]\right)\{U(s)\}=\{0\} \qquad (4.196)$$

with $[M]$, $[C]$, and $[K]$ defined in Equation (4.193) and $\{U(s)\}$ being the Laplace transform of $\{u(t)\}$. By using $j\omega$ for s in Equation (4.196) and by requiring that nontrivial solutions $\{U(s)\}$ be determined, Equation (4.194) is retrieved, which is the characteristic equation. The characteristic equation is also found by equating the polynomial in the denominator of the transfer function to zero.

Example 4.20
    Find the resonant frequencies of the system shown in Figure 4.32 by applying the transfer function approach. Consider the system behaves as a two DOF one with the coordinates being the displacements x_1 and x_2. Also consider the input is the force acting on the body of mass m_1.

Solution:
    The dynamic equations of the two DOF system are:

$$\begin{cases} m_1\ddot{x}_1+2kx_1+k\left(x_1-x_2\right)=f \\ m_2\ddot{x}_2+kx_2+k\left(x_2-x_1\right)=0 \end{cases} \qquad (4.197)$$

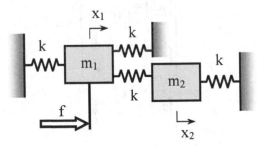

Figure 4.32 Two DOF mechanical microsystem

By applying the Laplace transform with zero initial conditions to Equation (4.197), the following transfer functions can be formulated:

$$G_1(s)=\frac{X_1(s)}{F(s)}=\frac{m_2s^2+2k}{\left(m_1s^2+3k\right)\left(m_2s^2+2k\right)-k^2} \qquad (4.198)$$

$$G_2(s) = \frac{X_2(s)}{F(s)} = \frac{k}{\left(m_1 s^2 + 3k\right)\left(m_2 s^2 + 2k\right) - k^2} \qquad (4.199)$$

It can be checked that both transfer functions have the same denominators and, consequently, the characteristic equation is:

$$\left(-m_1\omega^2 + 3k\right)\left(-m_2\omega^2 + 2k\right) - k^2 = 0 \qquad (4.200)$$

The eigenvalues resulting from Equation (4.200) are:

$$\omega_{1,2}^2 = \frac{k}{2m_1 m_2}\left(2m_1 + 3m_2 \pm \sqrt{4m_1^2 + 7m_1 m_2 + 9m_2^2}\right) \qquad (4.201)$$

4.2.8 Tuning Forks

Another example of the Coriolis and Euler effects in MEMS is the tuning fork, which is used to sense angular velocity and acceleration in non-inertial reference frames.

Figure 4.33 indicates the Coriolis effect for two common situations: in the case of Figure 4.33 (a), the relative velocity is achieved through out-of-the-plane

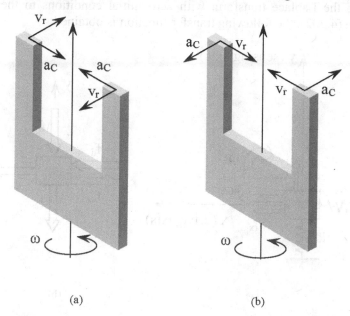

(a)   (b)

Figure 4.33 Tuning-fork microgyroscope: (a) out-of-the-plane driving and in-plane sensing; (b) in-plane driving and out-of-the-plane sensing

driving of the two tines that will bend in opposite directions. The superposition between the tines motions and the external angular velocity ω will generate Coriolis accelerations, which are directed towards the rotation axis and within the plane of the tines. The case pictured in Figure 4.33 (b) applies in-plane driving and the Coriolis accelerations of the two tines is sensed as out-of-the-plane bending.

By considering one tine can be modeled by using lumped parameters (a mass m and two springs, k_x and k_y, located at the tine's free end), Figure 4.34 (a) can be used to indicate the velocity and accelerations that occur. It is also considered that the longitudinal axis of tine coincides with the input rotation axis, such that no centrifugal force acts on the tine. The interaction between the drive velocity v_{rx} and the external angular velocity ω generates the Coriolis acceleration a_{Cy}, whereas the resulting vibration along the y-axis and the corresponding relative velocity v_{ry} interacting with the angular velocity will generate the Coriolis acceleration about the x-axis. The forces acting on the mass m are shown in Figure 4.34 (b). By assuming small displacements and a linear system, the equations of motion about the x- and y-axes are:

$$\begin{cases} m\ddot{x} + k_x x - 2m\omega\dot{y} = f_d \\ m\ddot{y} + k_y y + 2m\omega\dot{x} = 0 \end{cases} \quad (4.202)$$

The forcing (drive) term is f_d, as shown in the first Equation (4.202). By applying the Laplace transform with zero initial conditions to the second Equation (4.202), the following transfer function is obtained:

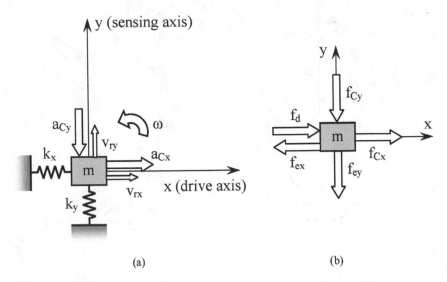

(a)   (b)

Figure 4.34 Lumped-parameter model of one tine: (a) physical model; (b) free-body diagram

$$G(s) = \frac{Y(s)}{X(s)} = -\frac{2m\omega s}{ms^2 + k_y} \tag{4.203}$$

Equation (4.203) enables finding the sensed amplitude from the drive amplitude and the modulus of the cosinusoidal transfer function as:

$$Y = \frac{2m\omega\omega_d}{k_y - m\omega_d^2} X \tag{4.204}$$

where ω_d is the drive frequency. When the tuning fork is used as a sensor, ω
can be found from Equation (4.204) as:

$$\omega = \frac{k_y - m\omega_d^2}{2m\omega_d} \times \frac{Y}{X} \tag{4.205}$$

Equation (4.203) also shows that the signal y is $\pi/2$ behind the x signal and is
calculated as:

$$y = Y\cos\left(\omega_d t - \frac{\pi}{2}\right) = Y\sin(\omega_d t) = \frac{2m\omega\omega_d}{k_y - m\omega_d^2} X \sin(\omega_d t) \tag{4.206}$$

By also applying the Laplace transform to the first Equation (4.202) under
zero initial conditions, by eliminating $X(s)$ and by using Equation (4.203), the
following transfer function is obtained, which combines the sensed displacement to the input force, namely:

$$\frac{Y(s)}{F_d(s)} = -\frac{2m\omega s}{4m^2\omega^2 s^2 + (ms^2 + k_x)(ms^2 + k_y)} \tag{4.207}$$

The characteristic equation corresponding to the above transfer function is:

$$-4m^2\omega^2\omega_r^2 + (-m\omega_r^2 + k_x)(-m\omega_r^2 + k_y) = 0 \tag{4.208}$$

whose roots are the resonant frequencies of the two DOF lumped-parameter
system of Figure 4.34 (a):

$$\omega_{r,1,2}^2 = \frac{k_x + k_y + 4m\omega^2 \pm \sqrt{(k_x + k_y + 4m\omega^2) - 4k_x k_y}}{2m} \tag{4.209}$$

Example 4.21

A tuning fork is used to measure an external angular velocity by measuring the sensed and drive displacements capacitively. Knowing that a tine has a length of l and a rectangular cross-section defined by w and t ($w < t$ and t is parallel to the y-axis), study the influence of the tine's dimensions, as well as of the drive and sensed amplitudes X and Y on the monitored ω.

Solution:

Equation (4.205) can be written as:

$$\omega = \frac{\dfrac{k_y}{m} - \omega_d^2}{2\omega_d} \times \frac{Y}{X} \tag{4.210}$$

and the tine stiffness is:

$$k_y = \frac{3EI_x}{l^3} = \frac{Ewt^3}{4l^3} \tag{4.211}$$

The lumped-parameter mass is equivalent to the total cantilever mass, namely:

$$m = \frac{33}{140}\rho lwt \tag{4.212}$$

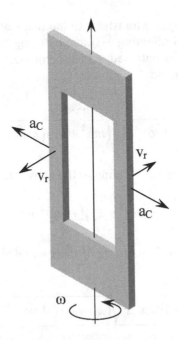

Figure 4.35 Tuning fork with clamped-clamped tines

By combining Equations (4.210), (4.211), and (4.212), the angular frequency can be expressed as:

$$\omega = \frac{\dfrac{35Et^2}{33\rho l^4} - \omega_d^2}{2\omega_d} \times \frac{Y}{X} \tag{4.213}$$

Equation (4.213) indicates that by increasing the thickness t of the tine and by decreasing its length l, it is possible to reduce the amplitude ratio Y/X to detect the external angular frequency ω.

Another tuning fork is shown in Figure 4.35 where the two tines are actually clamped at both ends instead of being clamped-free as they were with the first tuning fork analyzed. The model that has been developed previously for one clamped-free tine aligned with the external axis of rotation remains valid. Only the equivalent mass and stiffnesses about the drive and sense directions will change because the lumped mass is placed at the middle of the tine (instead of being located at its free end in the previous tuning fork configuration).

Example 4.22

Compare the sensed amplitude that is detected by a clamped-clamped tuning fork with the one captured by a clamped-free configuration.

Solution:

Equation (4.204) can be formulated as:

$$Y = \frac{2\omega\omega_d}{\dfrac{k_y}{m} - \omega_d^2} X \tag{4.214}$$

By considering a half-length model for a clamped-clamped beam, the lumped-parameter equivalent mass and stiffness are (as shown in Lobontiu [4]):

$$\begin{cases} k_y = \dfrac{96EI_x}{l^3} \\[2mm] m = \dfrac{13}{70}\rho lwt \end{cases} \tag{4.215}$$

The following signal ratio can be formulated, by also using the previous example:

$$\frac{X}{Y_{c\text{-}e}} - \frac{X}{Y_{c\text{-}f}} \approx 21\frac{Et^2}{\rho\omega\omega_d l^4} \tag{4.216}$$

where c-c means clamped-clamped and c-f stands for clamped-free. Consider, for instance, the following numerical values: $E = 150$ GPa, $\rho = 2400$ kg/m^3, $t = 1$ μm, $l = 100$ μm, $\omega = 10,000$ rad/s, $\omega_d = 20,000$ rad/s, which result in a value of approximately 4.5 for the signal ratio difference of Equation (4.216). This result indicates that for the same drive amplitude, drive frequency and external frequency, the clamped-free tuning fork senses a larger signal than the clamped one.

4.3 TIME-DOMAIN RESPONSE OF MEMS

4.3.1 Time-Domain Laplace-Transform Modeling and Solution

As the case was with the frequency domain, the Laplace transform and the inverse Laplace transform can be used to solve a differential equation (or a differential equations system for multiple DOF system) in the time domain, provided the equations are linear and with time-unvarying coefficients. The dynamic Equation (4.47), which defines the motion of a multiple DOF system involving the inertia, damping stiffness matrices, as well as a forcing vector, can simply be solved for a linear system with constant matrices by first applying the Laplace transform to Equation (4.47). It is known that:

$$\begin{cases} \mathscr{L}\big[\{\ddot{u}(t)\}\big] = s^2 \{U(s)\} - s\{u(0)\} - \{\dot{u}(0)\} \\ \mathscr{L}\big[\{\dot{u}(t)\}\big] = s\{U(s)\} - \{u(0)\} \end{cases} \tag{4.217}$$

where $\{U(s)\}$ is the Laplace transform of $\{u(t)\}$, the unknown time-domain vector. By combining now Equation (4.217) with Equation (4.47), it can be shown that:

$$\{U(s)\} = \big[A(s)\big]^{-1}\{B(s)\} \tag{4.218}$$

with:

$$\begin{cases} \big[A(s)\big] = s^2 [M] + s[C] + [K] \\ \{B(s)\} = \{F(s)\} + [M]\big(s\{u(0)\} + \{\dot{u}(0)\}\big) + s[C]\{u(0)\} \end{cases} \tag{4.219}$$

The solution is determined by applying the inverse Laplace transform to Equation (4.218), which results in:

$$\{u(t)\} = \mathscr{L}^{-1}\big[\{U(s)\}\big] = \mathscr{L}^{-1}\big[[A(s)]^{-1}\{B(s)\}\big] \tag{4.220}$$

Example 4.23

Consider the tine of a tuning fork is located at a distance R from the axis of rotation, the longitudinal direction of the tine being parallel with the rotation axis. By using the lumped-parameter modeling and the Laplace transform, express the system's matrices and vectors corresponding to cosinusoidal excitation with zero initial displacement and velocity.

Solution:

It can be shown that in this situation, a new force, the centrifugal one acts on the lumped-mass about the x-axis as shown in Figure 4.36.

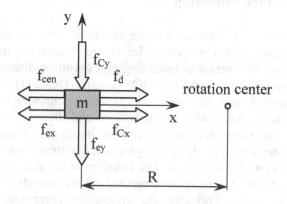

Figure 4.36 Lumped-parameter model of a tine from a tuning fork

The dynamic equations of motion for the two DOF system are of the matrix form shown in Equation (4.47) with the system's matrices being:

$$
\begin{cases}
[M] = \begin{bmatrix} m & 0 \\ 0 & m \end{bmatrix} \\[2mm]
[C] = \begin{bmatrix} 0 & -2m\omega \\ 2m\omega & 0 \end{bmatrix} \\[2mm]
[K] = \begin{bmatrix} k_x - m\omega^2 & 0 \\ 0 & k_y \end{bmatrix}
\end{cases}
\tag{4.221}
$$

The driving force and unknown coordinate vectors are:

$$
\begin{cases}
\{f\} = \{f_d - m\omega^2 R \quad 0\}^t \\
\{u(t)\} = \{x(t) \quad y(t)\}^t
\end{cases}
\tag{4.222}
$$

The vector $\{B\}$ of Equation (4.219) simplifies to:

$$\{B(s)\} = \{F(s)\} \qquad (4.223)$$

because the initial condition vectors are zero, and therefore:

$$[U(s)] = \left[s^2[M] + s[C] + [K] \right]^{-1} \{F(s)\} \qquad (4.224)$$

4.3.2 State-Space Modeling

An alternate method to representing the dynamic behavior of micro/nano systems is the *state-space* approach, which is widely used in control systems. This approach is instrumental in modeling cases with nonlinearities, as well as multiple-input multiple-output (MIMO) systems (Ogata [1], Nise [2]). The state-space approach uses the concepts of *state*, *state variables* and *state vectors* to formulate the *state-space equations* and then solve them for the *output vector*. The state variables are the minimum set of parameters that fully define the response of a dynamic system at any time moment t when the input is known at an initial time t_0. The state variables at a particular time t define the state of the system at that time and can be collected in one vector, named the state vector. Formally, the state-space representation uses the following two equations:

$$\begin{cases} \{\dot{x}(t)\} = [A]\{x(t)\} + [B]\{u(t)\} \\ \{y(t)\} = [C]\{x(t)\} + [D]\{u(t)\} \end{cases} \qquad (4.225)$$

The first of the two equations in Equation (4.226) is known as the *state equation* whereas the second is the *output equation*. The vector $\{x\}$ is the state vector, $\{u\}$ is the *input vector* and $\{y\}$ is the *output* (unknown) *vector*. Matrix $[A]$ in Equation (4.225) is the *state matrix*, $[B]$ is the *input matrix*, $[C]$ is the *output matrix*, and $[D]$ is the *direct transition matrix*. They are all assumed constant.

Application of the Laplace transform to Equation (4.225) with zero initial conditions produces the following equations:

$$\begin{cases} s\{X(s)\} = [A]\{X(s)\} + [B]\{U(s)\} \\ \{Y(s)\} = [C]\{X(s)\} + [D]\{U(s)\} \end{cases} \qquad (4.226)$$

where:

$$\begin{cases} \{X(s)\} = \mathcal{L}[x(t)] \\ \{U(s)\} = \mathcal{L}[u(t)] \\ \{Y(s)\} = \mathcal{L}[y(t)] \end{cases} \qquad (4.227)$$

By expressing $\{X(s)\}$ from the first Equation (4.226) and then substituting it into the second Equation (4.226), the Laplace-domain output vector is determined as:

$$\{Y(s)\} = [G(s)]\{U(s)\} \qquad (4.228)$$

where the matrix $[G(s)]$ is calculated as:

$$[G(s)] = [C][s[I] - [A]]^{-1}[B] + [D] \qquad (4.229)$$

The example of a single DOF cantilever will be studied next as an application of the state-space approach.

Example 4.24
Use the state-space approach to model the forced damped vibrations of a cantilever beam by considering the single DOF model and the following force acting on it: $u(t) = e^{-at}\cos(\omega t)$. Consider the following numerical values: $a = 4000$, the frequency of the excitation function is $f = 100$ Hz, the material parameters of the microcantilever's material are $E = 150$ GPa, $\rho = 2400$ kg/m³. The microcantilever dimensions are $l = 300$ μm, $w = 80$ μm, $t = 1$ μm, and the damping coefficient is $\varsigma = 0.01$.

Solution:
The equation of motion of the single DOF system can be written as:

$$m\ddot{y}(t) + c\dot{y}(t) + ky(t) = u(t) \qquad (4.230)$$

The following two state variables are selected:

$$\begin{cases} x_1 = y \\ x_2 = \dot{y} \end{cases} \qquad (4.231)$$

Using Equation (4.231) together with Equation (4.230) results in:

$$\dot{x}_2 = -\frac{k}{m}x_1 - \frac{c}{m}x_2 + \frac{1}{m}u \qquad (4.232)$$

Equations (4.231) indicate that:

$$\dot{x}_1 = x_2 \tag{4.233}$$

Equations (4.232) and (4.233) can now be written in vector-matrix form, which is the state equation expressed generically in the first Equation (4.225):

$$\left\{ \begin{array}{c} \dot{x}_1 \\ \dot{x}_2 \end{array} \right\} = \left[\begin{array}{cc} 0 & 1 \\ -\dfrac{k}{m} & -\dfrac{c}{m} \end{array} \right] \left\{ \begin{array}{c} x_1 \\ x_2 \end{array} \right\} + \left\{ \begin{array}{c} 0 \\ \dfrac{1}{m} \end{array} \right\} u \tag{4.234}$$

The matrices [A] and [B] of the first Equation (4.225) can be identified in Equation (4.234) and they are:

$$\begin{cases} [A] = \left[\begin{array}{cc} 0 & 1 \\ -\dfrac{k}{m} & -\dfrac{c}{m} \end{array} \right] \\ \{B\} = \left\{ 0 \quad \dfrac{1}{m} \right\}^t \end{cases} \tag{4.235}$$

Because the input vector has only one component, the excitation function $u(t)$, the matrix [B] reduces to a vector, as shown in Equation (4.235). The first Equation (4.231) can be written in vector-matrix form as:

$$y = \{1 \quad 0\} \left\{ \begin{array}{c} x_1 \\ x_2 \end{array} \right\} + 0 \times u \tag{4.236}$$

which is the output equation generically given in the second Equation (4.225). Inspection of Equation (4.236) shows that:

$$\begin{cases} \{C\} = \{1 \quad 0\} \\ D = 0 \end{cases} \tag{4.237}$$

Having determined the four matrices [A], [B], [C], and [D] allows calculation of the [G] matrix defined in Equation (4.229), which is actually a scalar in this particular example, namely:

$$G(s) = \frac{1}{ms^2 + cs + k} \tag{4.238}$$

The Laplace transform of the input function is:

$$U(s) = \frac{s+a}{(s+a)^2 + \omega^2}$$ (4.239)

The unknown $Y(s)$ is therefore:

$$Y(s) = G(s)U(s) = \frac{s+a}{\left(ms^2 + cs + k\right)\left[(s+a)^2 + \omega^2\right]}$$ (4.240)

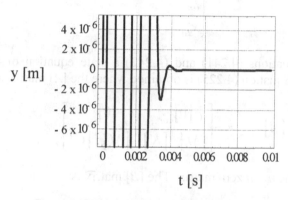

Figure 4.37 Time response of single DOF microcantilever

The lumped-parameter mass is given by Equation (4.66) and its numerical value is $m = 1.358 \times 10^{-11}$ kg. The stiffness is also given in Equation (4.66) and its numerical value is found to be $k = 0.333$ N/m. The damping coefficient is determined from the damping ratio, mass, and stiffness as shown in Chapter 3, and is found to be for the parameters of this problem $c = 0.425 \times 10^{-7}$ Ns/m. With these numerical values and by applying the inverse Laplace transform to Equation (4.240), the time-domain response of the single DOF microcantilever is obtained and plotted in Figure 4.37.

Consider now applying the state-space formalism to the dynamics of a system having n DOF, and whose behavior is described by the matrix-form equation:

$$[M]\{\ddot{y}(t)\} + [C]\{\dot{y}(t)\} + [K]\{y(t)\} = \{u(t)\}$$ (4.241)

where $[M]$, $[C]$, and $[K]$ are the system's regular mass, damping and stiffness matrices, respectively, all square and of $n \times n$ dimension. The vectors $\{y(t)\}$ and $\{u(t)\}$ are the output and input vectors, both having the dimension n. As shown in the previous example, for a single DOF system whose dynamics is defined by a second order differential equations two state variables were needed.

The system discussed here has n DOF and therefore $2n$ state variables are needed. One possible choice of the state vector is the following one:

$$\begin{cases} x_1 = y_1 \\ x_2 = y_2 \\ \dots \\ x_n = y_n \\ x_{n+1} = \dot{y}_1 \\ x_{n+2} = \dot{y}_2 \\ \dots \\ x_{2n} = \dot{y}_n \end{cases} \tag{4.242}$$

By combining Equations (4.241) and (4.242) a state equation of the type shown in the first Equation (4.225) is obtained where the $[A]$ matrix is:

$$[A]_{2nx2n} = -\begin{bmatrix} [0]_{nxn} & [0]_{nxn} \\ [M]^{-1}[K] & [M]^{-1}[C] \end{bmatrix} \tag{4.243}$$

with $[0]_{n \times n}$ being the $n \times n$ zero matrix. The $[B]$ matrix is:

$$[B]_{2nx2n} = -\begin{bmatrix} [0]_{nxn} & [0]_{nxn} \\ [0]_{nxn} & [M]^{-1} \end{bmatrix} \tag{4.244}$$

The derivative of the state vector is:

$$\{\dot{x}\}_{2nx1} = \begin{Bmatrix} \{0\}_{nx1} \\ \{\dot{x}_{n+1} \quad \dot{x}_{n+2} \quad \dots \quad \dot{x}_{2n}\}' \end{Bmatrix} \tag{4.245}$$

and the input vector is written as:

$$\{u\}_{2nx1} = \begin{Bmatrix} \{0\}_{nx1} \\ \{u_1 \quad u_2 \quad \dots \quad u_n\}' \end{Bmatrix} \tag{4.246}$$

where $\{0\}_{nx1}$ is the n-dimension zero vector.

Equation (4.242) allows formulating the output equation in the simplified form:

$$\{y\} = [C]\{x\} \tag{4.247}$$

where the [C] matrix is:

$$[C]_{2n \times 2n} = -\begin{bmatrix} [I]_{n \times n} & [0]_{n \times n} \\ [0]_{n \times n} & [0]_{n \times n} \end{bmatrix} \tag{4.248}$$

with $[I]_{n \times n}$ being the identity matrix.

With the $[A]$, $[B]$, $[C]$ matrices defined in Equations (4.243), (4.244), and (4.249), the $[G(s)]$ matrix of Equation (4.229) simplifies to:

$$[G(s)] = [C][s[I] - [A]]^{-1}[B] \tag{4.249}$$

which enables calculation of the Laplace-domain output vector $Y(s)$, by means of Equation (4.228). In the end, the time-domain output vector $y(t)$ is found by applying the inverse Laplace transform to $Y(s)$.

While the formalism derived here demonstrated that the dynamic equation of a MIMO microsystem can be expressed in state space form, care should be exercised as the matrices defined involve singularities and inversion by regular methods cannot be performed. There are matrix methods that can solve this aspect, as shown in more advanced matrix methods texts (Perlis [7]). An alternative formulation allowing to solve for $Y(s)$ is also shown next. Consider the state variables as being divided into two vectors, namely:

$$\begin{cases} \{x_a\} = \{x_1 \quad x_2 \quad \dots \quad x_n\}^t = \{y_1 \quad y_2 \quad \dots \quad y_n\}^t \\ \{x_b\} = \{x_{n+1} \quad x_{n+2} \quad \dots \quad x_{2n}\}^t = \{\dot{y}_1 \quad \dot{y}_2 \quad \dots \quad \dot{y}_n\}^t \end{cases} \tag{4.250}$$

By combining Equation (4.250) with Equation (4.241) the following relationships are obtained:

$$\begin{cases} [M]\{\dot{x}_b\} + [C]\{x_b\} + [K]\{x_a\} = \{u\} \\ \{\dot{x}_a\} = \{x_b\} \end{cases} \tag{4.251}$$

The Laplace transform is applied to the two equations in Equation (4.251) considering zero initial conditions, and then $X_b(s)$ is eliminated from the two resulting Laplace-domain equations and $X_a(s)$ is found:

$$\{X_a(s)\} = \{Y(s)\} = [G(s)]^{-1}\{U(s)\} \tag{4.252}$$

with $[G(s)]$ being calculated as:

$$[G(s)] = s^2[M] + s[C] + [K] \tag{4.253}$$

Application of the inverse Laplace transform to $Y(s)$ of Equation (4.252) yields the output vector $y(t)$.

It should also be pointed put that the approach described herein is also one that allows transforming a state-space formulation into a transfer function model if $[G(s)]$ is regarded as a transfer function for a MIMO system.

Example 4.25
Study the time response of the micro mechanism sketched in Figure 2.61 when the outer body is driven electrostatically and in-plane by a force defined as $u(t) = Ae^{-at}\sin(\omega t)$. Consider the system operates in vacuum and $l = 200$ μm, $m = 10^{-8}$ kg, $E = 165$ GPa, $\omega = 200$ rad/s, $a = 4000$, $A = 0.00001$ N.

Solution:
The system behaves as a four DOF one when planar motion is analyzed, as discussed in Example 2.17. The stiffness and mass matrices are given in Equations (2.193) and (2.194). Being a four DOF system, eight state variables are needed and they are selected as follows:

$$x_1 = u_{ay}; x_2 = \theta_{az}; x_3 = u_{by}; x_4 = \theta_{bz};$$
$$x_5 = \dot{u}_{ay}; x_6 = \dot{\theta}_{az}; x_7 = \dot{u}_{by}; x_8 = \dot{\theta}_{bz};$$

(4.254)

Consequently, the matrices $[A]$, $[B]$, and $[C]$ defined in Equations (4.243), (4.244), and (4.248) are of an 8×8 dimension. By following the approach developed previously, the time response of the system is obtained. Because of the specific type of driving, the two bodies undergo pure translations and therefore the rotational DOF are identically zero. Figure 4.38 shows the variations of y_a and y_b as functions of time.

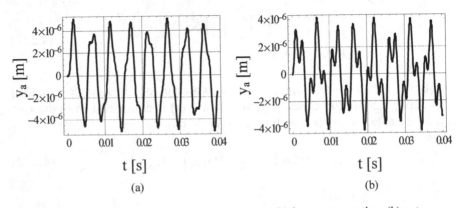

Figure 4.38 Time response of four DOF microsystem: (a) inner mass motion; (b) outer mass motion

4.3.3 Transfer-Function to State-Space Model Transformation

The transfer-function model can be transformed into a state-space model, as shown next. A single-input, single-output (SISO) system will be considered here, which is one of the most encountered cases in MEMS/NEMS. For such systems, and with particular application to dynamic systems that are modeled by means of second-order differential equations, the transfer function will involve second-degree polynomials in s of the form:

$$G(s) = \frac{Y(s)}{U(s)} = \frac{a_1 s^2 + a_2 s + a_3}{b_1 s^2 + b_2 s + b_3} \tag{4.255}$$

and is symbolically shown in Figure 4.39. Figure 4.40 suggests an alternative representation of the transfer function of Figure 4.39 by means of an intermediate function $Z(s)$, which allows separation between the numerator and denominator of the transfer function $G(s)$.

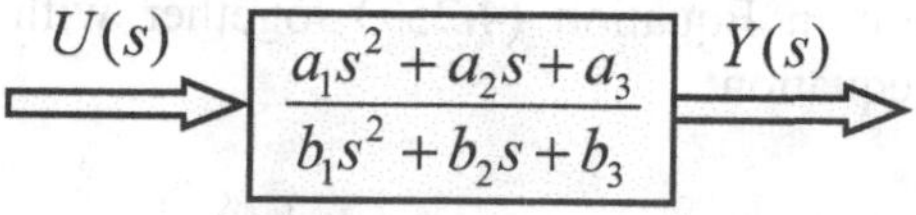

$$U(s) \longrightarrow \boxed{\dfrac{a_1 s^2 + a_2 s + a_3}{b_1 s^2 + b_2 s + b_3}} \longrightarrow Y(s)$$

Figure 4.39 Transfer function for a single-input single-output system

$$U(s) \longrightarrow \boxed{\dfrac{1}{b_1 s^2 + b_2 s + b_3}} \xrightarrow{Z(s)} \boxed{a_1 s^2 + a_2 s + a_3} \longrightarrow Y(s)$$

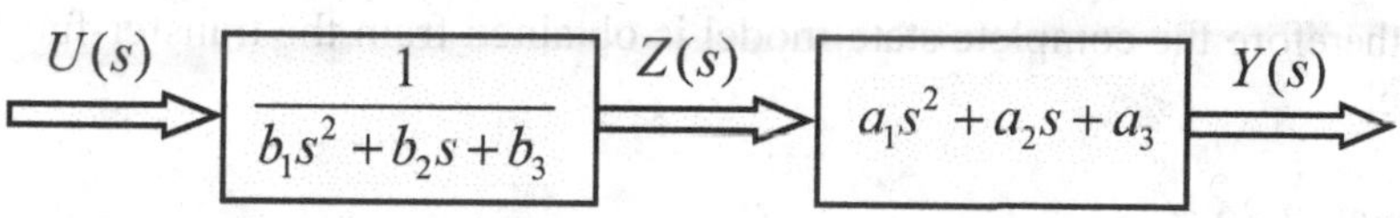

Figure 4.40 Alternative transfer function representation of a single-input single-output system

This coupling that is realized by means of $Z(s)$ allows defining the following transfer functions:

$$\begin{cases} \dfrac{Z(s)}{U(s)} = \dfrac{1}{b_1 s^2 + b_2 s + b_3} \\ \dfrac{Y(s)}{Z(s)} = a_1 s^2 + a_2 s + a_3 \end{cases} \tag{4.256}$$

Cross multiplication in Equation (4.256), followed by application of the inverse Laplace transform with zero initial conditions, produce the following equations in the time domain:

$$\begin{cases} b_1\ddot{z} + b_2\dot{z} + b_3z = u \\ a_1\ddot{z} + a_2\dot{z} + a_3z = y \end{cases} \tag{4.257}$$

The following selection is made for state variables:

$$\begin{cases} x_1 = z \\ x_2 = \dot{z} \end{cases} \tag{4.258}$$

by means of which the first of the equations in Equation (4.257) and the combination of the two of the equations in Equation (4.258) result in a state-form equation, namely:

$$\begin{Bmatrix} \dot{x}_1 \\ \dot{x}_2 \end{Bmatrix} = \begin{bmatrix} 0 & 1 \\ -\dfrac{b_3}{b_1} & -\dfrac{b_2}{b_1} \end{bmatrix} \begin{Bmatrix} x_1 \\ x_2 \end{Bmatrix} + \begin{Bmatrix} 0 \\ 1 \end{Bmatrix} u \tag{4.259}$$

The second equation in Equation (4.257) together with Equation (4.258) produce the output equation:

$$y = \begin{Bmatrix} a_3 - \dfrac{a_1 b_3}{b_1} & a_2 - \dfrac{a_1 b_1}{b_2} \end{Bmatrix} \begin{Bmatrix} x_1 \\ x_2 \end{Bmatrix} + a_1 u \tag{4.260}$$

and therefore the complete state-model is obtained from the transfer-function model.

Example 4.26

The transfer function corresponding to the excitation-response of a micro-cantilever is of the form: $G(s) = \dfrac{2s}{3s^2 + 5s + 2}$. Determine the equivalent state-space model for this SISO system.

Solution:

Equation (4.259) gives the following matrix [A]:

$$[A] = \begin{bmatrix} 0 & 1 \\ -\dfrac{2}{3} & -\dfrac{5}{3} \end{bmatrix} \tag{4.261}$$

whereas Equation (4.260) gives the row vector $\{C\}$ as:

$$\{C\} = \{0 \quad 2\} \tag{4.262}$$

The state and output equations, which define the state space model, are therefore:

$$\begin{cases} \begin{Bmatrix} \dot{x}_1 \\ \dot{x}_2 \end{Bmatrix} = \begin{bmatrix} 0 & 1 \\ -\dfrac{2}{3} & -\dfrac{5}{3} \end{bmatrix} \begin{Bmatrix} x_1 \\ x_2 \end{Bmatrix} + \begin{Bmatrix} 0 \\ 1 \end{Bmatrix} u \\ y = \{0 \quad 2\} \begin{Bmatrix} x_1 \\ x_2 \end{Bmatrix} \end{cases} \tag{4.263}$$

4.3.4 Time-Stepping Schemes

For single or multiple DOF systems, such as the ones analyzed thus far, it is convenient to use the so-called *time discretization* to solve (integrate) the dynamic equations of motion. For time discretization, the infinite time domain is partitioned into a finite number of time stations, separated by a time step, such that two consecutive time stations are connected by:

$$t_{n+1} = t_n + \Delta t \tag{4.264}$$

Based on this discretization, recurrence relationships have been derived enabling to determine the system variables at the moment t_{n+1} in terms of the known values of the system variables at the moment t_n and the forcing factors. One method of formulating a dynamic model in the time domain uses the truncated *Taylor series expansion* of the unknown time-dependent vector. This methodology results in a *time-stepping scheme* whereby a polynomial is selected to approximate the nodal vector in the following form:

$$\{u\} \approx \{u\}_n + \left\{ \frac{du}{dt} \right\}_n (t - t_n) + \frac{1}{2!} \left\{ \frac{d^2u}{dt^2} \right\}_n (t - t_n)^2 + \dots$$
$$+ \frac{1}{m!} \left\{ \frac{d^m u}{dt^m} \right\}_n (t - t_n)^m \tag{4.265}$$

Obviously, when the values of the polynomial and of its derivatives up to the order m are known at the moment n, similar values of the polynomial and its derivatives corresponding to the moment $n + 1$ can be computed by means of the equation above.

A time-stepping scheme needs to be *consistent* and *stable*. Consistency requires that the degree of the approximating polynomial is at least equal to the order of the differential equation modeling the system. For the dynamics of a structural microsystem, for instance, which is governed by a second-order differential equation, the approximating polynomial needs to be at least a second-degree one. To be stable, a time-stepping scheme cannot produce a solution that increases indefinitely with time (diverges) or a that is oscillatory. Certain algorithms comply with the stability requirement when the time step is less than a critical value:

$$\Delta t \le \Delta t_{cr} \tag{4.266}$$

Such algorithms are named *conditionally stable*, as contrasted to the *unconditionally stable* ones, which provide a stable solution independently of the time step.

Time-stepping schemes, as the one described by Equation (4.265) are named *single-step*, whereby the unknowns at time $n + 1$, which are $\{d^2 u_{n+1}/dt^2\}$, $\{du_{n+1}/dt\}$, and $\{u\}_{n+1}$, are expressed in terms of the same amounts at the previous time station, $\{d^2 u_n/dt^2\}$, $\{du_n/dt\}$, and $\{u\}_n$. Such algorithms are largely implemented in finite element software, for instance, and can be designed based on a constant or a variable time step to solve linear first- and second-order problems. Another class of algorithms for time-dependent dynamics problems is made up of *multiple-step algorithms* where the unknown $\{u\}_{n+1}$ is determined as a function of $\{u\}_n$, $\{u\}_{n-1}$, ..., $\{u\}_p$ and therefore the nodal vector derivatives are eliminated from the recurrence relationships. For *nonlinear problems*, special algorithms do exist, and more details can be found in specialized texts, such as the one by Wood [8]. The present text will briefly discuss single-step and nonlinear time-stepping schemes.

Another criterion of classifying the time-stepping schemes regards the algebraic equations system, which has to be solved at each time step. *Implicit algorithms* are solving coupled equations and this process involves more computational resources. *Explicit algorithms* result in models with the equations describing the system response at any given time moment that are decoupled, such as the situation is where lumped matrix formulation is used.

4.3.4.1 The Central Difference Method

One very popular time-stepping scheme, the *central difference method*, is presented first. This algorithm is a single-step one, is conditionally stable and can be formulated both implicitly and explicitly. The base equations are the following ones:

$$\begin{cases} \{\dot{u}\}_n = \dfrac{\{u\}_{n+1} - \{u\}_{n-1}}{2\Delta t} \\[4mm] \{\ddot{u}\}_n = \dfrac{\{u\}_{n+1} - 2\{u\}_n + \{u\}_{n-1}}{\Delta t^2} \end{cases} \qquad (4.267)$$

When the two definition equations in Equation (4.267) are used in conjunction with the lumped-parameter element equation corresponding to the time moment n:

$$[M]\{\ddot{u}\}_n + [C]\{\dot{u}\}_n + [K]\{u\}_n = \{F\}_n \qquad (4.268)$$

the following equation can be formulated that expresses the nodal variable at the time moment $n + 1$:

$$\left(\frac{1}{\Delta t^2}[M] + \frac{1}{2\Delta t}[C]\right)\{u\}_{n+1} = \left(\frac{2}{\Delta t^2}[M] - [K]\right)\{u\}_n$$

$$- \left(\frac{1}{\Delta t^2}[M] - \frac{1}{2\Delta t}[C]\right)\{u\}_{n-1} + \{F\}_n \qquad (4.269)$$

An alternative formulation to Equation (4.269) is:

$$\left(\frac{1}{\Delta t^2}[M] + \frac{1}{2\Delta t}[C]\right)\{u\}_{n+1} = \left(\frac{2}{\Delta t^2}[M] - [K]\right)\{u\}_n$$

$$- \left(\frac{1}{\Delta t^2}[M] - \frac{1}{2\Delta t}[C]\right)\left(\{u\}_n - \Delta t\{\dot{u}\}_n + \frac{\Delta t^2}{2}\{\ddot{u}\}_n\right) + \{F\}_n \qquad (4.270)$$

where the acceleration at moment n is:

$$\{\ddot{u}\}_n = -[M]^{-1}[C]\{\dot{u}\}_n - [M]^{-1}[K]\{u\}_n + [M]^{-1}\{F\}_n \qquad (4.271)$$

A time-stepping scheme starts based on initial conditions (at $t = 0$), which, for a second-order system, are known values of the displacement vector and of its first time derivative (velocity). Based on the equations presented herein, the central difference method progresses with evaluating the nodal vector, together with its velocity and acceleration at the second time moment and so on up to a moment n. Specifically, this time progression scheme involves the following calculations at time $n + 1$ based on known values of the nodal vector and its first time derivative at moment n:

(i)   Find the nodal vector $\{u\}_{n+1}$ by means of Equations (4.270) and (4.271).

(ii)   Find the nodal vector $\{u\}_{n+2}$ by means of Equation (4.269) when using $n + 1$ instead of n.

(iii)   Find the first time derivative of $\{u\}_{n+1}$ by means of the first Equation (4.267) when using $n + 1$ instead of n.

(iv)   Find the second time derivative of $\{u\}_{n+1}$ by means of Equation (4.271) when using $n + 1$ instead of n.

This algorithm is explicit only when the lumped-parameter inertia and damping matrices are diagonal. This assertion is demonstrated by inspection of the left-hand side of both Equations (4.269) and (4.270) for $[M]$ and $[C]$ in diagonal form. Consequently, the components of the nodal vector at moment $n + 1$, $\{u\}_{n+1}$, can be determined independently, as the n algebraic equations are decoupled.

As mentioned previously, the central difference algorithm is conditionally stable, and as Meirovitch [9] mentions, the critical time step is:

$$\Delta t_{cr} = \frac{2}{\omega_{max}} \qquad (4.272)$$

where ω_{max} is the maximum resonant frequency of the dynamic model.

Example 4.27
The proof mass of Figure 4.41 is supported by two identical folded beams and is electrostatically actuated and sensed in the x-direction by comb-type units. The driver applies a cosinusoidal force. Formulate an algorithm allowing evaluatation of the Couette damping between the moving plate and the above- and beneath-plate air. Assume the actuation frequency varies by a known time law.

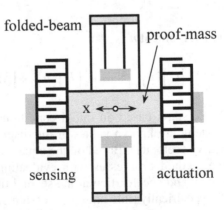

Figure 4.41 Comb-drive transduction of a proof mass supported by two identical folded beams

Solution:

When considering damping is produced through slide-film mechanisms and when accounting for friction between the plate and the surrounding air, the total damping coefficient is found by adding the two damping contributions given in Equations (3.173) and (3.155) as:

$$c = \mu \beta_1 A \left[1 + \frac{\sinh(2\beta_1 z_0) + \sin(2\beta_1 z_0)}{\cosh(2\beta_1 z_0) - \cos(2\beta_1 z_0)} \right] \tag{4.273}$$

with μ being the dynamic viscosity coefficient, A is the plate area, β_1 is given in Equation (3.169)—and it depends on the motion frequency ω, and z_0 is the constant gap between the plate and the substrate.

The dynamic equation of the proof mass is:

$$m\ddot{x} + c\dot{x} + kx = F\cos(\omega t) \tag{4.274}$$

Because c depends on ω through the coefficient β_1, and ω is variable, Equation (4.274) is not with constant coefficients and therefore regular methods, such as the transfer-function approach are not applicable. By taking into account that the proof mass displacements are known through electrostatic sensing, a time-stepping scheme that uses the displacements can be used in conjunction with Equation (4.274) to evaluate the damping coefficient. For a single DOF system, as the one of this example problem, Equation (4.267) simplifies to:

$$\begin{cases} \dot{x}_n = \dfrac{x_{n+1} - x_n}{2\Delta t} \\ \ddot{x}_n = \dfrac{x_{n+1} - 2x_n + x_{n-1}}{(\Delta t)^2} \end{cases} \tag{4.275}$$

Combination of Equations (4.274) and (4.275), the latter one written for the n-th time station, allows expressing the damping coefficient at the same time instant n as:

$$c_n = \frac{2\Delta t m \left(\dfrac{2x_n - x_{n-1} - x_{n+1}}{(\Delta t)^2} - \dfrac{k}{m} x_n \right)}{x_{n+1} - x_{n-1}} \tag{4.276}$$

4.3.4.2 The Newmark Scheme and Nonlinear Problems

By introducing a parameter β to account for various types of accelerations of the unknown nodal vector, Newmark [10] proposed the following single-step recurrence relationship:

$$\begin{cases} \{\dot{u}\}_{n+1} = \{\dot{u}\}_n + \dfrac{\Delta t}{2}\left(\{\ddot{u}\}_{n+1} + \{\ddot{u}\}_n\right) \\[2mm] \{u\}_{n+1} = \{u\}_n + \Delta t\{\dot{u}\}_n + \dfrac{\Delta t^2}{2}\{\ddot{u}\}_n + \beta\Delta t^2\left(\{\ddot{u}\}_{n+1} - \{\ddot{u}\}_n\right) \end{cases} \quad (4.277)$$

A generalization of the original Newmark algorithm is presented by Zienkiewicz and Taylor [11] who introduced two parameters, β_1 and β_2, in the form:

$$\begin{cases} \{u\}_{n+1} = \{u\}_n + \Delta t\{\dot{u}\}_n + (1-\beta_2)\dfrac{\Delta t^2}{2}\{\ddot{u}\}_n + \beta_2\dfrac{\Delta t^2}{2}\{\ddot{u}\}_{n+1} \\[2mm] \{\dot{u}\}_{n+1} = \{\dot{u}\}_n + (1-\beta_1)\Delta t\{\ddot{u}\}_n + \beta_1\Delta t\{\ddot{u}\}_{n+1} \end{cases} \quad (4.278)$$

Equations (4.278) are coupled with the dynamic equations written for the time stations n and $n + 1$, namely:

$$\begin{cases} [M]\{\ddot{u}\}_{n+1} + [C]\{\dot{u}\}_{n+1} + [K]\{u\}_{n+1} = \{F\}_{n+1} \\[2mm] [M]\{\ddot{u}\}_n + [C]\{\dot{u}\}_n + [K]\{u\}_n = \{F\}_n \end{cases} \quad (4.279)$$

By expressing the accelerations at time stations n and $n + 1$ from Equation (4.278) and substituting them into Equation (4.279), the nodal vector at time $n + 1$ can be expressed as:

$$\{u\}_{n+1} = [A]^{-1}\{F\}_{n,n+1} \quad (4.280)$$

where:

$$[A] = \left(b_1[M] + [C]\right)^{-1}\left(b_2[M] + [K]\right) - \dfrac{a_3}{a_1}[I] \quad (4.281)$$

and:

$$\begin{aligned} \{F\}_{n,n+1} &= \left(b_1[M] + [C]\right)^{-1}\left(\{F\}_{n+1} - a_2[M]\{\dot{u}\}_n + b_2[M]\{u\}_n\right) \\ &\quad - [M]^{-1}\left(\{F\}_n - \left(a_2[M] + [C]\right)\{\dot{u}\}_n + \left(a_3[M] - [K]\right)\{u\}_n\right) \end{aligned} \quad (4.282)$$

The new coefficients of Equations (4.281) and (4.282) are:

$$\left\{ \begin{aligned}
a_1 &= \frac{\beta_2}{(\beta_2 - \beta_1)\Delta t} \\[2mm]
a_2 &= \frac{2\beta_1 - \beta_2}{(\beta_2 - \beta_1)\Delta t} \\[2mm]
a_3 &= -\frac{2\beta_1}{(\beta_2 - \beta_1)\Delta t^2} \\[2mm]
b_1 &= \frac{\beta_2 - 1}{(\beta_2 - \beta_1)\Delta t} \\[2mm]
b_2 &= \frac{2(1 - \beta_1)}{(\beta_2 - \beta_1)\Delta t^2}
\end{aligned} \right. \qquad (4.283)$$

Equations (4.280), (4.281), and (4.282) indicate that the Newmark algorithm, unlike the central difference method, is self-starting, which means that successive calculation of $\{u\}_{n+1}$ only needs the initial values (at $t = 0$) of the nodal vector and of its first time derivative.

The algorithm is *unconditionally stable* when:

$$\beta_2 \geq \beta_1 \geq 0.5 \qquad (4.284)$$

as indicated by Zienkiewicz and Taylor [11], for instance. When $\beta_2 = 0$, the Newmark algorithm is explicit. It can also be shown that the Newmark algorithm reduces to the central difference scheme when $\beta_1 = 0.5$ and $\beta_2 = 0$.

In several categories of dynamic problems involving nonlinearities of material type (rate-dependent, for instance), geometric type (large deformations), or actuation, the system matrices or/and the load vector depend on the displacement vector and/or the velocity vector. In other cases, the damping coefficients are functions of the driving frequency. In a more generic case, it can be assumed that all the system matrices, as well as the forcing vector depend on both the displacement vector and its time derivative. A *collocation algorithm*, such as the Newmark scheme, can be employed to solve a nonlinear problem and the steps corresponding to one time station are the following ones:

(i)   Find the system matrices $[M(\{u\}_n, \{du/dt\}_n)]$, $[C(\{u\}_n, \{du/dt\}_n)]$, $[K(\{u\}_n, \{du/dt\}_n)]$, and the nodal vector $\{F(\{u\}_n, \{du/dt\}_n)\}$.

(ii)   Find the acceleration vector $\{d^2u/dt^2\}_n$ from the dynamic equation:

$$\left[M\left(\{\dot{u}\}_n,\{u\}_n\right)\right]\{\ddot{u}\}_n+\left[C\left(\{\dot{u}\}_n,\{u\}_n\right)\right]\{\dot{u}\}_n$$
$$+\left[K\left(\{\dot{u}\}_n,\{u\}_n\right)\right]\{u\}_n=\left\{F\left(\{\dot{u}\}_n,\{u\}_n\right)\right\}_n \qquad (4.285)$$

(iii)   Find the first time derivative of $\{u\}_{n+1}$ as well as $\{u\}_{n+1}$ by solving the equation system:

$$\begin{cases}
\{u\}_{n+1}=\{u\}_n+\Delta t\{\dot{u}\}_n+\left(1-\beta_2\right)\dfrac{\Delta t^2}{2}\{\ddot{u}\}_n \\[2mm]
+\dfrac{\beta_2\Delta t}{2\beta_1}\left[\{\dot{u}\}_{n+1}-\{u\}_n-\left(1-\beta_1\right)\Delta t\{\ddot{u}\}_n\right] \\[2mm]
\left[M\left(\{\dot{u}\}_n,\{u\}_n\right)\right]\left(a_1\{\dot{u}\}_{n+1}+a_2\{\dot{u}\}_n+a_3\{u\}_{n+1}-a_3\{u\}_n\right) \quad (4.286) \\[2mm]
+\left[C\left(\{\dot{u}\}_n,\{u\}_n\right)\right]\{\dot{u}\}_n+\left[K\left(\{\dot{u}\}_n,\{u\}_n\right)\right]\{u\}_n \\[2mm]
=\left\{F\left(\{\dot{u}\}_n,\{u\}_n\right)\right\}_n
\end{cases}$$

Example 4.28
Formulate the time-stepping algorithm that would solve the dynamics of the cosinusoidally excited undamped vibrations of the single DOF mass-spring system shown in Figure 4.42 where the electric actuation force is of plate type. Consider a lumped-parameter model where the moving plate has a mass m and the spring has a stiffness k. The initial gap is g, the plate area is A and the air permittivity is ε.

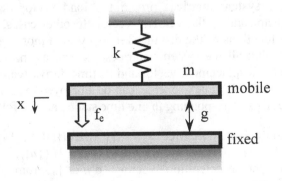

Figure 4.42 Plate-type electrostatic harmonic actuation of a single DOF mass-spring system

Solution:

Considering the voltage applied between the two plates is:

$$e = E \cos(\omega t) \tag{4.287}$$

the electrostatic attraction force is:

$$f_e = \frac{\varepsilon A}{2(g-x)^2} e^2 \tag{4.288}$$

By applying Newton's second law to the single DOF lumped-parameter system of Figure 4.42, and by also using Equations (4.287) and (4.288), the following equation is obtained:

$$m\ddot{x} + kx = \frac{\varepsilon A E^2}{2(g-x)^2} \cos^2(\omega t) \tag{4.289}$$

which has the driving force depending on the variable x.

The driving force at moment n can be expressed as:

$$f_n = \frac{\varepsilon A E^2}{2(g-x_n)^2} \cos^2(\omega n \Delta t) \tag{4.290}$$

The acceleration at the same moment n is determined from the dynamic equation of motion, namely:

$$\ddot{x}_n = \frac{f_n - kx_n}{m} \tag{4.291}$$

The following equations system needs to be solved, corresponding to the generic model (Equation (4.286)):

$$\begin{cases} m\left[a_1\dot{x}_{n+1} + a_2\dot{x}_n + a_3\left(x_{n+1} - x_n\right)\right] + kx_n = f_n \\ x_{n+1} = x_n + \Delta t\dot{x}_n + (1-\beta_2)\frac{\Delta t^2}{2}\ddot{x}_n + \frac{\beta_2\Delta t}{2\beta_1}\left[\dot{x}_{n+1} - \dot{x}_n - (1-\beta_1)\Delta t\ddot{x}_n\right] \end{cases} \tag{4.292}$$

After that, the algorithm is repeated for the next time station, $n + 1$.

4.3.5 Approximate Methods and Nonlinear MEMS Problems

Situations with relatively small nonlinearities can arise in MEMS due to large deformations, when the spring-type force is no longer linear, as shown next. It is known that nonlinear springs can be of a *hardening* type or of a *softening* type, as shown in Figure 4.43, where the characteristic of a linear spring is also included.

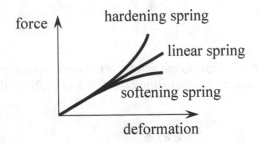

Figure 4.43 Linear and nonlinear spring characteristics

For a nonlinear hardening spring more force (for instance) is needed to obtain the same deformation when compared to a linear one, whereas less force from a softening spring will produce the same displacement as a linear spring. The nonlinear theory of vibrations (e.g., see Rao [12] or Thomson [13]), indicates that the free non-damped vibrations of a softening and a hardening spring, respectively, attached to a mass in a single DOF system are described by the equations:

$$\begin{cases} m\ddot{x} + kx - \mu m x^3 = 0 \\ m\ddot{x} + kx + \mu m x^3 = 0 \end{cases} \tag{4.293}$$

Equations (4.293) can be rearranged in the form known as the *Duffing oscillator's equations*, namely:

$$\begin{cases} \ddot{x} + \omega_l^2 x - \mu x^3 = 0 \\ \ddot{x} + \omega_l^2 x + \mu x^3 = 0 \end{cases} \tag{4.294}$$

with ω_l being the resonant frequency of the mass-linear spring system and μ is a small parameter. The solution to the Duffing's oscillator will be approached shortly but before that, an example demonstrating that a large-deformations microcantilever behaves as a softening-spring nonlinear free oscillator will be discussed. It is well-known in mechanics of materials that the nonlinear, large-deformation model of a cantilever produces tip deflections that are smaller than the ones predicted by the linear, small-deformation model (e.g., see Gere and Timoshenko [14]).

Example 4.29

Calculate the nonlinear coefficient μ corresponding to the free vibrations of a constant rectangular cross-section microcantilever, subjected to large deflections by considering it behaves according to the softening-spring Duffing oscillator. Express the coefficient for a tip angle of $75°$.

Solution:

Figure 4.44 shows the side view of a deformed microcantilever under the action of a tip force.

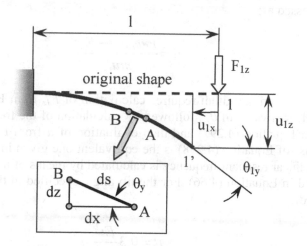

Figure 4.44 Microcantilever with tip force and large deformations

Under the assumption of large deformations, the free end 1 no longer moves only about the z-axis, it also has a motion about the longitudinal x-axis, as indicated in Figure 4.44. It can be shown (e.g., see Lobontiu and Garcia [3]) that the elementary arc of the inset of Figure 4.44 can be computed at an arbitrary position defined by an angle θ_y (not shown in the figure above) as:

$$ds = \frac{d\theta_y}{\sqrt{\dfrac{2F_{1z}}{EI_y}\left(\sin\theta_{1y} - \sin\theta_y\right)}} \tag{4.295}$$

The total length l of the deformed beam (which is assumed to remain constant) can therefore be found by integrating Equation (4.295) as:

$$l = \sqrt{\frac{EI_y}{2F_{1z}}} \int_0^{\theta_{1y}} \frac{d\theta_y}{\sqrt{\sin\theta_{1y} - \sin\theta_y}} \tag{4.296}$$

Similarly, the deflection about the z-axis is calculated as:

$$u_{1z} = \sqrt{\frac{EI_y}{2F_{1z}}} \int_0^{\theta_{1y}} \frac{\sin\theta_y \, d\theta_y}{\sqrt{\sin\theta_{1y} - \sin\theta_y}} \tag{4.297}$$

The force F_{1z} can be considered as the elastic force, a nonlinear one, and if it is further assumed this force is a nonlinear, softening-spring one, by inspection of the Duffing's oscillator (the second Equation (4.294)), the parameter μ can be expressed as:

$$\mu = \frac{m\omega_l^2 u_{1z} - F_{1z}}{mu_{1z}^3} \tag{4.298}$$

The calculation algorithm requires calculation of F_{1z} from Equation (4.296) for a selected value of θ_{1y}, followed by calculation of the free end deflection u_{1z} from Equation (4.297) and then evaluation of μ from Equation (4.298). The mass of Equation (4.298) is the equivalent one given in Equation (4.66) and the linear resonant frequency is calculated by means of it and the stiffness provided in Equation (4.66). For the tip angle specified in the example, it is obtained:

$$\mu = 0.3 \frac{Et^2}{\rho l^6} \tag{4.299}$$

For a microcantilever with $E = 165$ GPa, $\rho = 2300$ kg/m^3, $t = 1$ μm, and $l = 200$ μm, the coefficient is $\mu = 2.7 \times 10^{-6}$ m^{-2}-s^{-2}.

Solving the nonlinear Equation (4.294) can be performed by approximate analytic methods (Rao [12] or Thomson [13]) such as *perturbation*, *iteration* or *Ritz-Galerkin* (*finite element*-type) procedures. In what follows, the iteration method will be used to solve the following equation, which describes the forced undamped response:

$$\ddot{u}_z + \omega_l^2 u_z - \mu u_z^3 = F_{1z}\cos(\omega t) \tag{4.300}$$

where the subscript *1* has been dropped from both u_z and F_z.

The iteration method presupposes taking a first approximation to the real solution to Equation (4.300), and using that approximation in all applicable terms of Equation (4.300) except the one with the second time derivative, which is subsequently considered the second approximation. By selecting the first approximate solution as:

$$u_{z1} = U_z \cos(\omega t) \tag{4.301}$$

and substituting it in Equation (4.300), application of the next approximation, as mentioned previously leads to:

$$\ddot{u}_{z2} = -\left(U_z \omega_l^2 + \frac{3}{4}U_z^3 \mu - F_z\right)\cos(\omega t) + \frac{U_z^3 \mu}{4}\cos(3\omega t) \tag{4.302}$$

Equation (4.302) also took into consideration the following trigonometric relationship:

$$\cos^3(\omega t) = \frac{3}{4}\cos(\omega t) + \frac{1}{4}\cos(3\omega t) \tag{4.303}$$

Integration of Equation (4.302) with zero initial conditions produces:

$$u_{z2} = \frac{1}{\omega^2}\left(U_z \omega_l^2 - \frac{3}{4}U_z^3 \mu - F_z\right)\cos(\omega t) - \frac{U_z^3 \mu}{36\omega^2}\cos(3\omega t) \tag{4.304}$$

By halting the approximation series at this stage, and if the two approximations, u_{z1} and u_{z2} are valid choices, they should be similar and therefore the term in the higher harmonic of Equation (4.304) should not be considered. Consequently, by comparing Equations (4.304) and (4.303), it follows that the coefficients of $\cos(\omega t)$ in the two equations should be equal; this condition results in:

$$U_z = \frac{U_z \omega_l^2 - \frac{3}{4}\mu U_z^3 - F_z}{\omega^2} \tag{4.305}$$

For the free response, $F_z = 0$ in Equation (4.305), and therefore the frequency ω of the system's free response varies according to:

$$\omega = \omega_l \sqrt{1 - \frac{3\mu U_z^2}{4\omega_l^2}} \tag{4.306}$$

Equation (4.306) indicates that the response frequency of the softening spring decreases with the amplitude of vibration. It can be shown that for a hardening spring a plus sign should be taken in the same Equation (4.306), which shows the response frequency increases with the amplitude.

When damping is also taken into consideration, the equation of the Duffing oscillator is:

$$\ddot{u}_z + c\dot{u}_z + \omega_l^2 u_z - \mu u_z^3 = F_z \cos(\omega t) \tag{4.307}$$

It is known that damping introduces a phase difference between driving and response. If the approximation of Equation (4.301) is to be kept, than the excitation force should be of the following form, as suggested by Rao [12]:

$$f_z = F_z \cos(\omega t + \varphi) = F_1 \cos(\omega t) + F_2 \sin(\omega t) \tag{4.308}$$

By substituting the approximation solution of Equation (4.300) together with the force of Equation (4.308) and the trigonometric Equation (4.304) into Equation (4.307), the latter equation transforms into an equation with two terms, one in sin(ωt), the other one in cos(ωt), when, again, the term in cos($3\omega t$) is ignored. In order for that resulting equation to be identically equal to 0, the coefficients of the sine and of the cosine terms need to be 0, namely:

$$\begin{cases} c\omega U_z + F_2 = 0 \\ \left(\omega_l^2 - \omega^2\right)U_z - \dfrac{3}{4}\mu U_z^3 - F_1 = 0 \end{cases} \tag{4.309}$$

Equations (4.309) are now squared and then they are added up (by also considering that $(F_z)^2 = (F_1)^2 + (F_2)^2$), and the following algebraic equation is produced:

$$9\mu^2\left(U_z^2\right)^3 + 24\mu\left(\omega_l^2 - \omega^2\right)\left(U_z^2\right)^2$$
$$+16\left[c^2\omega^2 + \left(\omega_l^2 - \omega^2\right)^2\right]\left(U_z^2\right) - 16F_z^2 = 0 \tag{4.310}$$

For the $\omega_1 - \omega_2$ range, Equation (4.310) has three solutions U_z^2, as shown in Figure 4.45. The same figure represents the jump phenomenon, evident between points 2 and 5 and 3 and 6, respectively.

From Equation (4.309) one can also obtain the phase angle as:

$$\tan\varphi = \frac{F_1}{F_2} = \frac{-\left(\omega^2 - \omega_l^2\right)^2 + \dfrac{3}{4}\mu U_z^2}{c\omega} \tag{4.311}$$

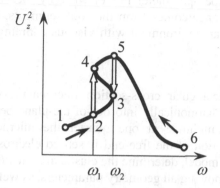

Figure 4.45 Jump phenomenon in a damped, harmonically driven Duffing oscillator

Problems

Problem 4.1

A harmonic force $f_z = F\cos(\omega t)$ is applied at the free end of a constant rectangular cross-section microcantilever. For the free end, compare the resulting deflection amplitude with the deflection produced by applying a force F statically at the same point. Consider the system has no energy losses.

Problem 4.2

A single DOF mass-dashpot translational MEMS is driven by a sinusoidal force. Its motion equation is of the form: $\ddot{x} + 3a\dot{x} + 2a^2 x = b\sin(\omega t)$ where a and b are positive real constants. Determine the system's response $x(t)$ for zero initial conditions by using the Laplace-transform approach.

Problem 4.3

By using the Laplace-transform approach, determine the frequency response of the system shown in Figure 2.64 of Problem 2.14 (Chapter 2) by considering a cosinusoidal force is applied about the motion direction to the proof mass. Known are the amounts defining the excitation, the proof mass, the dimensions, and material properties of the two identical beams of constant cross-section, as well as the equivalent viscous damping ratio. The initial conditions are zero displacement and zero velocity.

Problem 4.4

A paddle microbridge of known geometry and material parameters is actuated torsionally by a cosinusoidal electrostatic torque at a frequency $\omega = 1.5\,\omega_r$ (the torsion resonant frequency). Determine the amplitude of the actuation torque knowing that the maximum angle of rotation of the microbridge is θ_{max}. Use a lumped-parameter model whereby inertia is provided by the paddle and compliance comes from the root segments. Consider the micro-bridge vibrates in an environment with viscous damping of known damping ratio ζ.

Problem 4.5

A constant rectangular cross-section microcantilever that is placed in vacuum is excited harmonically into out-of-the-plane bending vibration by a force of known amplitude at one quarter the microcantilever's resonant frequency. The motion of the free-end is sensed electrostatically. By using a lumped-parameter model, determine the elastic modulus E of microcantilever's material, by also knowing all geometry parameters, as well as the mass density.

Problem 4.6

A microdevice is formed of a shuttle mass and for identical beams, as shown in Figure 4.46. The microdevice can sense the out-of-the-plane motion of the mass electrostatically and is placed on a support that vibrates harmo-nically. Determine the motion of the support from the sensed motion of the shuttle mass. Known are the dimensions and material properties of the constant cross-section beams, as well as the mass of the shuttle mass and the viscous damping ratio.

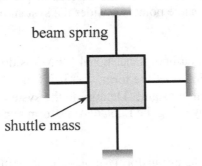

Figure 4.46 Shuttle mass supported by four identical beam springs

Problem 4.7

Determine the steady-state response of a translational single DOF microdevice consisting of a rigid plate of mass m, a spring of stiffness k, and a damper of damping coefficient c. A force $f = F\sin(\omega t)$ acts on the rigid plate.

Problem 4.8

The plate shown in Figure 4.47 has a mass $m = 1.5 \times 10^{-8}$ kg, is actuated electrostatically by a comb drive with a harmonic voltage $e = 20 \cos(\omega t)$ Volts and is suspended elastically above the substrate by two identical inclined beams. Derive the equation of motion of the plate in terms of the actuation frequency knowing the length of a beam is $l_b = 100$ μm, its cross-section is square with a side of 1 μm and the inclination angle is $\alpha = 5°$. The gap is $g = 1.5$ μm, the air permittivity is $\varepsilon = 8.8 \times 10^{-12}$ F/m, and $l_z = 5$ μm (comb thickness) and $E = 150$ GPa. Consider there are 15 comb drive actuation pairs.

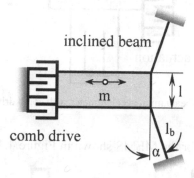

Figure 4.47 Single DOF proof mass with comb drive actuation and inclined beams suspension

Problem 4.9

Design a mechanical microsystem whose transfer function is of the form:

(a) $\dfrac{a}{bs^2 + cs + d}$

(b) $\dfrac{as + b}{cs^2 + ds + e}$

and identify the constants $a, b, c, d,$ and e with physical amounts.

Problem 4.10

Determine the electrical analogues of the mechanical microsystems of Problem 4.9.

Problem 4.11

Determine the lumped-parameter model of the microsystem shown in Figure 4.48, which consists of a shuttle mass and two folded-beam suspensions (the geometry of a folded beam is shown in Figure 2.23). Comb-drive actuation of cosinusoidal type is applied in a vacuum environment. Find the transfer function by considering the input is the electrostatic force and the output is the shuttle mass displacement.

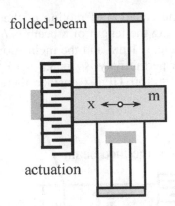

folded-beam

actuation

Figure 4.48 Folded-beam microaccelerometer with comb drive actuation

Problem 4.12

Solve Problem 4.11 for the MEMS shown in Figure 4.49.

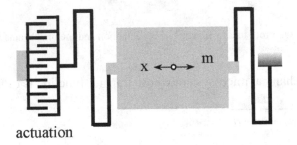

actuation

Figure 4.49 Spiral-spring microaccelerometer with comb drive actuation

The geometry of the serpentine spring, as the two identical ones that are connected to the proof mass, can be seen in Figure 2.25.

Problem 4.13

Evaluate the parameter μ of the nonlinear spring force corresponding to large-deformation bending vibrations of a paddle microcantilever by considering its root segment is compliant and massless, whereas the paddle section is rigid and provides the inertia of the microsystem. Consider the free end deformation angle is 50°. The geometric and material parameters are: $l_1 = 250$ μm, $l_2 = 100$ μm, $w_1 = 80$ μm, $w_2 = 10$ μm, $t = 1$ μm, $E = 150$ GPa, $\rho = 2400$ kg/m³.

Problem 4.14

A paddle microbridge is constructed from a material with parameters of uncertain values. By using a lumped-parameter model and cosinusoidal force excitation applied to the paddle, determine the shear modulus G and mass density ρ. Known are the geometric parameters $l_1 = 100$ μm, $l_2 = 200$ μm, $w_1 = 20$ μm, $w_2 = 200$ μm, $t = 1$ μm. The resonant frequency is $f_r = 25,000$ Hz, the excitation frequency is $\omega = 120,000$ rad/s, and the deflection amplitude is $U_z = 2$ μm. The equivalent damping ratio is $\zeta = 0.1$.

Problem 4.15

Find the maximum deflection of the constant cross-section microcantilever shown in Figure 4.50 resulting from the interaction between the current $i = I \cos(\omega t)$ passing through the circular loop and the external magnetic field B. Known are the microbridge dimensions: length l, width w, thickness t, as well as Young's modulus E, mass density ρ, the driving frequency ω, and current amplitude I. Consider the system incurs no energy losses.

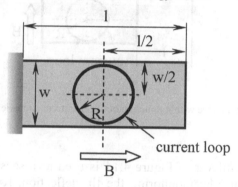

Figure 4.50 Microbridge with current loop and magnetic field interaction

Problem 4.16

The microdevice shown in Figure 4.51 is used to evaluate the overall damping ratio of the vibrating structure produced by the interaction of the current $i = I \cos(\omega t)$ passing through the circular loop and the external magnetic field B. Knowing B, I, R, l, w, t, m, ω, and the amplitude U_z of the out-of-the-plane vibration of the paddle free end, determine the damping ratio that corresponds to equivalent viscous damping.

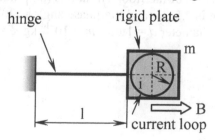

Figure 4.51 Paddle microcantilever with current loop and external magnetic field

Problem 4.17
Analyze the vibration produced by the interaction between the external magnetic field B and the harmonic current $i = I \cos(\omega t)$ passing through the circular loop shown in Figure 4.52. Considering that known are B, I, l, R ($R = l/2$), d (the diameter of the circular hinge), J_c (plate's central mechanical moment of inertia), m (plate's mass), ω (operating frequency), G (shear modulus) and τ_Y (yield strength), verify whether the flexures will yield under dynamic load.

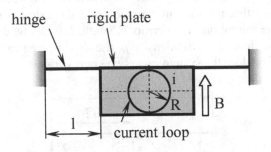

Figure 4.52 Microdevice with interaction between current and external magnetic field

Problem 4.18
The microcantilever of Figure 4.14 is used as a sensor to determine an external vibration y by monitoring the tip deflection. Knowing the overall Q-factor of the detection system Q as well as the excitation frequency ω, together with the microcantilever's geometric and material properties, calculate the amplitude of the external vibration Y. Also known is the amplitude U_z.

Problem 4.19
The microdevice of Figure 4.53 is used to gauge the floor vibrations that are known to be of sinusoidal nature by analyzing the frequency response. Evaluate the characteristics of the floor vibrations (amplitude and frequency) if the sensed amplitude is 2.5 μm and the phase angle is 20°. Known are also $l = 200$ μm, the beams diameter $d = 1$ μm, $m = 10^{-10}$ kg, $\zeta = 0.6$, and $E = 165$ GPa.

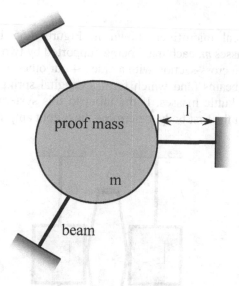

Figure 4.53 Three-beam microaccelerometer

Problem 4.20

The bent-beam (see Figure 2.30) microaccelerometer sketched in Figure 4.54 is used as a sensor being placed on a platform that rotates with an angle $\theta_i = \Theta_i \cos(\omega t)$. The relative rotation of the sensor is monitored electrostatically such that its amplitude, frequency and phase angle are known. By also knowing the beams dimensions, (equal-length legs of square cross-section) material properties, and by neglecting damping losses, evaluate the platform angular amplitude.

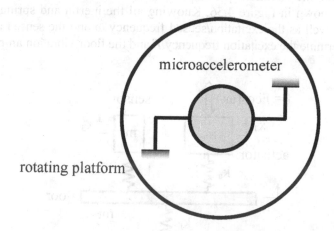

Figure 4.54 Rotating platform with microaccelerometer

Problem 4.21
The mechanical microfilter shown in Figure 4.55 is formed of two identical proof masses m, each mass being supported by two identical beams of length l and square cross-section with a side t. Four other beams, also identical with the support beams (and which form a saggital spring; see Figure 2.18) connect the two shuttle masses. If the input to this system is a force $f = F \cos(\omega t)$ applied on the first body, determine the output amplitude X_2.

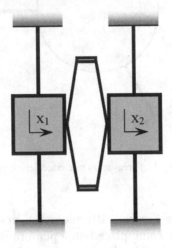

Figure 4.55 Two-mass mechanical microfilter

Problem 4.22
To measure the vibrations produced by an actuator, a sensor is placed on the floor, as shown in Figure 4.56. Knowing all the inertia and spring characteristics, as well as the excitation/sensed frequency ω and the sensed ampliztude X_3, determine the excitation frequency F and the floor vibration amplitude X_2.

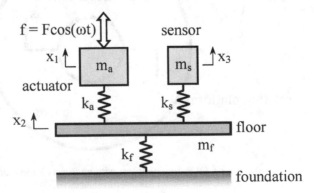

Figure 4.56 Microtransduction setup for monitoring floor vibration

Problem 4.23

For the electrical circuit shown in Figure 4.57, determine the output voltage amplitude E_o when the input voltage is $e_i = E_i \cos(\omega t)$.

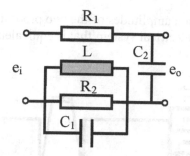

Figure 4.57 Electrical circuit under harmonic voltage excitation

Problem 4.24

Propose an electrical system that is analogous to the mechanical micro-filter of Figure 4.55 in Problem 4.21.

Problem 4.25

Determine an electrical system analogous to the mechanical microsystem shown in Figure 4.56 in Problem 4.22.

Problem 4.26

Determine a mechanical microsystem to be analogous to the electrical system sketched in Figure 4.58.

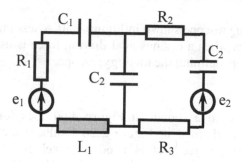

Figure 4.58 Two-loop electrical system

Problem 4.27

The MEMS of Figure 4.59 is actuated electrostatically by means of a comb drive that produces a cosinusoidal force f_e. Consider the two proof masses have each a mass m, and that the saggital spring is as the one defined in Figure 2.18, each of the four identical beams having a length of l and a circular cross-section of diameter d.

(a) Find the vibration amplitudes of the two proof masses.
(b) Determine the electrical system that is equivalent to the MEMS;

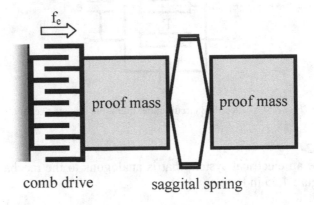

Figure 4.59 Two-mass microresonator with saggital spring and comb drive excitation

Problem 4.28
    Answer the two questions of Problem 4.27 for the MEMS shown in Figure 4.49 of Problem 4.12.

Problem 4.29
    For the microgyroscope shown in Figure 4.30, assess the external constant angular velocity ω when a cosinusoidal driving force is used. Known are all physical parameters defining the microgyroscope.

Problem 4.30
    The microgyroscope shown in Figure 4.60 uses four identical spiral springs as the one defined in Figure 2.26 Evaluate the amplitude of the cosinusoidal driving force that is needed to achieve a prescribed sense amplitude for a given external angular velocity. (Hint: Use $K_{Fx\text{-}ux}$ of Equation (2.41) and $K_{Fy\text{-}uy} = \dfrac{3EI_z}{2l_2^2\left(9l_1 + 4l_2\right)}$ (Lobontiu and Garcia [3]))

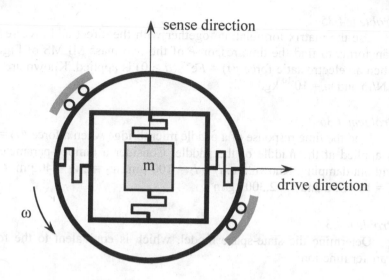

Figure 4.60 Microgyroscope with four spiral springs

Problem 4.31

For a tuning fork with out-of-the-plane cosinusoidal driving and in-plane sensing, as the one shown in Figure 4.33 (a), determine the driving frequency that will realize an amplitude ratio between the sensed and driving amplitudes (Y/X) of 2 for a given external angular velocity. A tine is defined by a length l and has square cross-section of side t.

Problem 4.32

Compare the Y/X amplitude ratios of the two tuning forks of Figure 4.33 when a cosinusoidal driving force is applied under external constant angular velocity and when a tine has a rectangular cross-section with the width w larger than the thickness t (the width is the dimension in the plane of the fork).

Problem 4.33

Calculate the resonant frequencies of the mechanical system of Figure 4.56 by applying the transfer function approach when considering the input is the force generated by the actuator and the output is the sensor displacement. Consider $k_f = 5k_a = 10k_s$, $m_f = 1,000m_a = 10,000m_s$, $k_s = 1$ N/m and $m_s = 10^{-10}$ kg.

Problem 4.34

Answer the same question of Problem 4.33 for the mechanical microfilter of Figure 4.55. The input is a force applied to the body on the left and the output is x_2.

Problem 4.35

Use the matrix formulation together with the direct and inverse Laplace transforms to find the time response of the two-mass MEMS of Figure 4.59 when an electrostatic force $f(t) = Fe^{-at}$ $(a > 0)$ is applied. Known are also $k = 1$ N/m and $m_s = 10^{-10}$ kg.

Problem 4.36

Find the time response of a paddle microbridge when a force $f(t) = 10^{-6}te^{-at}$ is applied at the middle of the paddle. Consider a lumped-parameter model without damping. Known are: $l_1 = l_2 = 100$ μm, $w_2 = 2w_1 = 40$ μm, $t = 1$ μm, $E = 160$ GPa and $\rho = 2{,}300$ kg/m^3.

Problem 4.37

Determine the state-space model, which is equivalent to the following transfer function:

$$G(s) = \frac{2s+3}{s^2+s+1}$$

Problem 4.38

Solve Problem 4.37 for the transfer function:

$$G(s) = \frac{4s}{s^2+5s+4}$$

Problem 4.39

The plate in Figure 4.48 translates due to an actuation force of the type $f(t) = Fte^{-at}\cos(\omega t)$. Consider there is no damping and find the time response of the system by using the Newmark algorithm.

Problem 4.40

Solve Problem 4.39 by using the central difference method. Consider slide-film damping defined by a coefficient c occurs between the plate and the substrate.

Problem 4.41

A force $f = \cos(\omega t)$ is generated electrostatically by using the micromechanism shown in Figure 4.61. Determine the time response of the shuttle mass in the presence of squeeze-film damping, which is defined by a constant damping coefficient, c.

folded-beam

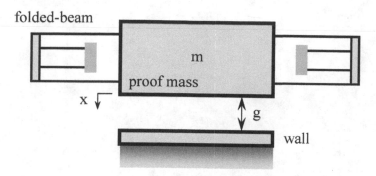

Figure 4.61 Proof mass with folded springs and plate-type actuation

References

1. K. Ogata, *System Dynamics*, Fourth Edition, Pearson Prentice Hall, Upper Saddle River, NJ, 2004.
2. N.S. Nise, *Control Systems Engineering*, Fourth Edition, John Wiley & Sons, Inc., New York, 2004.
3. N. Lobontiu, E. Garcia, *Mechanics of Microelectromechanical Systems*, Kluwer Academic Press, New York, 2004.
4. N. Lobontiu, *Mechanical Design of Microresonators: Modeling and Applications*, McGraw-Hill, New York, 2005.
5. L. Lin, R.T. Howe, A.P. Pisano, Microelectromechanical filters for signal processing, *Journal of Microelectromechanical Systems*, 7 (3), 1998, pp. 286–294.
6. F.D. Bannon III, J.R. Clark, C.T.-C. Nguyen, High-Q HF microelectromechanical filters, *IEEE Journal of Solid-State Circuits*, 35 (4), 2000, pp. 512–526.
7. S. Perlis, *Theory of Matrices*, Dover, Toronto, 1975.
8. W.L. Wood, *Practical Time Stepping Schemes*, Clarendon Press, Oxford, 1990.
9. L. Meirovitch, *Analytical Methods in Vibrations*, MacMillan, New York, 1967.
10. N.M. Newmark, A method for computation of structural dynamics, *Journal of ASCE*, 85, 1959, pp. 67–94.
11. O.C. Zienkiweicz, R.L. Taylor, *The Finite Element Method: Volume 2, Solid Mechanics*, Fifth Edition, Butterworth-Heinemann, 2000.
12. S.S. Rao, *Mechanical Vibrations*, Second Edition, Addison-Wesley Publishing Company, Reading, 1990.
13. W.T. Thomson, *Theory of Vibrations with Applications*, Third Edition, Prentice Hall, Englewood Cliffs, 1988.
14. J.M. Gere, S.P. Timoshenko, *Mechanics of Materials*, Third Edition, PWS Publishing Company, 1990.

INDEX

A

actuation, 18, 143, 247, 248, 329, 422
anelastic solid, 312
angular speed, 258

B

beam, 6, 10, 11, 14, 15, 20, 22, 42, 70, 79, 86, 113, 116, 119, 129, 130, 131, 143, 165, 167, 168, 171, 175, 176, 180, 181, 182, 183, 192, 194, 200, 201, 202, 204, 206, 207, 210, 211, 227, 230, 231, 237, 243, 246, 247, 248, 249, 251, 252, 253, 273, 284, 312, 319, 329, 330, 351, 378, 393, 397, 417, 422, 423, 424, 427
bending, 1, 3, 4, 5, 6, 8, 9, 10, 12, 14, 15, 17, 18, 19, 20, 21, 22, 24, 25, 28, 29, 30, 34, 36, 37, 38, 41, 42, 44, 45, 46, 47, 48, 49, 50, 51, 53, 54, 55, 57, 58, 60, 61, 62, 63, 64, 65, 68, 70, 71, 72, 73, 74, 75, 76, 77, 78, 79, 80, 82, 83, 85, 86, 87, 88, 90, 91, 94, 95, 96, 98, 100, 101, 102, 103, 104, 105, 106, 107, 108, 109, 110, 111, 112, 113, 114, 115, 116, 117, 118, 119, 120, 121, 122, 123, 125, 126, 127, 128, 129, 130, 131, 132, 133, 134, 135, 136, 137, 138, 139, 141, 142, 143, 144, 145, 146, 147, 148, 149, 150, 151, 152, 153, 155, 156, 158, 159, 160, 161, 162, 163, 164, 165, 169, 178, 180, 192, 193, 197,

199, 200, 207, 215, 225, 229, 232, 236, 237, 244, 245, 249, 251, 279, 280, 281, 318, 323, 324, 325, 348, 349, 351, 357, 389, 421, 424, 433
bending stiffness, 42, 101
bent beam spiral spring, 211
bent beam-spring, 180
bimorph, 68, 69, 71, 72, 114, 115, 150, 151
boundary conditions, 1, 4, 8, 75, 76, 78, 113, 160, 171, 172, 174, 218, 286, 294, 299, 300, 301, 303, 304, 305, 306, 307, 308, 309, 311, 328

C

Cartesian, 53, 54, 82, 214, 258
central difference method, 407, 408, 412, 433
characteristic equation, 186, 187, 189, 195, 204, 223, 386, 387, 388, 391
circularly filleted microcantilever, 27, 28, 143, 149
circularly notched microcantilever, 40, 41
complex, 4, 25, 44, 49, 98, 106, 154, 155, 183, 198, 257, 258, 259, 271, 301, 305, 308, 309, 312, 313, 321, 322, 330, 333, 334, 336, 339, 347, 348, 356, 364, 368, 370, 380, 381
complex stiffness, 271
compliance, 15, 148, 159, 160, 163, 185, 186, 187, 194, 195, 200, 207, 211, 218, 231, 244, 245, 250, 297, 324, 338, 345, 421